A History of the

Personal Computer

Blank page.

A History of the

Personal Computer

The People and the Technology

Roy A. Allan

Allan Publishing
London, Ontario, Canada.

First Edition 1.0

National Library of Canada Cataloguing in Publication Data

Allan, Roy A., 1931-
A history of the personal computer: the people and the technology

Includes bibliographical references and index.
ISBN 0-9689108-0-7

1. Microcomputers—History. I. Title.

QA76.17.A45 2001   004.16'09   C2001-901709-X

Additional copies or more information on the book are available from: Allan Publishing, 1624 Louise Blvd., London, Ontario, Canada. N6G 2R3

Printed and bound in Canada.

This book is dedicated to my dear wife Ann Louise for her constant care and support.

Blank page.

Contents

Preface xii

Acknowledgments xiv

Part I .... The Historical Background 1/1

Chapter 1 Development of the Computer 1/3

1.1 -- Original Digital Computers 1/3
1.2 -- IBM 1/6
1.3 -- Technology 1/8
1.4 -- Software 1/12
1.5 -- Other Developments 1/14
1.6 -- Small Computer Systems 1/15

Chapter 2 Personal Computing in the 1960's 2/3

2.1 -- Time sharing 2/3
2.2 -- Dartmouth DTSS and BASIC 2/5
2.3 -- The First Personal Computer 2/7
2.4 -- Small Computer Systems 2/8
2.5 -- Graphics and the User Interface 2/9
2.6 -- Software 2/11
2.7 -- Hobby & Amateur Computing 2/14

Part II .... 1970's -- The Altair/Apple Era 3/1

Chapter 3 Microprocessors in the 1970's 3/5

3.1 -- Intel 3/5
3.2 -- Motorola 3/10
3.3 -- Texas Instruments 3/12
3.4 -- Other Companies 3/13
3.5 -- Miscellaneous 3/16

Chapter 4 Transition to Microcomputers 4/1
4.1 -- The 1970-74 Transition 4/2
4.2 -- MITS Altair 4/8
4.3 -- Other Computers --1975-76 4/11
4.4 -- Commodore 4/15
4.5 -- Tandy/Radio Shack 4/16
4.6 -- Atari 4/17
4.7 -- Other Computers --1977-79 4/18

Chapter 5 Apple Computer in the 1970's 5/1
5.1 -- Wozniak/Jobs Early Years 5/1
5.2 -- Apple I Board 5/4
5.3 -- Founding of Apple Computer 5/9
5.4 -- Apple II 5/10
5.5 -- Apple Disk II Drive 5/12
5.6 -- 1978/79 Activities 5/13

Chapter 6 Microsoft in the 1970's 6/1
6.1 -- Gates/Allen Early Years 6/1
6.2 -- Altair/Basic 6/5
6.3 -- The Albuquerque Years 6/8
6.4 -- Relocation to Seattle 6/13

Chapter 7 Other Software in the 1970's 7/1
7.1 -- Operating Systems 7/1
7.2 -- Programming Languages 7/3
7.3 -- Word Processors 7/6
7.4 -- Spreadsheets 7/8
7.5 -- Databases 7/9
7.6 -- Miscellaneous 7/10

Part III ... 1980's - The IBM/Macintosh Era 8/1

Chapter 8 Microprocessors in the 1980's 8/3
8.1 -- Intel 8/3
8.2 -- Motorola 8/6
8.3 -- Other Microprocessors 8/7
8.4 -- Other Corporate Developments 8/8

Chapter 9 The IBM Corporation 9/1
9.1 -- Introduction 9/1
9.2 -- PC Approval and Development 9/4
9.3 -- The Original PC 9/6
9.4 -- The Following Models 9/11
9.5 -- Software 9/19
9.6 -- Corporate Activities 9/23

Chapter 10 Apple Computer in the 1980's **10/1**
10.1 -- Corporate & Other Activities 10/1
10.2 -- Apple III 10/9
10.3 -- Apple II's 10/11
10.4 -- Lisa 10/14
10.5 -- Macintoshes 10/18

Chapter 11 Competitive Computers **11/1**
11.1 -- Tandy/Radio Shack 11/1
11.2 -- Commodore 11/3
11.3 -- Osborne 11/6
11.4 -- Kaypro 11/8
11.5 -- Compaq 11/9
11.6 -- NeXT 11/11
11.7 -- Miscellaneous 11/12

Chapter 12 Microsoft in the 1980's **12/1**
12.1 -- Corporate & Other Activities 12/1
12.2 -- The IBM PC Software 12/11
12.3 -- Operating Systems 12/15
12.4 -- Windows 12/17
12.5 -- Languages 12/21
12.6 -- Application Programs 12/22

Chapter 13 Other Software in the 1980's **13/1**
13.1 -- Operating Systems 13/1
13.2 -- Programming Languages 13/7
13.3 -- Word Processors 13/9
13.4 -- Spreadsheets 13/13
13.5 -- Databases 13/16
13.6 -- Integrated Programs 13/19
13.7 -- Miscellaneous 13/22

Part IV .... 1990's -- Current Technology **14/1**

Chapter 14 Hardware in the 1990's **14/3**
14.1 -- Microprocessors 14/3
14.2 -- IBM Computers 14/8
14.3 -- Apple Computers 14/10
14.4 -- Other Computers 14/11

Chapter 15 Software in the 1990's 15/1

15.1 -- Microsoft 15/1
15.2 -- Apple Computer and IBM 15/7
15.3 -- Other Software 15/9
15.4 -- The Road Ahead 15/15

Chapter 16 Corporate Activities in the 1990's 16/1

Part V .... Bits and Bytes 17/1

Chapter 17 Hardware and Peripherals 17/3

17.1 -- Memory 17/3
17.2 -- Storage Devices 17/4
17.3 -- Input/Output Devices 17/10
17.4 -- Displays 17/11
17.5 -- Printers 17/12
17.6 -- Peripheral Cards 17/16
17.7 -- Modems 17/19
17.8 -- Miscellaneous 17/20

Chapter 18 Magazines and Newsletters 18/1

18.1 -- The Beginning 18/1
18.2 -- Apple Publications 18/4
18.3 -- PC Publications 18/6
18.4 -- Other Publications 18/7
18.5 -- Reference 18/9

Chapter 19 Other Companies, Organizations and People 19/1

19.1 -- Early Organizations 19/1
19.2 -- Conventions, Fairs and Shows 19/4
19.3 -- Historical Organizations 19/6
19.4 -- Retailers and Software Distributors 19/10
19.5 -- Networks and Services 19/13
19.6 -- Associations 19/17
19.7 -- Other Companies and People 19/20

Chapter 20 Miscellaneous Items 20/1

20.1 -- Bits and Bytes 20/1
20.2 -- Reference Sources 20/2
20.3 -- Standards and Specifications 20/3
20.4 -- Terminology: Clarification and Origins 20/6

Preface

This book has been compiled to fill a gap in personal computer literature. There are many biographical books about key individuals such as Bill Gates of Microsoft or John Sculley of Apple. Other books are also available providing details of certain companies and their products. These books quite naturally focus primarily on products associated with that particular individual or company.

The intent of this book is to provide a consolidated coverage of the significant developments in the evolution of the personal computer and related products. The book has some emphasis on the technical and commercial aspects of the developments as compared to the social details of the participants.

Part I of the book provides a historical background on the beginning of digital computer technology. It is a cursory overview of early developments in both hardware and software from the late 1930's to the late 1950's. It also describes the start of personal computing in the 1960's. Starting with time-sharing, then simpler programming languages, the first personal computer and finally significant improvements to the user interface.

Part II is devoted to the beginning of the microcomputer: This is "The Altair/Apple Era". It covers the period of the 1970's when the original microprocessors gave birth to microcomputers such as the Altair in 1975 and to the Apple II in 1977. This is the exciting period during which the *Byte* magazine started, the Homebrew Computer Club was founded, VisiCalc was created and many other entrepreneurs helped to create the microcomputer industry.

Part III is "The IBM/Macintosh Era" and describes the corporate commercialism of the industry. It is the period of the 1980's which began with the introduction of the IBM personal computer in 1981, followed by the release of the Apple Macintosh computer in 1984. This was another exciting period as the industry evolved from small entrepreneurial companies into participation by large corporations. The basis of the personal computer market had changed from the "hacker" of hardware and software, to the utilization by business and the non-technical home user.

Part IV is a brief overview of the hardware, software and corporate activities in the 1990's.

Part V of the book is called "Bit's and Bytes" and provides details of the peripherals, magazines, people, companies and other organizations associated with the personal computer. One chapter also discusses such items as reference sources, standards and terminology origins.

An extensive bibliography and two appendixes have been provided. The bibliography has a section on books and another section on periodical articles that describe initial product releases and other items of significance. These two sources provide extensive reference material for those interested in further study of personal computer history.

There are limitations on the amount and diversity of historical information that can be included in a book of this size. The amount of detail on a particular subject has therefore been limited to items of historical and commercial significance. As regards hardware, this has intentionally resulted in more detail on the significant early developments from Apple and IBM. In software there is greater coverage of details on the Microsoft Corporation and its significant products. Also the focus has been on North America, where most of the development in microprocessor and microcomputer technology has occurred. The references cited in the bibliography will extend each subject area as required.

A few comments on qualification of dates and dollar figures for prices are appropriate. Dates are sometimes termed announced, introduced, launched, released and shipped. The dollar figures will vary depending on the manufacturer's list price, the price in advertising, the street price and the date of publication. There also tends to be some inconsistency in both dates and prices depending on the source. These variations in dates and prices tend to create some ambiguity. It is hoped that the reader will understand this and take the dates and prices in a relative sense within a historical context.

I have been involved with computers for close to thirty years. However, My modest start was an assignment by General Motors to do the critical path planning on a vehicle prototype using an IBM 1130 computer. That humble beginning initiated my education and fascination with the technology. I do hope that you find this book as interesting and informative to read as it was to write.

Roy A. Allan,
London, Ontario, Canada
June, 2001.

Acknowledgements

My thanks to Geoffrey R. Pendrill for his partial review of the manuscript and his valued suggestions.

Photographs on the book cover are courtesy of:
Apple Computer, Inc., Compaq Computer Corporation, Intel Corporation and International Business Machines Corporation.

Finally my appreciation for access to the extensive library holdings at the University of Western Ontario.

Part I

The Historical Background.

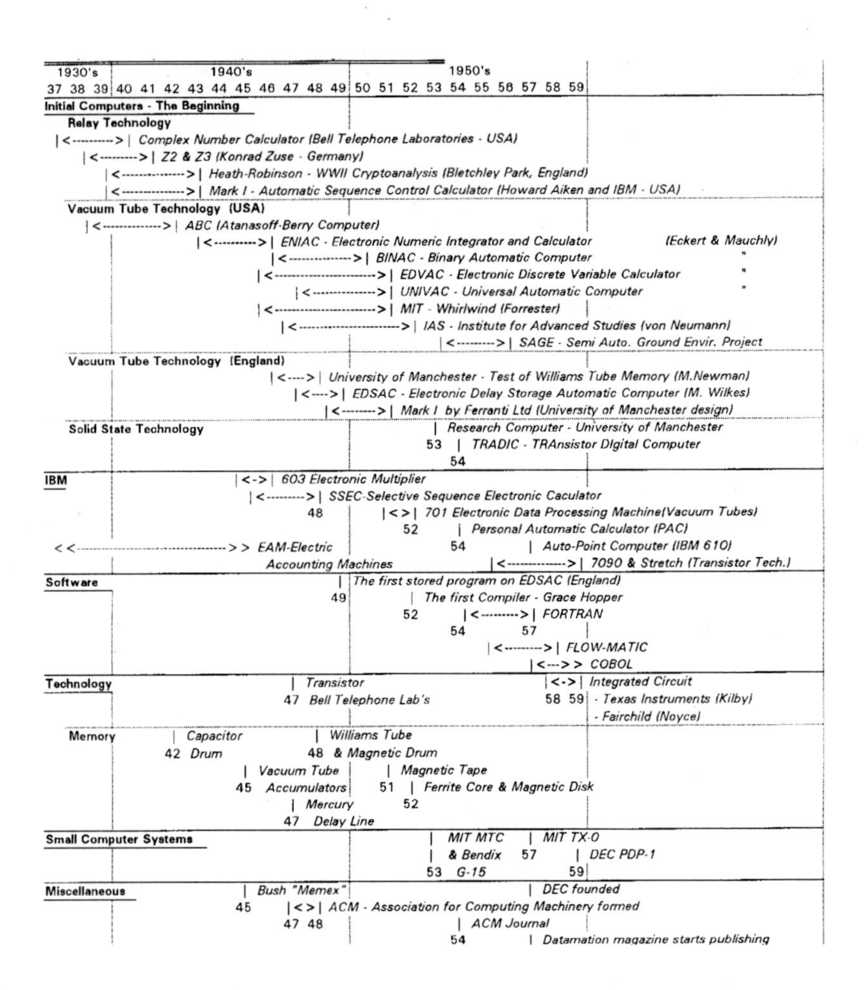

Figure 1.1: A graphical history of early computer technology (1937-1959).

Chapter 1 Development of the Computer

This chapter describes the beginning of digital computer technology. It is a cursory overview of the early significant developments. It starts with the beginning of relay technology in the late 1930s and concludes with the integrated circuit technology of 1959. The review of prior technology sets the stage for a discussion of personal computing and the microcomputer. It also forms the basis for a historic appreciation of the power and capabilities of today's personal computer, as compared to those early digital computers.

1.1 ... Original Digital Computers

Relay Technology

The original computers used mainly relay technology. Examples of these are those built by Howard Aiken in the USA, Bletchley Park in England, Konrad Zuse in Germany and those at Bell Telephone Laboratories in the USA.

Bell Telephone Laboratories developed a number of relay computers starting with the Complex Number Calculator. George Stibitz and S. B. Williams designed the calculator using 450 telephone relays and 10 crossbar switches. The machine became operational in January 1940 and Stibitz demonstrated it from a remote location, Dartmouth College in Hanover, New Hampshire in September 1940. Following this Bell Laboratories developed the Models III, IV, V and VI computers. Bell completed the Model V, also known as the Bell Laboratories General Purpose Relay Calculator in 1946. It contained over 9,000 relays. An addition or subtraction required 0.3 seconds, a multiplication 0.8 seconds and a division 2.2 seconds.

Konrad Zuse designed a completely mechanical computer, the Z1 in 1938. Zuse then developed a hybrid relay/mechanical unit named the Z2. A successful demonstration of this machine resulted in financing from the German Aeronautical Research Institute for a more advance Z3 machine. Zuse completed the Z3 computer in December 1941 using 2,300 relays. After the Second World War Zuse formed his own company that produced relay, vacuum tube and transistor computers.

Bletchley Park in England built a number of electro-mechanical machines at the beginning of the second World War for cryptoanalysis security work. M. H. A. Newman and C. E. Wynn-Williams completed a hybrid relay-vacuum tube machine called the Heath-Robinson in April 1943. The British Secrets Act has restricted the availability of technical information for the computing machines developed at Bletchley Park.

Howard H. Aiken and IBM developed the Automatic Sequence Control Calculator (ASCC). IBM built the computer at its Endicott laboratory during the period of 1939-43. It became operational in January 1943 and IBM presented the computer to Harvard University in August 1944. It had 2,200 counter wheels, 3,300 relays, was 51 feet long by 8 feet high and had a weight of approximately 5 tons. Addition or subtraction took about 0.3 seconds, multiplication 6 seconds and division could require 16 seconds. Harvard renamed it the Mark I and during the next decade Aiken developed the Mark II, III and IV computers.

Mechanical relay technology had inherent speed limitations as applied to computation. This led to the use of vacuum tube technology that provided significant speed improvements.

Vacuum Tube Technology (USA)

John V. Atanasoff developed the first electronic computer to use vacuum tube technology at the Iowa State College during the period of 1938-42. Atanasoff confirmed the design concepts but did not use the computer in a practical sense. He developed the design to mechanize the calculation of large systems of linear algebraic equations. Atanasoff subsequently named the computer the ABC (Atanasoff-Berry Computer) in recognition of the contributions by an associate Clifford E. Berry.

The first electronic computer used for calculations and solving practical problems, was ENIAC (Electronic Numeric Integrator and Calculator). J. Presper Eckert and John W. Mauchly developed the computer at the Moore School of Electrical Engineering, University of Pennsylvania during the period of 1943-46. ENIAC was a huge machine, having over 18,000 vacuum tubes and 1,500 relays that consumed 174 kilowatts of power. An addition or subtraction calculation took 200 microseconds, a multiplication 2,800 microseconds and a division 24,000 microseconds. The clock rate was 60-125 kHz. The U.S. Army Ordnance Department funded the computer development for use in calculating ballistic tables.

Other early computers developed and built during the late 1940s and early 1950s using vacuum tube technol-ogy were EDVAC, BINAC, UNIVAC, Whirlwind and IAS. During the early development of the digital computer there was a close liaison between research institutions in the USA and the United Kingdom. This resulted in significant research and development of vacuum tube computers in England.

Vacuum Tube Technology (England)

Between 1946 and 1948 the University of Manchester built an electronic computer to test the concept of electrostatic memory storage. F. C. Williams developed this memory concept using the cathode ray tube.

The first general purpose electronic computer in the United Kingdom was the Electronic Delay Storage Automatic Calculator (EDSAC). Maurice V. Wilkes developed EDSAC at Cambridge University between 1947 and 1949. EDSAC was the first computer to use the stored program concept.

Ferranti Ltd., did additional developmental work on the University of Manchester computer for commercial production. They completed the first computer, the MARK I in 1951.

The invention of the transistor significantly affected computer development in England and the USA. Bell Telephone Laboratories developed the transistor in 1947/48. Utilization of solid state technology resulted in significant cost savings and reliability improvements.

Solid State Technology

The first transistor computer to operate was in England. T. Kilburn of the University of Manchester designed the experimental computer that was operational in November 1953. In the early 1950s, Bell Telephone Laboratories received an Air Force contract to build a special computer called TRADIC (**TRA**nsistor **DI**gital **C**omputer). Bell built TRADIC and had it operating in early 1954.

Most early computer development had occurred in the USA and England at university research institutions. However a major USA corporation providing tabulating equipment worldwide entered the computer field. That company was IBM and they became the dominant supplier of computers.

1.2 ... IBM -- International Business Machines

Thomas J. Watson Sr., left National Cash Register Company and became General Manager of the Computing-Tabulating-Recording Company in May 1914. The company name changed to International Business Machines Corporation (IBM) in 1924. Tabulating machines were the basis for the initial growth of the corporation. These evolved from simple mechanical card punching, counting and sorting machines to electric accounting machines. These machines met the diverse needs of business, industry and some fields of science into the 1940s.

The first entry of IBM into the field of computers was the collaborative effort of the company with Howard Aiken of Harvard University. IBM approved construction of the Automatic Sequence Control Calculator (ASCC) in 1939 and completed it as described in Section 1.1 in January 1943.

The second world war created a significant growth in electronic technology. Utilization of this new technology resulted in the development of the 603/4 Electronic Multiplier in September 1946. IBM initiated construction of the "Super Calculator," the SSEC (Selective Sequence Electronic Calculator) in 1945 and dedicated it in January 1948. It had 21,400 relays, 12,500 vacuum tubes and operated until January 1952.

With the start of the Korean war in June 1950, IBM initiated steps to assist in the war effort with the development of the Defense Calculator. This new computer started in 1951, utilized a number of concepts new to IBM's existing products. Some of these were binary notation, magnetic drum storage, electrostatic cathode ray tube storage, magnetic tape storage and utilization of germanium diodes. The system became known as the IBM Electronic Data Processing Machine and the individual units had 700 series numbers assigned. The first production 701 system shipped in December 1952. Another important project was the joint development of the SAGE (Semi-Automatic Ground Environment) computer with MIT between 1952 and 1956. The company announced their last vacuum tube computer system, the 705 Model III, in September 1957. In October 1957 the company issued a memo stating a policy to use solid-state technology on all new computer developments.

IBM demonstrated a transistorized version of the Type 604 calculator in October 1954 and the first transistorized product was the 608 calculator that shipped in December 1957. These two machines did not have a stored program. Consequently IBM called them calculators, not computers. IBM initiated a project in late 1955 for a supercomputer utilizing transistors and the latest technology that became known as the Stretch system. Then in October 1958 they announced the 7090 Data Processing System using the Stretch technology. The first delivery of a 7090 system was in November 1959.

Computer development at IBM and other institutions was very dependent on technology advances. Of particular importance were the advances in memory capabilities and solid state technology.

1.3 ... Technology

Memory

The following is relative to the computer internal memory and not auxiliary storage units such as magnetic drums, disks or tape drives. Memory was a crucial technology in the expanding use of the computer.

The first electronic digital computer developed by Atanasoff during the period of 1938-42 used capacitors on a rotating drum for memory. The computer had two drums and each drum had 32 bands of 50 capacitors around the circumference. This is a memory capacity of 1,600 bits, or in more familiar terms 200 bytes.

The ENIAC computer used vacuum tubes for memory storage. It used the decimal system and had 20 accumulators for storage of variables. Each accumulator could hold a 10 place number. Each place number was a 10 stage ring counter corresponding to the digits 1 to 9. Therefore it required 100 unique flip-flop vacuum tubes (plus 2 for sign) in each accumulator to store a 10 place number. By taking certain liberties we could say the system had 100 unique bits in each of the 20 accumulators, for a total of 2,000 bits or 250 bytes in our more familiar terms.

The acoustic delay line concept increased the capacity and significantly reduced the number of vacuum tubes used for memory. A typical delay line used piezoelectric quartz crystals at each end of a mercury filled tube and could store 1,000 bits. Eckert and Mauchly applied the concept on the EDVAC computer during the period of 1945-51. Other early computers such as EDSAC, BINAC for Northrop Aircraft, Inc., and the initial UNIVAC computers also used the acoustic delay line concept

The Princeton University, Institute for Advance Studies (IAS) considered using electrostatic memory utilizing the cathode ray tube for the IAS computer in 1945. However F. C. Williams developed the first functional implementation of electrostatic memory on the University of Manchester test computer between 1946-48. This was the first high speed random access memory. By the middle of 1948 Williams was able to demonstrate a unit with a capacity of several thousand bits. Subsequently it was operational on the IAS computer and the IBM 701 System in 1952.

Jay W. Forrestor evaluated magnetic core memory at the Massachusetts Institute of Technology (MIT) between 1949 and 1951. Ferrite cores were operating at MIT and on the IBM 405 Accounting Machine in 1952. The first computer application of ferrite cores was on the Whirlwind computer at MIT in 1953. The preceding computer evolved into the SAGE (Semi Automatic Ground Environment) computers for a USA national air defense system. MIT selected IBM to build the computers. This defense program added significant impetus to the development of ferrite core technology.

Each of these stages of memory development resulted in lower costs and greater reliability. This also enabled the use of larger and more complex software. Associated developments in secondary memory storage complemented the advances in internal memory.

Secondary Memory (Storage)

The first electronic computers used paper cards for storage. Atanasoff's computer used 8.5-by-11-inch cards and ENIAC used IBM cards.

A. D. Booth of the United Kingdom was an early developer of various forms of memory storage. Booth experimented with thermal, rotating disk-pin and magnetic drum memories after the Second World War. By May of 1948 he had installed and demonstrated a working magnetic drum memory in the Automatic Relay Computer (ARC). In the USA a company called Engineering Research Associates Inc., (ERA) presented papers describing magnetic tape and magnetic drum storage research in 1947 and 1948.

The concept of magnetic disk storage was conceived by Jacob Rabinow at the National Bureau of Standards (NBS) in 1952. In early 1952, IBM established an advanced research development laboratory in San Jose, California that began looking at magnetic disk storage as an inexpensive fast data retrieval system. The requirements evolved from advance development of card-related applications and a United States Air Force request for a large random access storage device. Also customers for inventory and accounting applications were requesting a change from batch to random access in the method of updating file systems. Extensive research was conducted on disk materials and coatings, various types of movable heads and electronic control systems for data storage.

IBM demonstrated the first magnetic disk drive assembly with movable read-write-heads in May 1955 and described the concept at the Western Joint Computer Conference in February 1956. The assembly consisted of a stack of fifty coated aluminum disks, 24 inches in diameter, rotating at 1,200 revolutions per minute. Each disk surface contained 100 concentric recording tracks that provided a total storage capacity of 5 million characters (the equivalent of 50,000 IBM cards). The system was subsequently announced as the IBM 350 Disk Storage system. The preceding and a scheme for automatic addressing of data, was incorporated into the IBM 305 RAMAC (Random Access Method of Accounting and Control) system announced in September 1956. IBM announced a 14 inch diameter hard disk drive with a removable disk-pack in October 1962.

Improvements in memory technology were important in advancing computer capabilities. However solid state research that created the transistor and integrated circuit would also provide significant improvements in computer technology.

Transistor and Integrated Circuit

William B. Shockley was the director of a team of researchers that developed the transistor at Bell Telephone Laboratories in New Jersey, USA. The first transistor was operational in December 1947. John Bardeen and Walter Brattain received a patent for the transistor in 1948.

The concept of including a number of components on a single semiconductor chip was first developed by Jack St. Clair Kilby at Texas Instruments, Inc., in 1958. Kilby constructed the integrated circuit in germanium. Each component on the integrated circuit required precise and laborious interconnection by hand.

Independently in 1959, Robert N. Noyce at Fairchild Semiconductor Corporation also developed a monolithic concept of integrating components on a single silicon chip. However, Noyce had also developed a practical method to interconnect the co mponents. This new technology deposited an insulating layer on the silicon semiconductor that could be selectively removed by a photo resist technique. Then by depositing a vaporized metal layer, interconnection of the components was achieved in a practical manner. Noyce's concept had been facilitated by the invention of the planar process by Jean Hoerni, a technique to make a flat, or planar surface for transistors. The founding of Fairchild Semiconductor in 1957, had been financed by Fairchild Camera and Instrument Corporation, who completed its ownership of the new company in 1959.

For commercial viability, the practical method of interconnecting the components and the planar process were just as important as the integrated circuit concept. This would be crucial in the ensuing patent litigation between the two inventors and respective companies. In resolving the coverage of the patent applications between the two companies, an agreement resulted in Jack Kilby and Robert Noyce being declared as co-inventors of the integrated circuit.

Technology improvements and memory advances in particular resulted in increased utilization of the computer. However, it is software that facilitated the practical application of the computer.

1.4 ... Software

Development of the stored program concept had a direct relationship to advances in memory technology. The size, speed of access, reliability and cost of memory were crucial factors in the evolution of software.

The Beginning

The initial computers did not have a stored program, or what became known later as software. The first practical electronic computer ENIAC, used forty plug-boards, that required a configuration of wire connections for each instruction. Later changes utilized three panels, each containing 1,200 ten-position switches to control the instruction sequence.

John von Neumann described the concept of storing a program and data in 1945 during the construction of ENIAC. However, the first computer to function with a stored program was EDSAC at Cambridge University, England in 1949.

Grace M. Hopper was an early pioneer in the development of initial programming languages. Her initial work started on the Mark 1 at Harvard University in 1944. Then at the Eckert-Mauchly Computer Corporation. The first compiler was A-0 and ran on the UNIVAC computer in 1952. This compiler formed the basis for other variations such as ARITH-MATIC, MATH-MATIC and FLOW-MATIC (1955-58) which assisted in the later development of COBOL.

Graphics

One of the earliest applications of computer graphics was on the MIT Whirlwind computer, that was operational by 1953. This was an interactive system that was used to display radar information. It was then applied to the SAGE defense system built by IBM.

FORTRAN

In 1954 IBM established a project directed by John W. Backus to develop a compiler for the Model 704 computer. This project resulted in the creation of the language FORTRAN (FORmula TRANslation) that IBM finished in April 1957. The language had a notation oriented to mathematicians and scientists. Improvements in the language followed with the subsequent release of versions II, III and IV.

COBOL

By 1957 a number of people had become concerned that a common programming language designed for commercial users was not available. The United States government also stated a concern about the proliferation of different compilers from the various computer manufacturers. In 1959 the Defense department initiated activities that resulted in an Executive Committee being formed called CODASYL (Committee On DAta SYstem Languages). The committee coordinated the development of a new language that became COBOL (COmmon Business Oriented Language). CODASYL formed other committees, to work on the definition and further development of the language. These developments resulted in the 1960 release of COBOL.

Games

Artificial intelligence researchers started using computers and developed game software such as chess and checkers between the late 1940s and the 1950s. There were many researchers in this field of artificial intelligence. Early chess playing programs were developed by Claude E. Shannon in 1949 and Alex Bernstein in 1957. Arthur L. Samuel of IBM developed various games, especially checkers in the 1950s.

William Higinbotham and associate Dave Potter developed the first video game at Brookhaven National Laboratory in Upton, New York in 1958. Called Merlin, it simulated a game of tennis using an analog computer, oscilloscope display and paddle-type controllers for rackets.

Software was facilitating and extending the utilization of computers. As the number of users increased, new periodicals were published and new associations formed to disseminate information and facilitate user interaction.

1.5 ... Other Developments

Associations

One of the first groups to be formed in the computer industry was the Association for Computing Machinery (ACM), founded during 1947/48. The Data Processing Management Association (DPMA) formed in 1949, is one of the largest groups with involvement in education and certification of data processing professionals. IBM customers were one of the first to form user groups. The SHARE group that formed in 1955 had an initial orientation to scientific users. An IBM commercial users group called GUIDE (Guidance of Users of Integrated Data-processing Equipment) formed in 1956. Then the International Federation for Information Processing (IFIP) formed in 1959.

Magazines

One of the earliest periodicals was the *Digital Computer Newsletter* started by the Office of Naval Research in 1949. Another early publication was the *IBM Technical Newsletter* series started in 1950. The ACM issued a quarterly *Journal* in 1954 that evolved into the *Communications of the ACM* in 1958.

A popular data processing magazine called *Datamation* started in October 1957 as *Research and Engineering (The Magazine of Datamation)*. It started as a semi-monthly publication, then became a monthly publication in 1967.

Memex

Vannevar Bush published a futuristic article entitled "As We May Think" in the July 1945 issue of *Atlantic Monthly* [12, pp. 47-59]. The article described a "future device for individual use" called "Memex" in which a person "stores all his books, records, and communications ... that may be consulted with exceeding speed and flexibility." Another essential feature of Memex was "associative indexing ... whereby any item may be caused at will to select immediately and automatically another." Bush's Memex was similar to future concepts of an interactive personal computer navigating a field of knowledge with hypertext links.

Most of the activities in computer development and use related to large mainframe computers. However various organizations were attempting to lower the cost of computers by introducing small computer systems.

1.6 ... Small Computer Systems

MIT

MIT built the **M**emory **T**est **C**omputer (MTC) between 1952 and 1953 to test magnetic core memory planes for the Whirlwind 1 computer. Harlan Anderson and Kenneth Olsen designed MTC and would later co-found Digital Equipment Corporation (DEC). The computer was a 16-bit unit built from standard Whirlwind plug-in circuit package forms.

Following the completion of the MTC computer Olsen and Wesley A. Clark proposed the construction of a large transistorized computer, the TX-1. Rejected by management, Clark designed a smaller 18-bit unit, the TX-0 in 1957. Then MIT developed the ARC (Average Response Computer) and a much larger computer, the TX-2 in 1958. One of the students using these computers was Charles E. Molnar. Molnar and Clark would later develop the first personal computer, the MIT LINC.

IBM

John L. Lentz worked on the development of a small Personal Automatic Calculator (PAC) project, starting in the late 1940s. He then described details of an engineering model of the PAC project in December 1954. This project evolved into the IBM 610 Auto-Point Computer announced in September 1957. The system consisted of three units. A floor-standing cabinet that enclosed the electronics, two paper-tape readers and punches, plugboard and magnetic drum. The other two units were an operator keyboard for control and data entry and an electric typewriter for printed output. It was one of the last two vacuum tube computer models built by IBM and the company considers this to be the first "Personal Computer." IBM built about 180 units at a purchase price of $55,000.

DEC

One of the earliest entrepreneurs in the minicomputer market was Kenneth H. Olsen. After graduating from MIT he worked at the Lincoln Laboratories, then on the Whirlwind and SAGE computer systems between 1950 and 1957. Olsen co-founded Digital Equipment Corporation (DEC) with Harlan E. Anderson in August 1957, to provide low cost logic modules and computers for engineers and scientists.

The first DEC minicomputer was the 18-bit PDP-1 (Programmed Data Processor - One). DEC demonstrated the PDP-1 prototype at the Eastern Joint Computer Conference in December 1959. It had a cathode-ray tube display, keyboard and was the first small commercial interactive general-purpose computer.

Conclusion

The general use of large commercial and scientific computers was firmly established by the end of the 1950s. However the major computer manufacturers did not commit sufficient resources to the development of small competitive computer systems. This resulted in other companies releasing small computers such as the Bendix Aviation Corporation G-15, the Librascope/General Precision LGP-30 in 1956 and the DEC PDP-1 announcement in 1959.

These small systems were lowering the cost of computer technology. Although they targeted at scientific users, they were a part of the evolution to personal computing and the first personal computer in the 1960s.

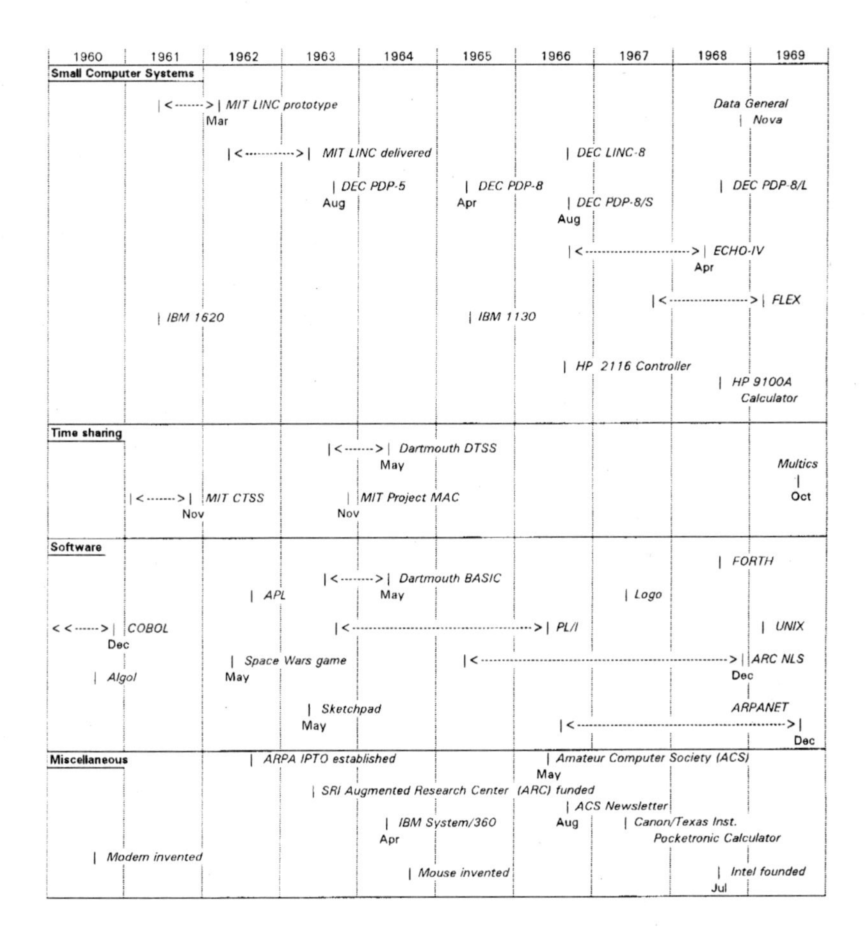

FIGURE 2.1:
A graphical history of personal computing in the 1960's.

Figure 2.2: Wesley A. Clark and the MIT LINC computer. A number of people consider LINC as being the first personal computer.

Photograph is reprinted with permission of MIT Lincoln Laboratory, Lexington, Massachusetts.

Chapter 2 Personal Computing in the 1960's

Most mainframe computers used a sequential batch process type of operation for computing tasks at the end of the 1950's. This mode of operation resulted in a slow and cumbersome interaction with the user.

The 1960's was a decade that saw many improvements to personalize the computer user interface. The government funded research in organizations such as the Massachusetts Institute of Technology (MIT) and the Information Processing Techniques Office (IPTO) by the Advanced Research Projects Agency (ARPA) of the Department of Defense. This resulted in innovations of significant importance for personal computing. Educational institutions such as MIT, Dartmouth College and Stanford Research Institute (SRI) created items such as the first personal computer, the "mouse," computer networks, BASIC programming language and time sharing.

2.1 ... Time sharing

A time sharing computer system is one that interacts with many simultaneous users through a number of remote consoles. An interleaving in time of two or more jobs on one processor gives what appears to each user, as the dedicated use of the computer. The first paper published describing time sharing was by Christopher Strachey at the Paris International Conference on Information Processing in June 1959.

MIT developed and tested the concepts of time sharing between December 1958 and early 1959 on an IBM 704 computer. Principals in this early development were John S. McCarthy, an early advocate of time sharing and Herbert M. Teager.

A Study Group investigated MIT's computational requirements for research and teaching in 1960 and made recommendations in early 1961. This resulted in the development of a time sharing system for an IBM 709 computer by Fernando J. Corbató and his staff that was first demonstrated in November 1961. This group developed the system further on an IBM 7090 computer and it became known as the Compatible Time Sharing System (CTSS). MIT presented a paper describing CTSS at the San Francisco Spring Joint Computer Conference in May 1962.

Jack B. Dennis wrote a proposal for a time sharing system on the MIT TX-0 computer in 1959. Then in 1961, MIT received a Digital Equipment Corporation (DEC) PDP-1 minicomputer. Under the direction of Dennis, a graduate student John E. Yates developed a time sharing system for the DEC PDP-1 that became operational in the spring of 1963.

Bolt Beranek and Newman Inc., (BB&N) in Cambridge, developed another time sharing system for the DEC PDP-1 in 1962. Two of the principals in the development were J.C.R. Licklider and John McCarthy. BB&N presented a paper entitled "A Time-Sharing Debugging System for a Small Computer" describing these developments at the Spring Joint Computer Conference of 1963.

The Advanced Research Products Agency (ARPA) appointed J.C.R. Licklider in 1962 to be in charge of the new Information Processing Techniques Office (IPTO). Convinced that time sharing would be an important computer technology, Licklider selected Robert M. Fano at MIT to head the development of a major new system. This new time sharing system became known as Project MAC (Multiple-Access Computer or Machine-Aided Cognition) and an early version was operational by November 1963. Initially it could serve 24 users simultaneously. In less than a year it was serving 200 users with 100 teletypewriter terminals. Improvements made to the system throughout the 1960's resulted in it becoming an important node in the ARPANET.

In late 1963 the Project MAC Group began a search for a more suitable time sharing computer. This resulted in the selection of the General Electric GE-645 computer in 1964. Shortly after, General Electric and Bell Telephone Laboratories joined MIT in the development of a new comprehensive time sharing system. This new system became known as Multics (Multiplexed Information and

Computing Service). It became operational at MIT in October 1969, and within two years was serving 500 users.

These early developments of time sharing technology were important in providing personal computing capabilities in a more friendly interactive mode at lower cost to an increasing number of users. Other institutions such as Dartmouth College started to evaluate these new concepts.

2.2 ... Dartmouth DTSS and BASIC

Between 1956 and 1962, a small university called Dartmouth College in Hanover, New Hampshire, started developing simple high-level programming languages on a small LGP-30 computer. Principals in these developments were John G. Kemeny and Thomas E. Kurtz. Kemeny and Kurtz had a conviction that, "...knowledge about computers and computing must become an essential part of a liberal education." To implement this objective and achieve acceptance by the students they had to simplify both the computer interface and the programming language used by the students.

Existing languages such as FORTRAN and ALGOL were too complex for the majority of students. The early languages developed at Dartmouth were: DARSIMCO (**DAR**tmouth **SIM**lified **CO**de), DART, ALGOL 30, SCALP (**S**elf **C**ontained **AL**gol **P**rocessor) and DOPE (**D**artmouth **O**versimplified **P**rogramming **E**xperiment). During this period Kurtz became aware of time sharing technology at MIT and BB&N. To meet their computer educational objectives they decided to develop and implement a new time sharing system and a new simple programming language for interactive computing.

In 1963, Dartmouth College selected a new hardware system consisting of a General Electric (GE) GE-225 computer for user programs, a GE Datanet-30 computer for communications and scheduling, a disk drive and other peripherals. Kurtz supervised undergraduate students in the development of the time sharing software, emphasizing simplicity of use for the novice. This software became known as the Dartmouth Time Sharing System (DTSS). At the same time Kemeny developed a compiler for the simple high-level programming language they had designed, named Beginner's All-purpose Symbolic Instruction Code (BASIC). Features incorporated in the BASIC programming language were influenced by the knowledge and experience of using ALGOL, FORTRAN and the early languages developed at Dartmouth. The College received the new computer system equipment in February 1964 (the GE-225 computer was changed to a GE-235 in the summer). The new time sharing system, BASIC compiler and the first test programs operated successfully on May 1, 1964. Subsequently, the College put the programming language in the public domain in order to improve its widespread acceptance.

Between 1964 and 1965, Dartmouth College began an association with GE that resulted in a joint effort to develop a time sharing system and a BASIC compiler for a new GE-635 computer. The College developed the BASIC compiler that became known as GE-BASIC. GE became a significant contributor in the dissemination of the BASIC programming language on larger computers. Other companies such as Digital Equipment Corporation (DEC) and Hewlett-Packard (HP) developed interpreter implementations of BASIC for smaller computers in the late 1960's.

The Dartmouth College development of a simple time sharing system and the BASIC programming language made a significant contribution to personal computing in the mid 1960's and following years. Continued improvements to the language were made with the release of versions two to five between 1965 and 1968. All implementations of BASIC at Dartmouth College have been compilers, whereas most of the later implementations on small personal computers were interpreters. Time sharing became widespread and BASIC became a popular programming language for personal computing.

However for intensive computational applications, time sharing could not provide the processor resources required. Also the convenience and flexibility of having one's own dedicated processor led to the development of the personal computer.

2.3 ... The First Personal Computer

To avoid confusion, one has to define the term "personal computer" as being a computer designed for use by one person. In the 1960's most computers were large mainframes, shared by many users.

A number of people consider the MIT LINC to be the first personal computer. The Massachusetts Institute of Technology (MIT) developed LINC to facilitate the use of computer technology in biomedical research laboratories. LINC is an acronym for **L**aboratory **IN**strument **C**omputer. Principal designers were Wesley A. Clark and Charles E. Molnar. MIT demonstrated a prototype in March 1962 at the Lincoln Laboratory and completed sixteen units in mid 1963. The scientific users assembled the units to improve their understanding of the system that cost about $32,000. Initial software was a text editor, an assembler and some utilities.

The LINC system had four console modules, an electronics cabinet and a keyboard. The electronics cabinet was about the size of a refrigerator. The processor logic circuits used transistorized system modules from Digital Equipment Corporation (DEC). Memory was a magnetic core type with a basic capacity of 1,024 twelve-bit words, expandable to 2,048 words.

The four console modules consisted of a control console, an oscilloscope module, a tape module with two magnetic-tape drives and a terminal module. The oscilloscope module could display a 512 by 512 point image.

A small number of scientific laboratories used the LINC computer in dedicated applications. In 1966, DEC released a refined version of the LINC computer that they named LINC-8 and sold for $43,000. During this period of time other companies were extending the low end of the market by developing small computer systems.

2.4 ... Small Computer Systems

Other organizations developed small computer systems or minicomputers during the 1960's for scientific and commercial users (See Chapter 1.6 for earlier small computer systems). This continued to lower the cost of computers and extend the concept of personal computing.

DEC delivered the first production version of the 18-bit PDP-1 minicomputer in early 1961. A minimum system cost $85,0000. This was followed by the PDP-4, then the 12-bit PDP-5 that cost $27,000 in August 1963. The 12-bit PDP-8, which became very successful, was announced in late 1964 and the first units delivered in April 1965. DEC marketed the PDP-8 as "the world's lowest-priced, fully programmable computer system." The computer used new technology such as small scale hybrid integrated circuits to reduce the price of the unit to $18,000. The PDP-8 had 4K words of core memory, expandable to 32K words. Then in August 1966, DEC introduced the low cost PDP-8/S system with a console teletype for $10,000. In 1967, DEC developed the PDP-8/I computer that incorporated TTL technology in place of the hybrid integrated circuits. This was followed in the summer of 1968 by a similar machine with fewer options called the PDP-8/L. The PDP-8/L with 4K of memory and a teletype sold for $8,500.

IBM introduced small systems for technical and professional users starting with the Model 1620 in 1961. This was a batch-oriented system for FORTRAN users. Then IBM released the Model 1130 computing system in 1965. The 1130 was a single-user system with an integrated disk-based operating system.

William Hewlett and David Packard founded Hewlett-Packard (HP) as a partnership in January 1939 and incorporated it as a company in 1947. It was initially a manufacturer of electronic test equipment. HP began evaluating computers to automate their instrument measurement systems in September 1964. This resulted in the development of the Model 2116 instrument controller in 1966. This was the company's first computer and HP sold many units as stand-alone minicomputers. The HP 2114A minicomputer was released shortly after at a price of $9,950. HP also started development of the 9100 series of electronic desktop calculators in 1966. The company released the HP 9100A in 1968, that has been described as a "computing calculator." It was a predecessor to the programmable calculators released in the early 1970's.

Other companies competed with DEC, HP and IBM for a share of the small system market. Some of these companies were: Computer Control Corporation (3C), Control Data Corporation (CDC), Data Machines, Honeywell, Scientific Control Systems (SCS), Scientific Data Systems (SDS) and Systems Engineering Laboratories. A group of people from DEC established Data General in April 1968 and announced their first 16-bit minicomputer called the Nova later that year.

Time sharing, simple high-level programming languages and small low cost computer systems were significantly increasing the number of users. However the interaction between the user and the computer was in most cases an awkward process through a teleprinter type of console. The 1960's was a decade of significant early improvements to the user interface

2.5 ... Graphics and the User Interface

Douglas C. Engelbart developed concepts for augmenting the human intellect at the Stanford Research Institute (SRI) in Menlo Park, California between 1957 and 1960. In March 1960, J.C.R. Licklider wrote an important paper entitled "Man-Computer Symbiosis" [12, pp. 306-318]. The paper contained ideas similar to Vannevar Bush's Memex concept, but extended them through the use of interactive computer technology. In late 1962, Engelbart presented his SRI report entitled "Augmenting Human Intellect: A Conceptual Framework" and a proposal to the newly formed Information Processing Techniques Office (IPTO). Licklider, who was the first IPTO director, approved the proposal that provided early government funding for an Augmented Research Center (ARC) at SRI in 1963.

Various graduate students at MIT and staff at General Motors Corporation did research on computer assisted drafting in the early 1960's. However a turning point occurred when Ivan E. Sutherland conducted significant research on computer graphics for a doctoral thesis in 1962 at MIT. Sutherland created an interactive graphics system, that enabled a user to create graphical figures on a video display using a light pen. The geometrical shapes could be copied, expanded, moved, rotated and shrunk. This also resulted in the development of the first user interface that incorporated a split screen (two-tiled windows), menus and the use of icons for such things as constraints to limit line lengths. Sutherland published the results of this research in an article entitled "Sketchpad: A Man-Machine Graphical Communications System" in May 1963. This became the basis for computer assisted drafting (CAD) and computer assisted engineering (CAE) software systems. In the late 1960's various companies such as Applicon, Calma and Computervision offered turnkey CAD systems.

The ARC research on facilitating the use of computers to extend human knowledge and intellect, resulted in the development of significant improvements to the user interface. Engelbart concentrated his research on an interactive graphics environment as compared to the then prevalent teletype communication interface. Engelbart evaluated various methods of interacting with the screen display. During this research he developed the "mouse" in 1964, for which he holds the patent. The interactive graphics system also used a small five-key keyset that supplemented the selection capabilities of the mouse. Another principal in the research was William K. English.

Engelbart's group also developed an on-line system with new capabilities called NLS. NLS was an acronym for "on-line system." He demonstrated the new system at the ACM/IEEE-CS Fall Joint Computer Conference in San Francisco, California in December 1968. The NLS system incorporated the capabilities for mixing text and graphics on the screen display. It also used a split screen (two-tiled windows) that led to the development of multiple tiled windows in 1969. The group also developed an electronic-mail (e-mail) system and incorporated it into NLS.

Between 1967 and 1969, Alan C. Kay and Edward Cheadle built a computer called FLEX at the University of Utah. Kay and Cheadle also developed a user interface that included multiple tiled windows and square icons representing data and programs.

The graphics and user interface research at ARC, MIT and the University of Utah led to significant developments later at Xerox PARC, Apple Computer and Microsoft. SRI transferred activities of the ARC research group to Tymshare, Inc., of Cupertino, California in 1977. Tymshare renamed the NLS system AUGMENT and provided marketing for the product. McDonnell Douglas Corporation acquired Tymshare in 1984.

2.6 ... Software

COBOL

The CODASYL group had started development of COBOL in 1959. In 1960 the United States government advised that they would not accept computer equipment without a COBOL compiler. The development groups defined the language and compilers were operating by December of 1960.

PL/I

IBM decided to create an advance common programming language that would meet the requirements of both the scientific, commercial and system users in 1963. The SHARE user's group assisted in the language development. IBM initially intended to release the language with IBM's new line of System/360 computers. The first description of the language was in March 1964.

The initial name assigned to the new language was NPL - New Programming Language. However in 1965 the name became PL/I - Programming Language /I. Some ambiguity exists in the suffix (I). The pronunciation is

"Programming Language One," however the suffix is the Roman character I and not the Arabic numeral 1. IBM released the PL/I compiler in 1966.

The American National Standards Institute (ANSI) subsequently defined a version of the language with a reduction in the number of features. ANSI named this version, the "G" Subset of PL/I.

Logo

Bolt Beranek and Newman Inc., (BB&N) in Cambridge, Massachusetts started developing Logo in 1966. They then pilot-tested the language in the summer of 1967. Principals in the development were Seymour Papert, Daniel Bobrow, Richard Grant and Wallace Feurzeig who gave Logo its name. Charles R. Morgan and Michael Levin developed an extended version of Logo during 1967-68. The language is designed for use in education by children.

FORTH

Charles H. Moore developed the FORTH programming language around 1968. The language was designed by Moore to improve programming productivity. Additional features were that it was easy to move to a different machine and it required a small amount of memory.

UNIX

UNIX is an operating system developed by Kenneth L. Thompson and Dennis M. Ritchie at AT&T's Bell Laboratories. The main feature of the operating system was its portability that enabled it to run on almost any computer. Bell Laboratories released UNIX in 1969 and provided a free license to educational users. This resulted in its widespread use at academic institutions.

ARPANET

In the 1960's most computers were large mainframes with restricted user access. Some had time sharing and remote terminals. However remote communication by a phone line and data transfer between different computers was difficult.

Paul Baron conducted distributed communications research at the RAND Corporation as early as 1962, and published details in 1964. Then between 1965 and 1966, Donald Davies from the National Physical Laboratory (NPL) in the United Kingdom wrote papers describing concepts of digital communications using short messages or "packets." J.C.R. Licklider, who was the first director of the Information Processing Techniques Office (IPTO) of the Advanced Research Projects Agency (ARPA), had also been promoting the concept of a "Intergalactic Computer Network." A concept that tried to define the benefits and problems of computer networking.

Around this time period the IPTO was funding computer research projects and wanted to improve data communications for time sharing and networking. Bob Taylor who was the current director of the IPTO and a proponent of computer communications, recruited Lawrence G. Roberts to lead this networking project.

Roberts began the experimental computer network research in 1966. He then received an appointment to manage the IPTO programs for the ARPA in 1967. This led to an initial plan for an ARPANET being published in October of that year. This network plan would enable load sharing, message service, data sharing to link university computers and researchers. It would also use the "packet" concept, interface message processors (IMP's) and leased telecommunication lines. ARPA awarded a contract to develop the network to Bolt Beranek and Newman (BB&N) in January 1969. Robert Kahn was a principal in the overall system design. By December of that year they had four nodes of the network installed and operating. This packet switching data communication system became highly successful in connecting many major universities, government organizations and research institutions. The ARPANET formed the basis for what subsequently became known as the Internet.

Games

Student hackers and academic staff developed some of the earliest games at the Massachusetts Institute of Technology (MIT) on the TX-0 and PDP-1 computers. One of the games was for a mouse, that would poke its way through a maze constructed by a light pen to find a blip in the shape of cheese.

Then in 1961 Stephen Russell who was a science fiction enthusiast and a student hacker created an interactive game called "Space Wars" on the PDP-1 computer. The game displayed rocket ships that could fire missiles in a celestial battlefield. MIT displayed the game at the annual Science Open House in May 1962. The software was free and received wide distribution. J. M. Graetz, Peter Samson and others contributed various enhancements to Space Wars that provided additional challenges to participants.

Ralph Baer of Sanders Associates in Nashua, New Hampshire received a patent for a ball-and-paddle video game using a TV set. Development started in September 1966. Baer had basic ball-and-paddle games working in early 1967 and a hockey game by September. Magnavox marketed the game system as Odyssey 100, the world's first home video game.

2.7 ... Hobby & Amateur Computing

This aspect of personal computing started from an interest by many enthusiasts in the building of their own computer. These people were both amateurs and professionals with a strong technical interest in hardware and software. Prior to the development of large scale integrated (LSI) memory chips and the microprocessor, it was not easy to build a computer. It required a knowledge of vacuum tube or transistor circuitry, digital logic, core memory, peripherals and other areas. It could also be a costly investment to create a complete system.

In May 1966 Stephen B. Gray founded the Amateur Computer Society (ACS). In August of that year he published the first *ACS Newsletter* devoted to hobby computing. The society and newsletter were a significant source of information for building a computer in the late 1960's and early 1970's.

The November 1967 issue of the *ACS Newsletter* included a survey requesting details of each member's computer. The January 1968 issue reported the following results. Clock speeds ranged from 500 kHz to 1 MHz, with the average 500 kHz. Instruction sets were small ranging from 11 to 34 instructions. The number of registers ranged from 2 to 11, with three being the most common. Word sizes were from 4 to 32 bits, with 12-bits being the average. Memory size ranged from 4 to 8K, all magnetic core. Most computers used discrete transistors and a few reported the use of integrated circuits. A Teletype terminal was the most common input/output device. Cost ranged from zero to $1,500, with an average of $650.

The April 1968 issue of *Popular Mechanics* included an article entitled "A Computer in the *Basement*?" The article described the ECHO IV, one of the few home-built computers actually completed. ECHO was an acronym for **E**lectronic **C**omputing **H**ome **O**perator. It had a designation of IV because it used surplus boards from a Westinghouse PRODAC IV computer. James F. Sutherland designed and built the computer between 1966 and 1968. ECHO IV was seven feet long, six feet high and 18 inches deep. A console desk included an electric typewriter keyboard, surplus teletype printer, 8-channel paper tape punch and an 8-channel paper tape reader. The unit had 120 circuit boards using 2N404 transistors and NOR logic elements. It had four registers, used 18 instructions with a clock speed of 160 kHz. Memory was an 8K surplus core unit. The computer is now located at The Computer Museum in Boston.

Many amateurs copied existing designs of small computer systems. Some based their designs on instruction sets from IBM 1620 or DEC PDP-8 computers. Enthusiasts started many computers. However only the most determined completed them.

Time sharing, small computer systems and to some extent, amateur computing started personal computing in the 1960's. Research had also improved the human interface to the computer. However the personal computer was still complex and expensive to construct. New integrated circuit developments that led to the creation of the microprocessor significantly reduced this complexity.

Part II

1970's -- The Altair/Apple Era.

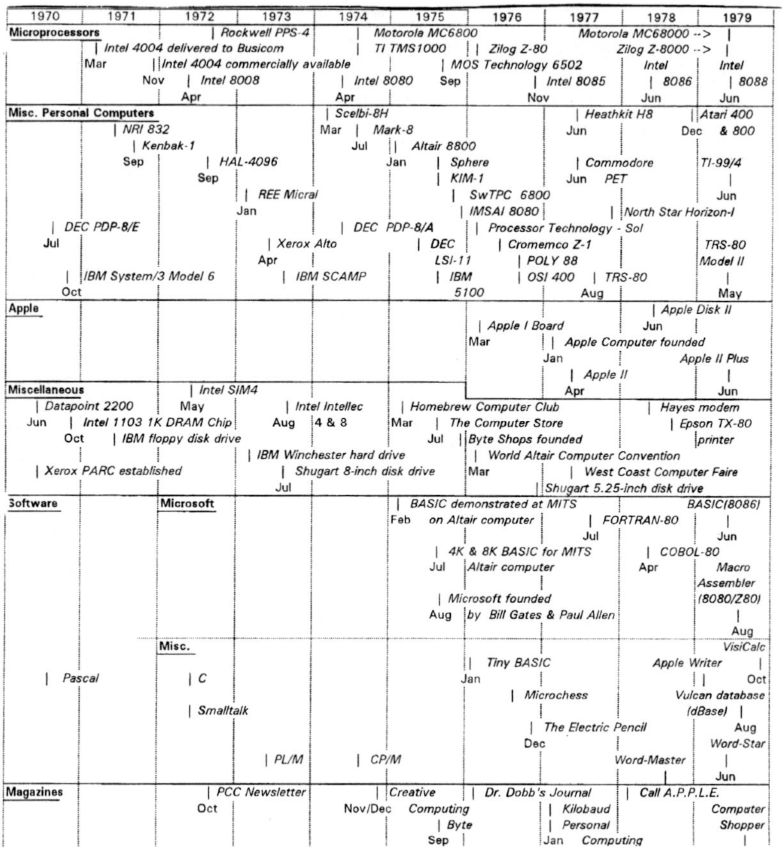

1970 | 1971 | 1972 | 1973 | 1974 | 1975 | 1976 | 1977 | 1978 | 1979

Microprocessors

Rockwell PPS-4
Motorola MC6800
Motorola MC68000 -->
Intel 4004 delivered to Busicom
TI TMS1000
Zilog Z-80
Zilog Z-8000 -->
Mar
Intel 4004 commercially available
MOS Technology 6502
Intel
Intel
Nov
Intel 8008
Intel 8080
Sep
Intel 8085
8086
8088
Apr
Apr
Nov
Jun
Jun

Misc. Personal Computers

Scelbi-8H
Heathkit H8
Atari 400
NRI 832
Mar
Mark-8
Jun
Dec
& 800
Kenbak-1
Jul
Altair 8800
Sep
HAL-4096
Jan
Sphere
Commodore
TI-99/4
Sep
KIM-1
Jun
PET
REE Micral
SwTPC 6800
Jun
Jan
IMSAI 8080
North Star Horizon-I
DEC PDP-8/E
DEC PDP-8/A
Processor Technology - Sol
Jul
Xerox Alto
DEC
Cromemco Z-1
TRS-80
Apr
LSI-11
POLY 88
Model II
IBM System/3 Model 6
IBM SCAMP
IBM
OSI 400
TRS-80
Oct
5100
Aug
May

Apple

Apple Disk II
Apple I Board
Jun
Mar
Apple Computer founded
Jan
Apple II Plus
Apple II
Apr
Jun

Miscellaneous

Intel SIM4
Datapoint 2200
May
Intel Intellec
Homebrew Computer Club
Hayes modem
Jun
Intel 1103 1K DRAM Chip
Aug
4 & 8
Mar
The Computer Store
Epson TX-80
Oct
IBM floppy disk drive
Jul
Byte Shops founded
printer
IBM Winchester hard drive
World Altair Computer Convention
Xerox PARC established
Shugart 8-inch disk drive
Mar
West Coast Computer Faire
Jul
Shugart 5.25-inch disk drive

Software

Microsoft

BASIC demonstrated at MITS
BASIC(8086)
Feb
on Altair computer
FORTRAN-80
Jul
Jun
4K & 8K BASIC for MITS
COBOL-80
Jul
Altair computer
Apr
Macro
Assembler
(8080/Z80)
Microsoft founded
Aug
by Bill Gates & Paul Allen
Aug

Misc.

VisiCalc
Tiny BASIC
Apple Writer
Pascal
C
Jan
Oct
Microchess
Vulcan database
Smalltalk
(dBase)
The Electric Pencil
Aug
Dec
Word-Star
PL/M
CP/M
Word-Master
Jun

Magazines

PCC Newsletter
Creative
Dr. Dobb's Journal
Call A.P.P.L.E.
Oct
Nov/Dec
Computing
Kilobaud
Computer
Byte
Personal
Shopper
Sep
Jan
Computing

Figure 3.1: A graphical history of personal computers in the 1970's, the MITS Altair and Apple Computer era.

Figure 3.2: Andrew S. Grove, Robert N. Noyce and Gordon E. Moore.

Figure 3.3: Marcian E. "Ted" Hoff.
Photographs are courtesy of Intel Corporation.

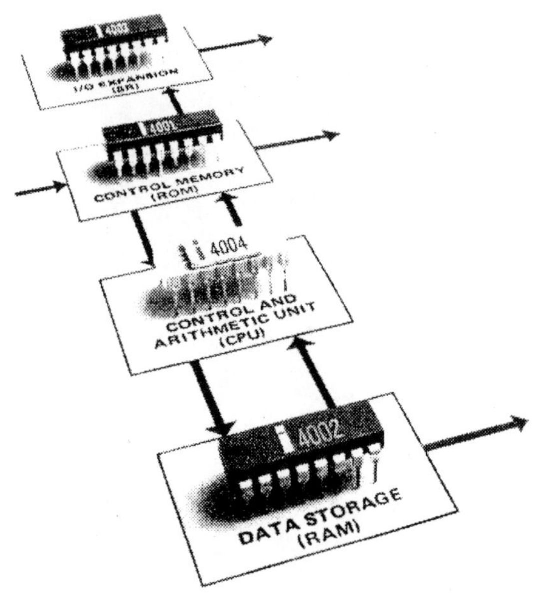

Figure 3.4: The Intel MCS-4 (Micro Computer System 4) basic system.

Figure 3.5: A photomicrograph of the Intel 4004 microprocessor.
Photographs are courtesy of Intel Corporation.

Chapter 3 Microprocessors in the 1970's

The creation of the transistor in 1947 and the development of the integrated circuit in 1958/59, is the technology that formed the basis for the microprocessor. Initially the technology only enabled a restricted number of components on a single chip. However this changed significantly in the following years. The technology evolved from Small Scale Integration (SSI) in the early 1960's to Medium Scale Integration (MSI) with a few hundred components in the mid 1960's. By the late 1960's LSI (Large Scale Integration) chips with thousands of components had occurred.

This rapid increase in the number of components in an integrated circuit led to what became known as Moore's Law. The concept of this law was described by Gordon Moore in an article entitled "Cramming More Components Onto Integrated Circuits" in the April 1965 issue of *Electronics* magazine [338]. Moore's Law initially stated that the number of transistors on a semiconductor chip would keep doubling every year. This was later changed to every 18 to 24 months. The "law" has held up remarkably well since 1965 and has had a profound effect on computer technology.

Advanced chip technology enabled Texas Instruments to develop the first electronic calculator in 1967. Then they worked with the Canon company to produce the worlds first pocket calculator, the Pocketronic in April 1971. During this period another company called Intel had entered the semiconductor market.

3.1 ... Intel

The Beginning

Gordon E. Moore and Robert N. Noyce resigned from Fairchild Semiconductor and founded a company called N M Electronics that they incorporated in July 1968. Noyce had been the general manager and was the co-inventor of the integrated circuit at Fairchild, Moore the director of the research and development laboratory. Arthur Rock and other investors provided venture capital for the company startup. Shortly after, they changed the company name to Intel Corporation, signifying *Int*egrated *el*ectronics. Andrew S. Grove who was assistant director of research and development at Fairchild joined Intel shortly after the founding as director of operations. The company became a public corporation in October 1971.

The initial focus of the company was to develop large scale integrated (LSI) memory chips to replace magnetic core memory. Intel introduced the 1101, a 256 bit MOS static random access memory in September 1969. Then Intel started producing the 1103 chip, the worlds first 1K bit dynamic random access memory (DRAM) in October 1970. This provided significant reductions in the cost of computer memory. However a request for logic chips would change the focus of the company to include microprocessors.

The First Microprocessor (4-Bit)

In April 1969, a Japanese calculator company called Busicom asked Intel to develop a set of at least twelve custom logic chips for a new low cost desk top calculator. Marcian E. "Ted" Hoff evaluated this request and determined that the design configuration proposed by the customer was too complex. Hoff had experience with the DEC PDP-8 architecture and proposed that Intel develop a four-chip set incorporating a general-purpose processor. The processor would accommodate the Busicom calculator BCD (Binary Coded Decimal) requirements and possibly other applications. Intel's executive staff approved the proposal and design proceeded with assistance from associate Stan Mazor. Busicom approved the Intel proposal for a general processor and associated chips in October 1969.

Busicom paid Intel $60,000 to produce the chip family configuration for use in their calculators in early 1970. Federico Faggin who joined Intel in April 1970, did the detailed circuit design and layout of the chips. The final configuration consisted of four 16 pin chips. Those chips were; a 4-bit central processing unit

(CPU), a 256 by 8-bit read only memory (ROM), a 320-bit random access memory (RAM) and a shift register for input/output (I/O). The CPU chip measured one-eighth of an inch wide by one-sixth of an inch long and had about 2,300 transistors. The CPU included a 4-bit parallel adder, sixteen 4-bit registers, an accumulator and a push-down stack on one chip. This CPU chip became known as the 4004 4-bit microprocessor. It had a set of forty five instructions, operated at 108 kilohertz and could execute 60,000 operations a second. Faggin produced working samples in nine months and Intel delivered sets of chips to Busicom in March 1971.

In early 1971, Busicom got into financial difficulties and Intel obtained the design rights for applications other than calculators in return for a lower price on the chips. In mid-November 1971, the first advertisement for a commercially available microprocessor, the Intel 4004 appeared in *Electronic News*. Intel advertised it as the MCS-4 (Micro Computer System 4-bit) family of four integrated chips. The advertisement stated it was "Announcing a new era of integrated electronics... A micro-programmable computer on a chip!" The 4004 CPU chip sold for $200. It was indeed the beginning of a new era, but the 8-bit microprocessor is what started the microcomputer industry.

8-Bit Microprocessors

In late 1969, concurrent with the Busicom development, Computer Terminal Corporation (CTC), which later became Datapoint, contracted with Intel and Texas Instruments to evaluate a set of chips for a new intelligent terminal. Ted Hoff and Stan Mazor analyzed the request and determined that a single 8-bit integrated chip could contain all the logic. Intel initially assigned the chip design to Hal Feeney. Then at the end of 1970 Federico Faggin directed the development of the chip that became the Intel 1201. During the summer of 1971 Datapoint agreed to let Intel use this architecture in exchange for a release from the development charges. Experience with the Intel 4004 microprocessor facilitated adaptation and development of the 8-bit chip. Intel then changed the number of the 1201 chip to 8008.

Intel introduced the first 8-bit 8008 microprocessor in April 1972 as a family of products, the MCS-8 (Micro Computer System 8-Bit). The 16 pin 8008 had 3,500 transistors and operated at about 0.2 MHz. The microprocessor had a set of 66 instructions, with 45 instructions oriented toward character string handling, six 8-bit general registers and could address 16K bytes of memory. The chip implemented interrupts, however the interrupts worked poorly. This chip with its 8-bit architecture suited the handling of data character processing as compared to the 4004 4-bit microprocessor. It was the first true general purpose microprocessor.

This is the microprocessor that started the microcomputer industry. The earliest microcomputers such as the French Micral in 1973 and the Scelbi-8H and Mark-8 in 1974 used the Intel 8008 microprocessor.

The 4004 and the 8008 chips had started the technology, but a new chip, the 8080 had a more significant impact on the microcomputer market. Federico Faggin proposed the new faster chip and its use of high-performance NMOS (N-channel Metal Oxide Semiconductor) technology as compared to PMOS (P-channel Metal Oxide Semiconductor) used on the 4004 and 8008 microprocessors. Masatoshi Shima who had moved from Busicom headed the design team. Intel approved development in the summer of 1972 and introduced the 8-bit 8080 chip in April 1974.

This new 40 pin chip with 5,000 transistors, operated at 2-3 MHz, could execute 290,000 operations a second, had 111 instructions and could address 64K bytes of memory. The chip offered a tenfold increase in throughput compared to the 8008. It also required only 6 support chips for operation as compared to 20 for the 8008. Intel had an initial price of $360 for the chip in quantities up to 24, with significant discounts on volume purchases.

MITS used this chip on the Altair 8800 in 1975. The microprocessor had a significant impact on the early development of microcomputers. However Motorola and Zilog had now entered the microprocessor marketplace.

The Motorola MC6800 had a single +5-volt power supply requirement and the Zilog Z-80 was faster with more features.

The competitive pressure from Motorola and Zilog resulted in the release of a faster 8-bit microprocessor, the 8085 in November 1976. It had more functions integrated, a 113 instruction set and only required a single 5-volt power supply.

The 8085 was faster but National Semiconductor Corporation had announced the 16-bit Pace microprocessor in 1974. Intel had started a 16-bit microprocessor project but was having problems that resulted in a conservative development schedule.

16-Bit Microprocessors

Management had approved development of a unique 16-bit architecture in late 1974. Then in mid 1975, William W. Lattin who had been the manager of memory and microprocessor products at Motorola, moved to Intel and was put in charge of the design team. The new microprocessor would incorporate a complex multiprocessing architecture that became the 32-bit iAPX 432 (Intel Advanced Processor Architecture) product. However due to product development delays and competitive pressures, Intel released the 8-bit 8085 in 1976 and began a second 16-bit project that became the 8086.

Development of the 8086 microprocessor with a 16-bit architecture started in 1976. Jean Claude Cornet was assigned to manage the project and William B. Pohlman who had managed the 8085 design was also the leader for the 8086 design team. Stephen P. Morse defined the microprocessor architecture and created it as an extension of the 8080 architecture to provide software compatibility. Intel introduced the 8086 with ten times the performance of the 8080 in June 1978.

The chip has 29,000 transistors with a minimum feature size of 3 microns and at a clock speed of 5 MHz it has a rating of 0.33 MIPS (million instructions per second). The instruction set is an expanded version of the 8080 with 133 instructions. Memory addressability is 1 megabyte and the microprocessor is available at clock frequencies of 5, 8 and 10 MHz. The price at launch was $360.

Intel introduced the 8088 microprocessor in June 1979. It has a 16-bit internal architecture with an 8-bit external data bus. This 8-bit external data bus allowed for a simpler interfacing to the rest of the system. The number of transistors and operating characteristics is similar to the 8086. The memory addressability is 1 megabyte and the microprocessor is available at clock frequencies of 5 and 8 MHz.

IBM selected this microprocessor for the PC computer announced in August 1981. This computer became a dominant product in the personal computer marketplace of the early 1980's, and was crucial in helping to position Intel as a major supplier of microprocessors.

An *IEEE Micro* article entitled "A History of Microprocessor Development at Intel" [342] and an Intel publication entitled "A Revolution in Progress: A History of Intel to Date" [48] provide additional details on Intel microprocessors and a history of the company. Intel had pioneered the introduction of microprocessors. However other companies had entered the marketplace with competing products. One of the other dominant companies was Motorola.

3.2 ... Motorola

Paul V. Galvin founded the Galvin Manufacturing Corporation in Chicago, in 1928. Then the company name changed to Motorola, Inc., in 1947. Following the name change and the invention of the transistor in 1947, the company opened a research laboratory in 1950, in Phoenix, Arizona to explore solid-state electronics. Motorola is now a major supplier of discreet semiconductors, integrated circuits and microprocessors.

Motorola introduced the 8-bit MC6800 MPU (Micro Processor Unit) in mid-1974. Principals in the design of the microprocessor and peripheral chips were Charles Melear and Chuck Peddle. The MC6800 used six-micron NMOS technology, contained 4,000 transistors and was the first microprocessor to require a single 5-volts power supply. This simplification of the power supply requirements and the related Motorola family of chips lowered product costs. Some early microcomputers using this chip were the MITS Altair 680B, Sphere and SwTPC 6800.

Motorola introduced the MC6801 in 1978, as "the world's first 35,000-transistor single-chip microcomputer." This microcomputer system had ROM/EPROM, RAM, I/O, timer, clock and CPU. The MC6809 was a microprocessor with an 8-bit external bus that processed data internally in 16-bit words. Motorola announced the microprocessor in 1978. Apple Computer selected this chip for the initial Macintosh computer design in 1979.

The MC68000 was the first of the MC68000 family of 16/32-bit microprocessors. Principals in the design were Tom Gunter and Gene Schriber. The MC68000 has 61 instructions and a capability of two million instructions per second (MIPS). The chip had 68,000 transistors. It can address 16 megabytes directly and 64 megabytes through functional segmentation. It has a 16-bit data bus that processed data internally in 32-bit words. Motorola introduced the MC68000 in 1979. Motorola also released an 8-bit external data bus version, the 68008.

This microprocessor had a significant impact in the marketplace. Apple Computer selected the MC68000 for the Lisa computer, with the processor operating at 5 MHz, and for the final design of the Macintosh personal computer, with the processor operating at 7.83 MHz.

Motorola and Intel became two dominant designers and suppliers of microprocessors for microcomputers in the 1970's. However the other company that played a significant role in the introduction of the microprocessor was Texas Instruments.

3.3 ... Texas Instruments

J. Clarence Karcher and Eugene McDermott founded Geophysical Service as a partnership in May 1930. The company name changed to Coronado Corporation, with Geophysical Service, Inc., (GSI) as a subsidiary in January 1939. Cecil H. Green, J. Erik Jonsson, Eugene McDermott, H. B. Peacock and others purchased the GSI subsidiary in 1941. The Coronado Corporation dissolved in 1945 and GSI's name changed to Texas Instruments, Inc., in 1951.

Texas Instruments (TI) obtained a license for the manufacture of transistors in May 1952 and aggressively expanded its semiconductor business during the 1950's. Jack St. Claire Kilby of TI co-developed the integrated circuit in 1958. During the 1960's TI was a significant supplier of chips for consumer products such as calculators, watches and toys.

Computer Terminal Corporation (CTC), which later became Datapoint, contracted with Texas Instruments (and Intel) to evaluate a set of chips for a new intelligent terminal in late 1969. This requirement resulted in TI developing and submitting a single-chip microprocessor to CTC.

TI has stated in company literature that it invented the single-chip microprocessor in 1970. The company applied for a patent in 1971 with the invention being credited to Gary W. Boone an Michael Cochran (the patent was approved in February 1978). The company described their invention as a "microcomputer --the first integrated circuit with all the elements of a complete computer on a single chip of silicon" [178]. TI advertised the integrated circuit developed for the Datapoint terminal in the *Electronics* magazine with the caption "CPU on a chip" in June 1971. This was the first announcement of a single-chip microprocessor. TI had functional problems with the chip and never marketed it. However, it did lead to the subsequent development of the TMS1000 microprocessor.

TI released the TMS1000 4-bit microprocessor in 1974. The chip had a set of 43 instructions and included a 256-bit RAM for data storage and an 8,192-bit ROM for program storage. The chip was used mainly in low-cost embedded applications. TI announced the TMS9000 series of 16-bit microprocessors in the spring of 1976, then the TI 9980 in the late 1970's. The TI 9980 had an 8-bit external bus but processed data internally in 16-bit words. Texas Instruments used the TMS9900 microprocessor in the TI-99/4 microcomputer released in 1979. However the computer was not successful and TI discontinued it in 1983.

Texas Instruments failed to penetrate the microcomputer market and concentrated in other applications. Texas Instruments is a leading supplier of 4-bit microprocessors for games, toys and low-end controller applications. However a number of other companies had entered the microprocessor market.

3.4 ... Other Companies

A number of other companies produced microprocessors in the 1970's. However two companies produced products that had a significant impact on the microcomputer market. Those two companies were MOS Technology and Zilog.

MOS Technology

In July 1975, MOS Technology, Inc., started advertising the 8-bit 6501 microprocessor that would be compatible with the Motorola MC6800 and would cost only $20. Chuck Peddle who had left Motorola and joined MOS Technology was a principal in the design of the microprocessor. Motorola subsequently forced the company to withdraw the 6501 because of technology infringements.

However MOS was also developing another chip, the 8-bit 6502 with additional features that would only cost $25. The chip had approximately 9,000 transistors. This pricing had a significant effect on the lowering of prices for microprocessors at both Intel and Motorola. This reduction in the cost of microprocessors resulted in additional impetus to the development of the microcomputer. MOS sold the new 6502 microprocessor at the WESCON show in September 1975.

The MOS 6502 microprocessor had a significant impact on the early microcomputer market. The Apple II, Atari 400 and 800 and Commodore PET microcomputers used the chip. Commodore International purchased MOS Technology in October 1976.

Zilog Inc.

In the summer of 1974, Federico Faggin and Ralph Ungermann decided to leave Intel and founded Zilog Inc., in November. Zilog is an acronym for the "last word (Z) in integrated (i) logic (log)." Shortly after the founding, the powerful oil corporation Exxon, invested $1.5 million in Zilog for a 51 percent controlling interest in the company. Faggin and Ungermann developed ideas for a new microprocessor and family of components that would be compatible and more powerful than the Intel 8080. Assisting in the design effort was Masatoshi Shima who had also left Intel and joined Zilog. It would be faster, have more features, registers and instructions than the 8080.

Zilog announced the Z-80 microprocessor in 1975 and it became available in February 1976. It operated at 2.5 MHz, could address 64K bytes of memory and incorporated the 8080's instruction set within 158 instructions. The chip had 8,500 transistors and was manufactured by the Mostek company. The Z-80 with a low price of $200, became a successful alternative to the Intel 8080. Zilog announced a faster 4 MHz version of the chip, the Z-80A in February 1977. A number of early microcomputers such as the Radio Shack TRS-80 used the Z-80.

The Z-8000 was a 16-bit microprocessor with 17,500 transistors and 110 instructions. The memory address capability was 48 megabytes in six segments of eight megabytes and the operating speed was 2.5 to 3.9 MHz. Zilog priced the microprocessor at $195 for small quantities and announced it in early 1979. The instruction set was not compatible with the Z-80 and other design problems prevented it from becoming successful.

Other Sources

A number of other companies became second sources for the major suppliers. The following are some of the more significant suppliers.

Walter Jeremiah (Jerry) Sanders was a principal in the founding of Advanced Micro Devices (AMD), Inc., with seven other Fairchild personnel in mid 1969. Sanders had worked for Motorola then Fairchild Semiconductor where he became sales and marketing director. AMD had an initial orientation as a second source for other manufacturers chips by creating an equivalent design or by obtaining a license agreement from the other company. Early license agreements were with Intel for the 8085 microprocessor and with Zilog for the Z-8000. AMD became a public company in September 1972 and the Siemans AG company made an investment of $30 million for nearly 20 percent of AMD in late 1977.

Hewlett-Packard developed a proprietary microprocessor in the late 1970's. This microprocessor was used in the company's HP-85 personal computer released in 1980.

Mostek was founded by L. J. Sevin and subsequently became a subsidiary of United Technologies. It was a major supplier of semiconductor memory chips and a second source for other company's microprocessors such as the Zilog Z-80.

National Semiconductor Corporation was founded in 1959. Charles E. Sporck who had been at Fairchild Semiconductor joined the company in 1968, became the president and a significant contributor to the success of the company. The company offered microprocessor systems on a 8.5 by 11-inch PC card. The IMP-8 was an 8-bit microprocessor system and the IMP-16 was a 16-bit microprocessor system available in 1973. Then National Semiconductor announced the first 16-bit single chip microprocessor, called the Pace in 1974. A faster microprocessor called the Super-Pace followed. The SC/MP was an early 8-bit microprocessor developed by National.

RCA introduced the 1802, the first complementary metal oxide semiconductor (CMOS) 8-bit microprocessor in 1974, that was followed by the 1804 microprocessor.

Rockwell introduced the PPS-4 that was a 4-bit parallel processor in late 1972. It had a 50 instruction set. Rockwell subsequently released the PPS-8.

Signetics Corporation was founded by four Fairchild Semiconductor executives in 1961. The company name is an acronym for "Signal Network Electronics." The company shipped an 8-bit Programmable Integrated Processor (PIP) in 1974 that had more than 64 instructions. The Signetics 2650, is another processor available from the company around 1978. Signetics was also a second source for microprocessors to companies such as Motorola.

3.5 ... Miscellaneous

Patent Controversy

During the 1960's and early 1970's, integrated circuit technology continued to improve the capability of integrating a larger number of circuit elements. This evolved to consideration by some in the industry of possibly creating a central processor unit (CPU) on a single chip. However the assignment of credit for the invention of the microprocessor and the awarding of patents for it have been controversial. Companies such as Intel and Texas Instruments and entrepreneur Gilbert Hyatt are principals in the controversy.

According to Gilbert Hyatt he built his first breadboard concept of a small computer in 1968, trademarked the term "microcomputer" and started a company called Micro Computer Inc (MCI). Hyatt made application for a broad patent with a title "Data Processing System" in 1969. The application was rejected by the U.S. Patent Office. This was appealed by Hyatt and in December 1970 the Patent Office accepted a new application with a title "Factored Data Processing System for Dedicated applications." Hyatt continued working on printed circuit prototypes but did not demonstrate or prove that his concept was practical. In question was Hyatt's ability to implement the patent on a single chip at the time of the patent application. At that time the technological capabilities were just evolving and Hyatt had not created his concepts on a single integrated chip. In September 1971, Hyatt's Micro Computer company went out of business. Hyatt continued to have problems with his patent application that resulted in further appeals and changes to the application. Then in July 1990, the U.S. Patent Office awarded Hyatt a patent that now had a title of "Single Chip Integrated Circuit Computer Architecture." This immediately created significant controversy in the industry. However in June 1996, the patent was overturned.

Between 1969 and March 1971, Intel conceived and developed the 4004 microprocessor as described in Section 3.1. Intel did not consider that they had developed a unique patentable device, but had merely extended existing technology. This rather casual approach resulted in Intel not filing a patent application for the microprocessor. Ted Hoff has stated that he published an article in March 1970 describing the feasibility of building a central processor on a chip. This in essence placed the concept of a microprocessor in the public domain, thereby precluding a viable patent application.

Texas Instruments has stated that it invented the single-chip microprocessor in 1970 as described in Section 3.3. However TI withdrew their microprocessor due to functional problems and the following products were not accepted by any major microcomputer manufacturer. As stated previously TI has been a major supplier of microprocessors for consumer products.

The Texas Instrument patent that was approved by the U.S. Patent Office in 1978 has not been successfully contested. However Hyatt's patent that was approved by the U.S. Patent Office in July 1990 was overturned in 1996. Neither of these patents has affected Intel which became the dominant supplier of microprocessors for personal computers.

An article in *Byte* magazine "Micro, Micro: Who Made the Micro?" details the patent controversy with interviews of Hyatt, Hoff and Faggin [328]. Some additional comments by Federico Faggin are in another *Byte* magazine article, "The Birth of the Microprocessor"

[331] and in a *Popular Science* article "Gilbert Who?" [324].

References

A further article in *Byte* magazine entitled "Evolution of the Microprocessor" provides an informal history [333]. The *IEEE Micro* article entitled "A History of Microprocessor Development at Intel" provides some details of developments at other companies [342]. The *Encyclopedia of Microcomputers*, "Architecture of Microprocessors," Vol. 1, pages 269-282 [236] contains an extensive bibliography.

Conclusion

A number of companies became second sources for different manufacturers. For instance Synertek and Rockwell International produced the 6500 series of microprocessors designed by MOS Technology. Solid state technology had made significant advances during the 1970's. The integrated circuit enabled the development of low-cost memory chips and the microprocessor. These are the developments that resulted in the personal computing transition to microcomputers.

Chapter 4 Transition to Microcomputers

The 1970's were a period of transition for personal computing. There was a change from the common usage of time sharing on large computers to the use of low-cost personal computers. This cost aspect changed our general understanding of what defined a computer as being personal. Previously the definition only required that a computer be designed for use by a single person, such as those developed by DEC, IBM and MIT in the 1960's. During the 1970's, the definition evolved to include a price level that was affordable to the average consumer. This change became possible by a period of hardware transition. The microprocessor and memory chips replaced discrete components and core memory. This reduced the complexity and cost of building a personal computer. Then factory-built "turn-key" units changed the market from the computer hobbyist and software enthusiast to the "appliance user" and a larger consumer market.

This was also a decade in which the main computer companies failed to develop personal computer products for the consumer market. Intel created systems for software development. However, they did not extend these products into a consumer computer. Digital Equipment Corporation (DEC) could also have adapted their PDP-8 and LSI-11 designs to a consumer product. A product was proposed by an employee David Ahl, but rejected by the company. Hewlett-Packard also rejected an offer by Stephen Wozniak to market what became the highly successful Apple II computer. Federal antitrust actions discouraged IBM from any product line expansion during the 1970's. Xerox also failed to take advantage of innovations they had created in the Alto computer. These dominant companies had the financial resources, major research facilities and marketing power. However, they did not develop any viable products for the larger consumer market. These companies allowed entrepreneurs to start what became the "microcomputer revolution."

4.1 ... The 1970-74 Transition

The first microprocessor became available from Intel in 1971. However, commercial use of the new processors by companies other than Intel, did not occur in North America until 1974. Companies did use integrated circuits in calculators and computer products, but most of the early computer products had limited exposure. Though, the CTC "smart" terminal received some interesting exposure when used as a computer by some customers.

Phil Ray and Gus Roche founded Computer Terminal Corporation (CTC) in July 1968 to make terminals for other companies. In 1969, the company decided to develop a "smart" terminal with a processor. Vic Poor, the technical director, hired Jack Frassanito to help with the design. CTC designed the unit as a replacement for IBM keypunch machines and other vendors terminals. The company contracted with Intel and Texas Instruments to integrate a number of functions on a single large-scale chip. However, CTC decided to complete the design using discrete components. This design became the Datapoint 2200 computer terminal that CTC introduced in June 1970. The terminal had a cost of about $5,000. In 1972, the company name changed to Datapoint Corporation. An article in *Invention & Technology* [276] credited Frassanito as being "The Man Who Invented the PC." The article states that a company had "written a payroll program on the 2200 itself, using machine language" and that "Other buyers were applying the Datapoint to accounting and process control."

The Datapoint unit was not a computer; but in 1970, DEC released the small PDP-8/E and PDP-11 computers. The PDP-11 was a 16-bit computer with a price of $20,000 that shipped in the spring. Then in July, DEC introduced the low-cost PDP-8/E 12-bit computer with a price of $4,990. Sales of the PDP-8 and PDP-11 series of computers were outstanding. This helped to position DEC, as the major supplier in the minicomputer market.

The market for small computers had also attracted another major company. IBM announced the System/3 Model

6 BASIC computer system in October 1970. This was a small personal disk based system, with a BASIC language user interface for a display unit. A typical system cost of $48,250 excluded the computer from the personal consumer market. However in 1971, a shift to low cost products began with the National Radio Institute computer kit and John Blankenbaker's Kenbak-1 computer.

In 1971, the National Radio Institute (NRI) offered one of the earliest low cost computer kits for a computer electronics course. Louis E. Frenzel designed the NRI 832 kit that cost $503.

The Kenbak Corporation, advertised one of the earliest personal computers, called the Kenbak-1 in the September 1971 issue of the *Scientific American* magazine. John V. Blankenbaker who had founded the company, priced the computer at $750, but sold only 40 units. An Early Model Personal Computer Contest, sponsored by The Computer Museum of Boston, ComputerLand and CW Communications in 1986, selected the Kenbak-1 as the winner. The criterion for selection of the winner was "interest, significance and date of each model" [235, page 288].

The Kenbak-1 was the first commercially assembled, low cost personal computer. However, computer enthusiasts had been building their own low cost computers. An example of such a computer was the HAL-4096 designed by Hal Chamberlain. The computer was described in the September 1972 issue of the *ACS Newsletter*, published by the Amateur Computer Society (ACS). Chamberlain offered construction plans for $2.

Low cost computers started to evolve from a different direction as some companies developed more powerful calculators. An example of this was the 9800 series of products from Hewlett-Packard. HP extended the capabilities of their desktop calculator product line with the release of the HP 9830A calculator in 1973, which had a BASIC programming capability. HP had also been working on a low-cost 16-bit general-purpose computer with the code name of Alpha, which it released in 1972 as the HP 3000 minicomputer.

A significantly smaller company than HP, the EPD Company, advertised a computer kit called the System One at a price of $695 in May 1973.

Also during the first half of 1973, IBM developed the first portable computer called SCAMP (Special Computer APL Machine Portable). The IBM General Systems Division in Atlanta proposed the product to raise the visibility of APL (A Programming Language). Paul Friedl developed the computer design at the IBM Scientific Center in Palo Alto, California. The processor was an IBM Palm microcontroller. Although not marketed commercially, it was the basis for the development of the IBM 5100 computer in 1975.

Unlike IBM, the Digital Equipment Corporation (DEC) continued to develop and market products with lower prices. In 1973, DEC began marketing the EduSystem series of low-cost PDP-8 computers for educational users. Then in May 1974, DEC announced the PDP-8/A computer with a Model 100 priced at $1,835. The PDP-8/A had a lower cost due to the use of medium scale integrated (MSI) chips and smaller circuit boards. Although DEC had achieved a pricing level whereby they could have entered the personal consumer market, they chose not to personalize their products. However, a new research center of the Xerox Corporation would personalize a computer that significantly affected the future personal computer market.

Xerox Alto

Xerox, a copier manufacturing company, decided to diversify and enter the computer market in 1970. It wanted to offer a "system of interrelated products to manage information in the office" and establish itself as a leader in "the architecture of information" [144]. In 1970, it founded the Palo Alto Research Center (PARC) in Northern California and acquired Scientific Data Systems (SDS). SDS was a manufacturer of computers for the engineering and scientific market. Xerox intended to use SDS as a basis for its entry into the general computer market.

PARC started development of a personal computer called Alto as a research project. It evolved from human interface ideas for the computer developed by Ivan Sutherland, Douglas Engelbart, and Alan Kay in the 1960's.

In September 1969, Kay described in a doctoral thesis *The Reactive Engine;* an interactive programming language and a computer called FLEX. By 1972, Kay had extended the concepts of FLEX to a proposed computer he named Dynabook. He envisaged a portable notebook computer that would contain a knowledge manipulator with extensive personalized reference materials, which could also be a dynamic interactive medium for creative thought. It was beyond the hardware capabilities then, so the Alto computer became an interim design.

Xerox started the Alto design in late 1972 and completed the first unit in April 1973. Although not produced commercially, the technology was of historical significance. Charles "Chuck" P. Thacker was a principal in the Alto hardware design; with significant design input from Alan Kay, Butler Lampson and Bob Taylor. The Alto hardware consisted of four units: the processor and disk storage cabinet, bit-mapped graphics display, keyboard and a mouse to facilitate graphics manipulation and control. The CPU used a 16-bit custom-made processor similar to the Data General Nova 1220. The system had estimated costs ranging from $20,000 to $32,000.

The research group introduced the desktop metaphor and extended Engelbart's windows graphical environment to include an intuitive user interface. By 1976, the group had created overlapping windows, pop-up menus and icons. Dan Ingalls facilitated this by developing a procedure known as BitBlt, for manipulating a bit-block of the screen display. Software system designers used BCPL, a language similar to C, to create the Alto operating system. Later, the MESA programming language, which was similar to Pascal, replaced BCPL. Butler Lampson and Charles Simonyi developed a word processing program called Bravo, which used a new idea called "what you see is what you get." Alan Kay developed a new interactive object-oriented programming language called Smalltalk. Another new system was the Ethernet network used to connect the Alto computers and peripherals. Robert Metcalfe designed the initial communications software called PUP (PARC Universal Packet) based on ARPANET concepts. Other principals in software design at PARC were William K. English and Lawrence G. Tesler.

Xerox PARC had created significant innovations in personal computer technology. The synthesis of hardware and software to facilitate a friendly user interface was unique. Subsequently, Apple Computer and Microsoft would use various concepts developed by PARC in their technology for the new microcomputers.

Figure 4.1: Xerox PARC Alto computer.

Photograph is courtesy of Xerox Corporation.

The Early Microcomputers

Intel developed the first 1K dynamic random access memory (DRAM) chip in 1970. Then in 1971, Intel created the 4-bit 4004 microprocessor, followed by the 8-bit 8008 microprocessor in 1972. The first application of the Intel 4-bit microprocessor was in the Japanese Busicom calculator. This was the technology that changed personal computing -- large scale integrated (LSI) memory chips and the early Intel microprocessors.

Intel Corporation released the SIM4 simulator board that used the 4004 processor in May 1972, and shortly after the SIM8 that used the 8008 processor. The company then introduced the Intellec 4 and 8 Development Systems in August 1973. These computers facilitated the development of software for new microprocessor products. Initially, Intel made significantly more money from the Development Systems, than from the sale of microprocessor chips. The simulator board was the first application of a microprocessor to a computer product. However, Intel designed the board as a development aid, not as a personal computer.

The first personal computer to use a microprocessor was the French built Micral developed by Realisations Études Électroniques S.A. (REE), also known as R2E. Truong Trong Thi, a French-Vietnamese engineer, cofounded the company and is credited with being inventor of the first microcomputer. Francois Gernelle was a principal in the design. The computer used an Intel 8008 processor and was released in January 1973 at a price of $1,900. Poor sales resulted in the company filing for bankruptcy in October 1975. However, France had the distinction of preceding the USA in producing the first microcomputer.

In the USA, the first personal computer to use a microprocessor was the Scelbi-8H from Scelbi Computer Consulting, Inc. Scelbi was an acronym for Scientific, electronic and biological. Nat Wadsworth formed the company in August 1973 and advertised the computer in the March 1974 issue of *QST*, an amateur radio magazine. Scelbi described it as "The totally new and the very first Mini-Computer." Wadsworth and Robert Findley designed the Scelbi-8H using the Intel 8008 processor. A kit sold for "as low as $440." In April 1975, Scelbi introduced the Scelbi-8B computer for business applications. Subsequently the company changed its business orientation to software and book publishing. The Scelbi computers were not successful. However, a computer called the Mark-8 obtained a certain amount of success when details of it appeared in a national magazine.

Jonathan A. Titus described how to build the Mark-8 Minicomputer in the July 1974 issue of *Radio-Electronics* magazine [282]. The computer used the Intel 8008 processor. The article advertised the 8008 processor as being available from Intel for $120.00 and a set of six boards for $47.50. The popularity of the computer resulted in Mark-8 clubs being formed and clubs' newsletters being published.

The Mark-8 and the other early microcomputers were important, but 1974 was significant for another reason. With the release of the Motorola MC6800 microprocessor, Intel now had a strong competitor that would affect future personal computer products. However, the personal computer industry really started in 1975 with the introduction of the MITS Altair computer.

4.2 ... MITS Altair

H. Edward Roberts, Forrest M. Mims and two other officers from the Effects Branch of the Air Force Weapons Laboratory in Albuquerque, New Mexico, founded MITS (Micro Instrumentation and Telemetry Systems), Inc., in 1969. Their initial interest was in telemetry systems for amateur rockets and an infrared voice communicator, but these were not a commercial success. In November 1970, Roberts bought out the other founders and started working on an electronic calculator kit. The November 1971 issue of *Popular Electronics* featured the new MITS 816 calculator. MITS sold thousands of the calculator kits, but by early 1974 competition from low-priced ready-made products forced the company out of the market. MITS required a new product. After evaluating the Intel microprocessors, especially the new 8080 chip, Roberts decided to create a computer kit.

After the article describing the Mark-8 computer appeared in the July 1974 issue of *Radio-Electronics*, the editors at *Popular Electronics* were eager to have an article on a superior computer for their magazine. The timing was right for Roberts, who agreed to provide an article on a design that used the Intel 8080 processor for the January 1975 issue. The magazine cover headline for an article written by Roberts and William Yates read, "World's First Minicomputer Kit to Rival Commercial Models ... Altair 8800" [301]. Yates and Jim Bybe assisted Roberts in the engineering design. Roberts initially selected the name PE-8 (Popular Electronics 8-bit) for the computer. However, Leslie Solomon, who was the technical editor of the *Popular Electronic* magazine, thought the name PE-8 dull, and selected the celestial name "Altair" for the computer. The introduction of the Altair 8800 became a major event in the start of the personal computing industry.

Figure 4.2: MITS Altair 8800 microcomputer.
Photograph is courtesy of Intel Corporation.

The computer had a front panel with toggle switches for control and input, and light emitting diodes (LED's) for output display. Roberts had negotiated with Intel to obtain a special high volume price of $75 for the 8080 microprocessor. The computer sold as a kit for $397, or completely assembled for $498. MITS hoped to sell about eight hundred computers a year. However a flood of orders overwhelmed MITS and created delivery problems due to inadequate production capacity.

A significant design decision by MITS was the use of an open bus architecture similar to that used in minicomputers. This provided expansion capabilities, for additional memory and peripherals. Originally called the "Altair Bus," it later became known as the "S-100 Bus" because the bus had 100 connection points. Memory cards were available as a kit or assembled at prices ranging from $90 to $400. The 4K memory card was the minimum size required to run Altair BASIC. However, the MITS memory cards used dynamic RAM chips that were unreliable. Providing reliable replacements for these cards and extending the capabilities of the Altair 8800 led to the founding of a number of other companies such as Cromemco and Processor Technology. MITS released an improved version, the Altair 8800b in April 1977.

To compete with companies that used the Motorola MC6800 processor, MITS announced a new computer called the Altair 680 in October 1975. The Altair *Computer Notes* newsletter announced that it would be less expensive than the Altair 8800. It would use the MC6800 processor, be smaller and have a limited expansion capability. However, a kit would only cost $293. Design problems delayed the release of the computer until 1976, by which time MITS had also changed the designation to the Altair 680b and the price to $466 as a kit or $625 assembled.

William Gates and Paul Allen developed the BASIC interpreter for the Altair. This development resulted in Roberts hiring Paul Allen, who became the director of software for MITS. During the development of the Altair 680b, Allen hired Mark Chamberlain to work on software for the new computer. In late 1976 Allen left MITS to work full time at Microsoft and Chamberlain became the new director of software.

Roberts started to have misgivings about the competition and his company in 1976. This resulted in a letter of intent being signed for the purchase of MITS, by the Pertec Computer Corporation in December. Pertec completed the purchase in May 1977. However, MITS was bankrupt by 1979.

MITS did not survive the transition from electronic hobbyist to the plug-and-play user market. However it left a significant legacy. It created a popular product, established the personal computer industry and started the "personal computer revolution."

4.3 ... Other Computers - - 1975-76

The years 1975 and 1976 were significant for personal computing. In 1975 MITS released the Altair 8800, the Homebrew Computer Club started in California, William Gates and Paul Allen founded Microsoft and *Byte* magazine published its first issue. In 1976 *Dr. Dobb's Journal* appeared, MITS held the first microcomputer convention and Intel announced the 8085 microprocessor. These were just a few of the more significant events that accompanied innovations by various companies and numerous microcomputer entrepreneurs. This was a period of significant growth. The following describes some of the more dominant companies and the computers they introduced.

The larger computer companies were slow to incorporate the microprocessor in their products. However, in April 1975, Digital Equipment Corporation (DEC) released its first microprocessor based product, the LSI-11 microcomputer board. The LSI-11 used a 16-bit processor designed and manufactured by DEC that incorporated features of the DEC PDP-11 minicomputer. Early applications of the board were in the DEC LSI-11/23, DEC PDP-11/03 microcomputer systems and the Heathkit H11 computer.

IBM announced the IBM 5100; the first commercially produced portable computer in September

1975. This evolved from the SCAMP computer designed in 1973. The unit weighed 50 pounds and the company described it as being "slightly larger than an IBM typewriter." Targeted primarily at the scientific market, it was capable of running APL and BASIC programming languages. The price at introduction ranged from $8,975 to $19,975 depending on the memory and programming language configuration. In 1977, the company introduced the IBM 5110 model with two diskette drives for the commercial market.

Mike Wise developed the Sphere microcomputers and advertised them in the first issue of the *Byte* magazine (September 1975). The Sphere Corporation advertised three models that used the Motorola MC6800 processor. The Hobbiest computer kit sold for $650, the Intelligent computer kit had additional features with a price of $750 and a computer named BASIC sold for $1,345. Sphere then released the System 340 that had an integrated monitor and keyboard. However, Sphere had a number of problems with their computers that resulted in the failure of the company.

In the fall of 1975, MOS Technology (who manufactured the 6502 processor) announced the KIM-1 computer designed by Chuck Peddle. The name was an acronym for Keyboard Input Monitor. The computer used a MOS 6502 processor and cost $250 fully assembled.

In November 1975, an established low-cost electronic kit company called Southwest Technical Products Corporation (SwTPC), founded by Dan Meyer, introduced the SwTPC 6800 Computer System. Before releasing their computer they had produced a Digital Logic Microlab for digital experimentation, a KBD-2 Keyboard and an Encoder Kit. Gary Kay, Don Lancaster and Meyer designed the SwTPC 6800 computer that used the Motorola MC6800 processor. The kit sold for $395. The SwTPC 6800 was one of the first computers to incorporate the Motorola mini-operating system called Mikbug. The unit used an SS-50 bus for expansion boards. It had no front panel switches for input; the system automatically started (booted) on power-up or by pushing the reset button. SwTPC was primarily a kit supplier and was not successful in the transition to complete computer systems for a larger consumer oriented market.

William H. Millard was a principal in the 1973 founding of a computer consulting company called IMS Associates, Inc. Two other personnel involved in the early development of IMS were physicist Joseph Killian and a computer science graduate, Bruce Van Natta. The company entered the microcomputer market through various attempts to network Wang computers for an automobile dealer. IMS considered the Altair for use in the network then discarded it due to financial and delivery problems with MITS. This resulted in Killian designing an improved version of the Altair 8800. IMS called the new computer the IMSAI 8080 and released it in December 1975. It used the Intel 8080A processor and a kit had a price of $499. The IMSAI 8080 became a successful alternative to the Altair 8800 and is notable for being the first clone! In October 1976, Millard made IMS Associates Inc., a holding company for IMSAI Manufacturing Corporation and the ComputerLand retail store organization.

Robert Marsh and Gary Ingram founded Processor Technology Corporation in April 1975. The initial products were input/output and memory expansion boards for the MITS Altair 8800 computer. Then in mid 1975, Leslie Solomon, who was the technical editor of *Popular Electronics*, approached Processor Technology with a request to develop a computer terminal suitable for an article in the magazine. By February 1976, Marsh and Lee Felsenstein developed not just a terminal, but a complete computer design that appeared in the July 1976 issue of *Popular Electronics*. The company named the unit "Sol Terminal Computer." Sol was an abbreviation of Solomon's name. The Sol-10 Terminal Computer kit used the Intel 8080A processor. The computer also had an attractive low profile style cabinet with wooden sides made of walnut. The ROM memory featured a "Personality Module" that facilitated an easy change of the operating system. Steve Dompier created the minimal operating system called CONSOL. The company introduced the factory assembled Sol-20 model at the Personal Computing 76 show in Atlantic City. The Sol-20 used an Intel 8080 processor and an advanced operating system called SOLOS. Processor Technology released later a dual eight-inch disk drive named Helios with a PT-DOS disk operating system. However the company did not maintain a competitive product line and terminated business in May 1979.

Two Stanford University professors, Harry Garland and Roger Melen, started a business called Cromemco Inc. in 1975. The company name was derived from their dormitory called Crothers Memorial Hall, at Stanford University. The first products were add-on boards for the MITS Altair 8800 computer. The company introduced the Cromemco Z-1 computer that used the Zilog Z-80 processor in 1976. The Z-2 Computer System followed in March 1977 and the System Zero, One, Two and Three computers later. Cromemco subsequently changed its market orientation to engineers and scientists, but it was not successful, which resulted in it being sold in 1986.

Mike and Charity Cheiky founded Ohio Scientific Instruments (OSI) in 1975. The company announced the OSI 400 Superboard computer in September 1976, that used either a Motorola MC6800, MOS 6501 or 6502 processors. OSI also released a Model 300 computer training board based on the MOS 6502 processor.

Rich Peterson, John Stephensen and Brian Wilcox established PolyMorphic Systems in December 1975. The company announced their first computer in October 1976. PolyMorphic initially named the computer the "Micro-Altair." However after objections from Ed Roberts of MITS, PolyMorphic changed the computer name to "POLY 88." It used the Intel 8080A processor, and sold for $685 as a kit. PolyMorphic then released the System 8813 that consisted of a main unit with a walnut cabinet, which used an Intel 8080 processor and had a detached keyboard. The system price started at $2,795. PolyMorphic developed its own operating system and Disk BASIC. The company ceased operations in the early 1980's.

John Ellenby headed the design group at Xerox PARC that started working on the design of the Alto II computer in June 1975. Company personnel received the

Alto II's in mid 1976. Ellenby proposed the design and manufacture of Alto III's in late 1976. However, Xerox executive staff did not support the future potential of the Alto computer and rejected the proposal.

Numerous other small companies released microcomputers between 1975 and 1979. *A Collectors Guide to Personal Computers and Pocket Calculators* [197] provides details of these computers.

The personal computer market changed in 1977. The dominant pioneering manufacturers such as Cromemco, IMSAI, MITS, OSI, PolyMorphic Systems, Processor Technology and SwTPC had attracted the electronic hobbyist. However they failed to develop products that were more applicable to a non-technical type of general user. New companies started offering products that were "plug in" systems for personal or business use. These new systems did not require any assembly or technical training. The Apple II computer as detailed in Chapter 5 was one of those new systems. The other dominant manufacturers in this new second generation of personal computers would be Atari, Commodore and Tandy/Radio Shack.

4.4 ... Commodore

In 1954, Jack Tramiel and a friend founded a typewriter repair business in New York called Commodore Portable Typewriter. The company moved to Toronto, Canada and became known as the Commodore Business Machines Company in 1956. In the 1960's the company sold adding machines, then hand-held calculators in the early 1970's. The company went public and became Commodore International Inc., in May 1974. However, competitive problems in the calculator market created financial difficulties for Commodore.

After corporate refinancing, Commodore bought MOS Technology (who supplied their chips) in October 1976. Chuck Peddle, who was the designer of the MOS 6502 processor, impressed Tramiel with the potential of the chip. Tramiel decided that the MOS 6502 could be "...the makings of the first low-cost personal computer" [180]. Commodore announced the PET 2001 computer in December 1976 and introduced it at the National Computer Conference (NCC) in June 1977. The name PET denoted "Personal Electronic Transacter" officially, or "Programmable Educational Terminal" at other times.

The PET 2001 used a MOS 6502 processor and had an integrated CPU, video display, keyboard and cassette tape drive. The ROM chip contained the cassette operating system and a Microsoft BASIC interpreter. Commodore priced the computer at $595 with 4K bytes of memory and $795 with 8K. Initially the company had an interesting policy of "pay now, delivery later."

Commodore then introduced the PET 4000 series in 1978 for educational and scientific users. The system had a MOS 6502 processor, an improved keyboard, but omitted the integrated cassette drive that had been on the original PET.

4.5 ... Tandy/Radio Shack

Dave Tandy and Norton Hinckley founded the Hinckley-Tandy Leather Company in 1927 as a retailer of leathercraft products. In 1961, they changed the company name to Tandy Corporation. Then Charles Tandy, a son of the founder, decided to enter the consumer electronics market in 1962. Tandy acquired control of Radio Shack, a distributor of electronic parts and products, by 1965. Donald H. French who was a buyer for the company and a computer hobbyist, promoted the release of a computer product. This resulted in Tandy deciding to enter the personal computer market in 1976.

Radio Shack announced the TRS-80 Micro Computer System designed by Steven Leininger in August 1977. TRS is an acronym for Tandy Radio Shack. Although the computer was not a single physical unit like the Commodore PET, it was a complete system with keyboard, video display and cassette tape recorder. The computer CPU and keyboard were a single unit. The power supply, video monitor and cassette recorder were separate units of the system. The computer system used a Zilog Z-80 processor and the ROM chip contained a Level-I BASIC interpreter. Leininger developed the BASIC interpreter by adaptation of a public domain program called "Tiny BASIC." Later, the company released a vastly improved Level II BASIC, developed by Microsoft. The complete system with processor and keyboard unit, power supply, video monitor and cassette recorder sold for $599.95. The processor/keyboard unit was available separately at a price of $399.95 with 4K bytes of RAM. With orders greater than initially expected, the TRS-80 was a success.

Radio Shack subsequently released an Expansion Unit with a TRS-80 Mini Disk System. Radio Shack developed a TRSDOS operating system that was not compatible with the popular CP/M operating system. The company then designed the TRS-80 Model II for business applications and announced the computer in May 1979. It used the faster Zilog Z-80A processor. It also had a floppy disk drive, additional functions, more memory and sold for $3,450 as a state-of-the-art computer. With the introduction of the Model II, the company changed the name of original TRS-80 to the TRS-80 Model I.

4.6 ... Atari

Nolan K. Bushnell and Ted Dabney founded the Atari Corporation in June 1972. They introduced a video table tennis game called "Pong" in November that was a huge success. The co-founders of Apple Computer, Steven Jobs and Stephen Wozniak designed a computer game called "Breakout" for Atari. Bushnell sold Atari to Warner Communications Inc., for $26 million in 1976 and shortly after left the company. Warner then appointed Raymond Kassar to replace Bushnell and manage the company. Atari announced the Atari 400 and Atari 800 computers in December 1978, although the units did not become available until late 1979.

The Atari 400 used a MOS 6502 processor and the housing included a plastic membrane keyboard. The Atari 400 had a price of $499 with 8K bytes of memory and a price of $630 with 16K bytes. The Atari 800 model also used the MOS 6502 processor, an improved keyboard and had a price of $1,000.

Atari had a restrictive policy in providing system information to software companies. This limited the development of application programs for the Atari computers that adversely affected sales.

4.7 ... Other Computers -- 1977-79

This section describes other computers developed from 1977 to 1979. Two major companies were the Heath Company and Texas Instruments.

Heath Company

The Heath Company was a subsidiary of Schlumberger Ltd., and was the dominant manufacturer of electronic kits in the 1970's. The company entered the personal computer market in June 1977 with the announcement of the H8 and H11 computer kits.

The H8 computer used the Intel 8080A processor, had a front panel with a 16-key keypad for octal input and a LED display. The price of the computer kit was $375. The H11 computer was an early 16-bit computer using the DEC LSI-11 microcomputer board. The LSI-11 had operating characteristics of a DEC PDP-11 16-bit minicomputer. This computer kit had a price of $1,295. Peripheral products included the H9 video terminal kit that had a price of $530 and the H10 paper tape reader and punch kit had a price of $350. Memory, parallel and serial interface cards were also available.

A Benton Harbor BASIC interpreter was available in 8K and 12K versions. The HDOS operating system was also available for a H7 floppy disk drive assembly. Gordon Letwin developed the software.

Zenith Radio Corporation acquired the Heath Company for $64.5 million in 1979. Following the acquisition, Zenith introduced a completely new computer called the Heath/Zenith-89. Zenith designated the factory assembled version the Z-89 and the Heathkit version the H89. It was an integrated desktop unit that included the CRT, keyboard and a single 5.25-inch floppy disk drive. The Heath/Zenith-89 used a Zilog Z-80 processor. A kit sold for $1,595 or an assembled unit for $2,295. Heath/Zenith provided a Text Editor, Assembler, HDOS operating system and an Extended Benton Harbor BASIC interpreter.

Texas Instruments (TI)

An early 1978 headline for the Texas Instruments SR-60A read "TI's First Personal Computer?" TI described the SR-60A as a "personal computer/calculator" having "the power of a computer with the simplicity of use and low cost of a calculator." It cost $1,995.

Texas Instruments introduced the TI-99/4 computer in June 1979. TI described it as "featuring easy-to-use computing power for personal finance, home management, family entertainment and education." The basic system consisted of a TI TMS9900 16-bit processor, an integrated keyboard within the housing and a separate 13-inch color monitor with 16 colors. TI provided BASIC software that was a full floating point, expanded version compatible with the ANSI standard. The price for the TI-99/4 system was $1,150. Also available was a range of Solid State Software command modules at prices ranging from $20 to $70 and a Solid State Speech synthesizer accessory for $150.

Other Computer Releases

Hewlett-Packard (HP) introduced a new generation of the HP 9800 series of computers between 1976 and 1979. This new generation changed the processor from TTL logic to a proprietary 16-bit NMOS processor called the "BPC." An example of the computers in this series was the HP 9831A desktop computer, introduced in mid 1977. The computer had a BASIC interpreter and a price of $7,200.

Noval Inc., introduced the Noval 760 computer that used an Intel 8080A processor in June 1977. The computer system was a unique package in a desk console with drawers, retractable keyboard and a fold-down top. The console had a 12-inch monitor, cassette drive and sold for $2,995.

In July 1977, the Digital Equipment Corporation introduced the DECstation, also known as the VT78 that was compatible with PDP-8 computers. DEC physically configured the computer similar to the VT52 terminal. The VT78 used an Intersil 6100 CPU, had 16K bytes of RAM and was advertised at a price of $7,995.

The Digital Group Inc., was a company that developed a unique computer design that enabled one to change the processor by inserting the appropriate CPU board. Robert Suding designed the Digital Computer System in the mid 1970's. It was available with one of four processors: AMD 8080A, MOS 6502, Mostek 6800 or Zilog Z-80. An 8080 computer system with 10K bytes of memory had a cost of $645.

Charles Grant and Mark Greenberg founded North Star Computers, Inc. They had also founded Applied Computer Technology, G&G Systems and Kentucky Fried Computers. The initial products were plug-in boards for the S-100 Bus and floppy disk systems. North Star released the Horizon-I computer that used a Zilog Z-80A processor in October 1977. The company developed their disk operating system and BASIC interpreter. North Star subsequently released the Horizon-II computer with higher capacity disk drives.

IMSAI Manufacturing Corporation introduced a desktop business computer called the VDP-80 in late 1977. The computer was an integrated unit; with a video monitor, 8-inch disk drives and keyboard. However, the computer had a number of technical problems. IMSAI replaced it with the VDP-40 that used 5.25-inch disk drives, but the product was too late. IMSAI then failed to respond to a market change from electronic hacker to the mass consumer user. William Millard, a founder of IMSAI, had also shifted corporate resources to his ComputerLand retail organization. This resulted in IMSAI going bankrupt in March 1979.

Xerox established a Systems Development Division (SDD) in California to advance PARC inventions into commercial products. SDD then started development of an Alto based office automation computer with the project code name of Janus. The code name changed to Star in late 1977. Then in January 1978, Xerox established an Advanced Systems Division (ASD) that implemented a market evaluation program to determine the feasibility of personal computing and the Alto computer. Xerox selected four test sites that included the White House and the U.S. House of Representatives. By the end of 1978, over 1,500 Alto computers were in use. However, Xerox terminated manufacture of the Alto computer in January 1979.

Ohio Scientific Instruments introduced various models in the Challenger series between 1977 and 1979. The Challenger III model was unique in having multiple processors: a 6502, Z-80 and a MC6800 on the same CPU board. OSI was an early developer of floppy-disk drives for their computers that used an operating system called OS 65. OSI had problems competing with other companies such as Apple Computer, Atari and Tandy/Radio Shack. This resulted in Ohio Scientific Instruments being acquired by MaCom, a large communications firm who sold OSI again to a company that went bankrupt.

The microprocessor and LSI memory chips enabled development of microcomputers, the electronic hobbyist and software enthusiast provided the market in the 1970's. However, most of the early dominant microcomputer manufacturers did not retain their market position. Other companies and entrepreneurs introduced personal computers for a new consumer market. Significant suppliers to that mass market were Apple Computer, Commodore and Tandy/Radio Shack. New processors such as the Intel 8088 and the Motorola MC68000 introduced in 1979 would start another new era in personal computing. The dominant producers would change from small entrepreneurs to larger corporations such as IBM. However, Apple Computer, as described in the following Chapter, was an exception. It started as a small entrepreneurial company, then attained a position as a leading supplier of personal computers.

Blank page.

Chapter 5 Apple Computer in the 1970's

Among the early entrepreneurial microcomputer companies, Apple Computer was an exception. Incorporated two years after the introduction of the MITS Altair computer, it became a dominant commercial success as compared to many other companies formed during that period. The company's initial product, the Apple II, facilitated the change from the hobbyist or technical hacker to the personal "appliance" user in the mass consumer market. The financial initiative of Mike Markkula enabled the aspirations of the two principal founders of company. Steven Jobs provided the entrepreneurial energy that complemented the innovative technical skills of Stephen Wozniak.

5.1 ... Wozniak/Jobs Early Years

Stephen Gary Wozniak was born on the 11th of August 1950 in San Jose, California. He was the first of two sons of three children of Jerry and Margaret Wozniak. The father was an electrical engineer and the mother was active in local politics.

Wozniak had an early interest in electronics and obtained his amateur radio (Ham) license in the sixth grade. He also designed and built electronic projects for his Homestead High School science fairs. One project named "A Parallel Digital Computer" received awards at the Cupertino School District Science Fair and the Bay Area Science Fair. Wozniak was also the president of the Electronics Club and secretary-treasurer of the Mathematics Club at high-school. Through Wozniak's high-school electronics teacher, he became a frequent visitor to the GTE Sylvania computer facility. For one year he and a friend Allen Baum would write FORTRAN programs for the IBM 1130 computer and get familiar with its technology. They also visited the Stanford Linear Accelerator Center's (SLAC) computer facility that had an IBM System/360 computer. Access to the SLAC library became a valued source for information on computer technology.

Wozniak enrolled at the University Of Colorado in Boulder in 1968. He now had access to the university computer and wrote programs in FORTRAN and ALGOL. However the year was a failure academically, so the next year he enrolled at the local De Anza Community College.

Wozniak and Allen Baum had been delving into the intricacies of Data General Nova computers. The analysis progressed to the design of their own versions of the computer. In 1969 Wozniak decided to build his own computer and enlisted the help of his neighbor and friend Bill Fernandez. They scavenged parts from surplus stores, Fairchild, Intel, Signetics and Hewlett-Packard. Wozniak worked on the logic design and Fernandez on the timing circuits and power supply. They called the machine the "Cream Soda Computer" on account of the amount of the drink they consumed during its construction.

Steven Paul Jobs was born on the 24th of February 1955 in San Francisco, California. He was the first of two adopted children of Paul and Clara Jobs. The father had several occupations such as machinist, finance company representative and real estate salesman. His mother had also worked at a number of jobs, including part-time in the payroll department at Varian Associates.

Jobs became interested in electronics during his elementary school years. At the age of twelve, a neighbor who worked at Hewlett-Packard, would take Jobs to visit the company and expose him to its technology. Through his interest in electronics Jobs obtained a summer job at Hewlett-Packard by calling one of the founders Bill Hewlett. He also obtained a part-time job at a surplus electronic parts retailer called Haltek. His familiarity with the parts enabled him to buy and sell parts to Haltek for a profit. Although Jobs had an interest in electronics, his expertise would tend to the commercial rather than the technical aspects. Bill Fernandez introduced Jobs to Wozniak in 1969. This was the beginning of the association and friendship between Jobs and Wozniak.

In 1971, Wozniak moved to Berkeley and enrolled at the University of California, Berkeley campus. About the same time, an article entitled "Secrets of the Little Blue Box" was published in the October 1971 issue of *Esquire* magazine, that initiated an interest by Wozniak and Jobs in telephone hacking. University library texts on phone systems and the infamous phone hacker John Draper, also known as "Captain Crunch" provided additional details. Wozniak developed a digital design to generate the audio tones required to place and route calls in the phone system. They incorporated the design components into a compact box and used it to hack the phone systems of North America and the world in early 1972. Jobs convinced Wozniak to sell the "blue boxes" to other hackers at prices varying from $80 to $300. However Draper's conviction in 1972 of phone fraud charges tempered the commercial activities of Jobs and Wozniak.

Wozniak joined Hewlett-Packard as an associate engineer in 1973. His first assignment was to work on refinements to the HP 35 calculator. It was during this period that Wozniak's interest in flying light aircraft developed.

On completion of high-school in 1972, Jobs enrolled at Reed College in Portland, Oregon. It is at Reed College that Jobs began Bohemian associations, an interest in Eastern mysticism and meditative religions such as Zen Buddhism. These associations and interests adversely affected his academic studies. Jobs left Reed College and started working for Atari engineering on their video games in early 1974. Jobs combined an Atari business trip to Europe with a personal mystic excursion to India. On his return from India he combined a Bohemian lifestyle, a renewed interest in Zen Buddhism and his association with Atari.

Wozniak attended the first meeting of what became the Homebrew Computer Club in March 1975 with Allen Baum. The meetings were an important forum for Wozniak to exchange information on the latest microcomputer technology. Jobs also attended a few of the club meetings with Wozniak. It was at the Homebrew Computer

Club that Wozniak met Chris Espinosa and Randy Wigginton who would later become employees of Apple Computer.

In mid 1975, Alex Kamradt enlisted the help of Wozniak to develop a convenient video terminal with a keyboard to replace the cumbersome Teletype terminals. Kamradt was the founder of Call Computer, a time-sharing company in Mountain View, California. Kamradt and Wozniak formed a subsidiary of Call Computer named Computer Converser. Kamradt provided the financing and Wozniak was to develop the design. Kamradt arranged for Jobs to run the business and develop the terminal for manufacture. The intent was for the initial terminal to evolve at a later stage into a computer. Wozniak designed and built a terminal, but he lost interest and withdrew from the project.

5.2 ... Apple I Board

During 1975 a number of factors occurred which motivated Wozniak to design a microcomputer. Hewlett-Packard was offering employees a large discount on the Motorola MC6800 microprocessor and related chips. Wozniak also had the experience of designing the Cream Soda Computer and the Computer Conserver terminal. Then the technology exposure and enthusiasm generated by the Homebrew Computer Club provided the synergy that resulted in a new microcomputer design by Wozniak. It would be an enhancement of the Computer Conserver terminal incorporating the MC6800 microprocessor.

Wozniak and Baum changed the design concept after attending the WESCON trade show in September 1975. Wozniak purchased a MOS Technology 6502 microprocessor that was a derivative of the Motorola MC6800 and only cost $25. The significantly lower cost compared to either the Motorola or Intel 8-bit microprocessors, resulted in Wozniak changing the microprocessor for his new design. He also decided to use dynamic rather than static memory chips. Prior to building the hardware Wozniak decided to write a version of BASIC for his new microcomputer. The number of changes required to convert the microcomputer design to the MOS 6502 microprocessor were minimal. Wozniak completed the computer board and interfaced it with his video terminal and keyboard in March 1976

Wozniak had discussions with Hewlett-Packard to determine their interest in producing a microcomputer. The company declined and provided Wozniak with a legal release. They advised him that "HP doesn't want to be in that kind of market."

Wozniak demonstrated the computer board at the Homebrew Computer Club in April. However the members did not have much enthusiasm for the board. Applications using the Intel 8080 microprocessor had greater popularity. Wozniak also handed out schematics of the microcomputer design at the club meetings.

Figure 5.1: Steven P. Jobs and Stephen G. Wozniak holding an Apple I board - a 10 year anniversary photograph.

Photograph is courtesy of Apple Computer, Inc.

In early 1976 Jobs convinced Wozniak that they should try and sell the microcomputer boards. The initial concept was that they would sell only a bare printed circuit board to friends, a few stores and to members of the Homebrew Computer Club. The customer would then purchase the components and insert them into the board to complete the microcomputer. To pay for the boards Wozniak sold his Hewlett-Packard calculator and Jobs his Volkswagen van.

Jobs and Wozniak decided to form a partnership and formalized it by placing an advertisement in a local newspaper. The advertisement required a name for the company and at Jobs suggestion they selected Apple Computer Company. Jobs obtained the services of Ron Wayne who was a field service engineer at Atari to draw the schematics of the board and design a motif for the company. Jobs also used Wayne to convince Wozniak of the commercial potential of the company. Jobs, Wozniak and Wayne signed a partnership agreement in April 1976. Jobs and Wozniak split ninety percent of the company and Wayne received the remaining ten percent. Wozniak would be responsible for engineering, Jobs for marketing and engineering and Wayne for documentation and mechanical engineering.

Apple Computer now called the microcomputer the Apple I. It included the MOS 6502 microprocessor, 256 bytes of ROM, 8K bytes of RAM and a partial power supply. Interface connectors provided for video output, a keyboard and an external card to connect a cassette recorder. The Apple I had no graphics or color, and no speaker. The video was 24 rows of 40 characters stored in shift registers and displayed at 60 characters per second. The program in ROM was a small monitor routine that enabled the input or output display of memory and a capability to start a program at a particular address. The BASIC interpreter required 4K bytes of memory, leaving 4K bytes for user programs.

Following the club demonstrations Jobs contacted Paul Terrell, the founder of Byte Shop's regarding the possible sale of the Apple I boards in his stores. Terrell agreed to purchase fifty boards with the component parts assembled on the board. He agreed to pay about $500 per board with cash on delivery. This was a significant breakthrough for Apple Computer. However it required extensive efforts by Jobs to arrange the procurement of parts with a "net 30 days credit." He also arranged to borrow $5,000 from Allen Baum and his father. Assembly of the boards was at Jobs parents home using his sister to insert the components on the boards. They then delivered the assembled boards to Terrell around July and received the cash as promised.

Jobs arranged for a mail-drop box and an answering service to give a proper company image. Jobs now hired his friend Bill Fernandez who became Apple's first employee. Jobs moved the assembly of a second batch of fifty boards from the home bedroom to the garage. With pressure from Terrell at Byte Shop's, Wozniak developed a small board for a cassette interface to facilitate the input of the BASIC interpreter software. The interface board with Wozniak's BASIC interpreter sold for $75. After input from the Homebrew Computer Club, Jobs and Wozniak established a price of $666.66 for the microcomputer board. Ron Wayne became concerned about his potential financial obligations to the company and terminated his share of the partnership in the summer of 1976. Jobs now arranged financing for a second batch of computer boards.

Wozniak started adding a number of improvements to the Apple microcomputer during 1976. A key improvement would be the addition of color capabilities. Cromemco had demonstrated color displays with their "Dazzler" machine at the Homebrew Computer Club. Innovative changes to the Apple I memory and video circuits provided for color with fewer chips and lower cost. Wozniak also wanted his computer to be capable of playing a game called Breakout that he and Jobs had developed for Atari. This resulted in circuits being added for game paddles and sound.

In August 1976 Jobs and Wozniak attended the Personal Computing 76 show in Atlantic City, New Jersey to demonstrate the Apple I board and the prototype of their improved design. Stan Veit of the New York Computer Mart retail store had agreed to provide them with space at his booth. Processor Technology demonstrated their new SOL microcomputer with its integrated keyboard and attractive enclosure at the Personal Computing Fair. The company also started advertising the Apple I board in *Dr. Dobb's Journal* and the September 1976 issue of *Interface Age*.

It was obvious to Jobs that the current prototype would require additional improvements to compete commercially. Consequently they decided to design a unit with an attractive case, built-in power supply and integrated keyboard. Wozniak also wanted to add expansion slots to facilitate expansion of the computer's capability. An expanded ROM had the Apple BASIC programming language and additional routines for items such as an improved video display, cassette recorder and a disassembler.

Wozniak had concentrated on the digital aspects of the computer and did not have a power supply to complete an integrated design. Jobs contacted Al Acorn, the chief engineer at Atari who recommended Rod Holt. Jobs wanted a power supply design that did not require a cooling fan. The computer would then be silent in operation compared to the competitors with their somewhat noisy fans. Apple hired Holt on a consulting basis to design the power supply.

In October 1976, Commodore International expressed an interest in the purchase of the Apple Computer Company. Commodore had just purchased MOS Technology and was planning to enter the microcomputer market. However the price of $100,000 and other conditions set by Jobs were not acceptable to Commodore. Apple Computer showed the new Apple computer to Stan Veit of the New York Computer Mart during a trip to the West Coast. Jobs offered Veit 10 percent of Apple Computer for an investment of $10,000. However Veit required all his available capital for his own store and rejected the offer.

5.3 ... Founding of Apple Computer

In June 1976, Jobs had contacted the advertising and public relations company founded by Regis McKenna. The company had Intel as a client and Jobs liked the unique advertising that the McKenna agency provided. The agency also handled the marketing activities for Paul Terrell's Byte Shop's and Atari. After discussions with Jobs, McKenna agreed to handle the Apple Computer Company.

Apple did not have the funds required for marketing and future development. In August, Nolan K. Bushnell of Atari advised Jobs to contact a venture capitalist Donald T. Valentine of Sequoia Capital, who in turn recommended A. C. "Mike" Markkula. Markkula was in his early thirties, a retired multi-millionaire from Intel. Markkula decided to finance Apple Computer by investing $91,000 of his own money and arranging and guaranteeing a $250,00 line of credit with the Bank of America. Markkula, Jobs and Wozniak had equal one-third shares of the business and Rod Holt received a small percentage. Markkula however insisted that Wozniak leave Hewlett-Packard and join Apple full time. The participants in the founding were the two initial Apple partners, Jobs and Wozniak, Holt the power supply designer and Markkula the financier. They founded the Apple Computer, Inc., in January 1977 and purchased the previous Apple Computer Company partnership in March.

To run the company Markkula selected Michael M. Scott who was an associate from his days at Fairchild Semiconductor. Markkula hired Scott as president of the company. He was an engineer who had worked at Beckman Instrument Systems, Fairchild Semiconductor and was director of National Semiconductor's manufacturing line. With the founding of the new company, Markkula had operations moved from the Jobs family garage to a building in Cupertino, California.

5.4 ... Apple II

The company decided to use the Apple II model designation for the improved prototype. They also decided to introduce the Apple II computer at the First West Coast Computer Faire. Jim Warren had scheduled the Faire for April 1977 at the Civic Auditorium in San Francisco, California. Apple also decided to develop a new logo to replace the original partnership motif. The Regis McKenna agency worked on a new design and developed the striped apple with six colors and a bite removed on the side.

Jobs was determined to have an attractive housing for the Apple II. A Hewlett-Packard designer Jerrold C. Manock had agreed to design the plastic case. The housing incorporated a 53-key Teletype-style keyboard. Holt had developed a digital design for the power supply. The unit was smaller, lighter and generated less heat than conventional power supplies. Once again Jobs obtained the services of Howard Cantin, an associate from Atari who had done the Apple I board, to do the artwork for the new motherboard. Wozniak had decided on eight 50 pin expansion slots. He also adjusted the BASIC interpreter to fit within the limitations of the ROM memory.

The computer used a MOS 6502 microprocessor and 4K bytes of dynamic RAM, expandable to 48K. The video system could display 24 rows of 40 characters in upper case only. The unit included an audio cassette interface and four analog game paddle inputs. Customers used their own TV sets as monitors by using a radio frequency modulator and stored programs on audio cassettes.

The ROM incorporated the system monitor program, utility routines and an Apple BASIC interpreter that also became known as Integer BASIC. Allen Baum, Chris Espinosa and Randy Wigginton assisted Wozniak in the software development.

Apple had a prime location at the front of the Faire hall due to an early commitment by Jobs. Markkula organized an attractive booth with a backdrop displaying the new company logo. Apple displayed all three of the only assembled computers with a large screen monitor for program demonstration. The Faire was a tremendous success with about thirteen thousand people attending. Apple received orders for about three hundred Apple II computers within a few weeks.

Markkula with the assistance of a consultant had developed a business plan. Michael Scott started hiring personnel for accounting, engineering, production and sales. He subcontracted anything that Apple could not produce at lower cost. Quality problems with the case resulted in the purchase of new mold tooling.

Some Apple II boards shipped around May 1977 and the Apple II computer was available to the general public by June at a price of $1,298. The company offered Apple I board owners the option of upgrading to the new Apple II computer. Apple sold about seven thousand Apple II's by the end of 1977.

Figure 5.2: Apple II computer, monitor and two disk drives.

Photograph is courtesy of Apple Computer, Inc.

An extended BASIC interpreter called Applesoft also became available. Apple had purchased a license in August 1977 for Microsoft BASIC that included functions for doing floating-point mathematics. Randy Wigginton then made changes to adapt the interpreter for the Apple II. This resulted in the hybrid product with the name of Applesoft I that was released in November. Loaded from a tape recorder, it required 16K bytes of memory. Then in the spring of 1978, Wigginton made improvements to the interpreter that was released as Applesoft II.

5.5 ... Apple Disk II Drive

Apple was under pressure from dealers and customers to develop an alternative to the slow and at times unreliable cassette tape storage system. Wozniak started evaluating the floppy disk drive controller technology of IBM, North Star Computers and Shugart Associates in late 1977. Jobs was also contacting Shugart who were one of the first manufacturers of minifloppy disk drives for the microcomputer market.

The electronic circuit design developed by Wozniak for the floppy disk-controller card was an innovative design using a minimum number of components. Shugart manufactured the 5.25-inch disk drive. The disk had 35 tracks with thirteen 256 byte sectors on each track. It had a storage capacity of 116K bytes per disk. Apple also released an operating system developed by Randy Wigginton and Wozniak called DOS 3.1.

The company introduced an early version of the disk drive at the Las Vegas, Consumer Electronics Show in January 1978. The final design of the Apple Disk II drive was released in June. The drive, controller card and DOS 3.1 operating system cost $495 at introduction. Apple had released the lowest priced floppy disk drive system offered by a microcomputer manufacturer. The Apple Disk II drive was a success and had a direct impact on further improvements to computer sales.

Fig. 5.3: Steven P. Jobs. Fig. 5.4: Stephen G. Wozniak.

Figure 5.5: A.C. (Mike) Markkula, Jr.

Photographs are courtesy of Apple Computer, Inc.

5.6 ... 1978/79 Activities

In January 1978 Apple obtained additional financial investment from venture capitalists Arthur Rock, Donald Valentine's Sequoia Capital and Venrock Associates. During 1978 the Apple II computer sales were overwhelming and an order backlog developed. Apple increased staff and by the summer of 1978 the company had about 60 employees. Chuck Peddle, formerly of MOS Technology and Commodore joined Apple during the summer of 1978. However differences between Apple and Peddle led to him returning to Commodore at the end of 1978. During 1978, Henry E. Singleton the chairman of Teledyne Inc., and a friend of Arthur Rock made a significant investment in Apple stock and became a company director. Apple sales for 1978 were 7.8 million dollars.

During 1978 the necessity of developing models to follow the Apple II became evident. This started the development of an interim model that became the Apple II Plus. Apple also started a project for a new computer with the code name of "Sara" in late 1978, that would become the Apple III. The success of the Apple Disk II led to delivery problems with the disk drive supplier. Apple established a second supplier called Alps Electric Company of Japan. It also resulted in a project with the code name of Twiggy for Apple to develop and produce a new floppy-disk drive. Apple named the drive Twiggy because of its thin design (a name associated with a female model of that era).

Apple introduced the Apple II Plus in June 1979 with a unit cost of $1,195. It had 48K bytes of memory, a new ROM with Applesoft Extended BASIC, an auto-start for easier startup and screen editing. Apple also introduced its first printer, the Silentype in June. Then they introduced a word processing program called Apple Writer 1.0 followed by the release of the Apple II Pascal programming language in August.

Personal Software, Inc., released the VisiCalc spreadsheet software developed by Dan Bricklin and Bob Frankston in October 1979. This program was only available for the Apple II initially and had a very significant impact on increasing computer sales.

Lisa Project

During 1979 Apple was looking at the development of a new product that would utilize the latest technology and target the office market. The Apple II product line would focus on the home and school user and the Apple III computer on the small business user.

The project had the code name of "Lisa." Writers have attributed the code name to a manager's daughter or to the name of Jobs alleged daughter. Apple based the new computer design on a 16-bit architecture using a new Motorola MC68000 microprocessor. The project plan targeted the computer as a $2,000 system to be shipped in 1981. Jobs wanted a movable keyboard to provide a more comfortable means of input. Ken Rothmueller was in charge of engineering and John D. Couch headed software development. Bill Atkinson who was a senior programmer was working on a bit-mapped graphics system for the new computer.

In the summer of 1979 a coincidence of interests developed between Xerox Corporation and Apple Computer. Xerox wanted the marketing/production expertise of Apple in the personal computer consumer market. Xerox was considering the possible introduction of a new computer product to implement the technology they had created at the Palo Alto Research Center (PARC) in California. Apple Computer was arranging additional financing and had an interest in the possible implementation of the new Xerox technology. This resulted in Xerox Corporation investing one million dollars in the shares of Apple Computer. Another factor was that Jef Raskin wanted Jobs to visit PARC and improve his understanding of graphic user interface concepts that Raskin had implemented on the Macintosh project. Lawrence "Larry" G. Tesler provided demonstrations of the Xerox systems to Atkinson, Couch, Jobs, Rothmueller and other Apple personnel in November and December of 1979.

The Xerox concept of a computer "desktop," graphical interface and the use of a mouse was the new vision of computer technology that Jobs had been searching for. Apple immediately changed the Lisa project to incorporate various Xerox technological concepts. Apple also decided to create a set of application programs for Lisa and to develop the software within the company.

Macintosh

In early 1979 Jef Raskin proposed a concept for an inexpensive computer for the masses. A "desktop appliance" with the screen, keyboard, storage, printer and all the software built-in. There would be no peripheral expansion slots. Raskin wanted to develop both the hardware design and the software together with the objective of creating a simple user friendly interface. He wanted to incorporate some aspects of the PARC graphics technology in a portable configuration that would sell for a price targeted at about $995. Raskin named the computer Macintosh. Although the name is similar to the McIntosh apple, he intentionally changed it to avoid a potential conflict with a high-fidelity audio equipment manufacturer called McIntosh Laboratories. Jobs opposed the new computer proposal. However Raskin received board approval to continue the research project in September 1979.

The design configuration changed as the interplay between Raskin's initial objectives and pricing constraints occurred. Principals in the initial computer design were Brian Howard and Burrell C. Smith. By late 1979 the design configuration had evolved to include the Motorola M6809E microprocessor, 32K bytes of ROM and 64K bytes of user memory. The video system had changed from using a television to a 7-inch built-in bit-mapped screen with a 256 by 256 pixel resolution that could display 25 lines with an average of over 80 characters per line.

Conclusion

Other companies such as Atari, Commodore, Radio Shack and Texas Instruments provided competing computers in a rapidly expanding market. However Apple computer sales had increased from about seven thousand units in the 1977 founding year, to an annual rate of 35,000 in December 1979. For the 1979 fiscal year, the company had net sales of 47.9 million dollars and employed 900 people. The combination of an easy to use color system, the Apple Disk II drive and the VisiCalc spreadsheet software gave Apple a significant advantage. Apple Computer, Inc. was now the dominant supplier of personal computers.

Blank page.

Chapter 6 Microsoft in the 1970's

6.1 ... Gates/Allen Early Years

William Henry Gates was born on the 28th. of October 1955 in Seattle, Washington. He was the only son and the second of three children by William and Mary Gates. The father was a prominent Seattle lawyer and the mother was active in community organizations such as the United Way.

Bill's parents enrolled him in Seattle's exclusive Lakeside School in 1967. This was a progressive private all male school with a disciplined approach to education. During the 1967/68 school year, the teaching staff recommended acquisition of computer facilities to expose the students to the technology. The school could not afford to purchase a computer. However the Lakeside Mothers Club agreed to finance the use of a time sharing service. In 1968 the school obtained an ASR-33 Teletype terminal and used a local access line to dial into a General Electric Mark II time sharing system. In his 1968/69 school year at the age of thirteen, Bill Gates started programming.

Another student interested in programming was Paul G. Allen, who was born in 1953. From their mutual enthusiasm for programming computers a friendship developed which continued through high-school and university. It is this association that eventually led to their founding of Microsoft.

Figure 6.1: Paul Allen and Bill Gates (standing) at the Lakeside School computer room teletype terminal, c1968. Photograph is courtesy of Microsoft Corporation.

The students' enthusiasm for programming quickly strained the schools' time sharing budget. Fortunately the parent of another student had become one of the founders of Computer Center Corporation (CCC). This corporation had installed a Digital Equipment Corporation PDP-10 computer to offer time sharing service in the Seattle area. The corporation had a shakedown agreement, whereby DEC did not require payment for the computer until the equipment and software operated to the customers' satisfaction. The company suggested to the school that the students help test the computer system for bugs. In return they would receive free computer time. This became an intensive period of programming and acquisition of computer technology knowledge for both Bill Gates and Paul Allen. Not only did they de-bug the DEC PDP-10 system, but they invaded the intricacies of the computer accounting files to access the passwords. This resulted in a temporary disciplinary period of exclusion from the system. However after a short period of assisting the company programmers, Computer Center Corporation encountered financial difficulties and went into receivership in March 1970.

The next source of computer access and time was the University of Washington where Paul Allen's father was the associate director of the libraries. The university had a Xerox Data Systems (XDS) Sigma V time sharing computer and a Control Data Corporation (CDC) Cyber 6400 computer. Gates and Allen used both computers for a period of time.

Lakeside now managed to obtain on loan a DEC PDP-8/L computer. Gates obtained the source code for a version of BASIC from DECUS, a DEC user's group. He then started working on his own first version of a BASIC interpreter. However before he had completed the software, Lakeside returned the DEC PDP-8/L. Lakeside then received a Data General Nova computer. However the school returned the computer before Gates could adapt his BASIC interpreter to the new machine architecture.

Then in the fall of 1970, Lakeside arranged with Information Sciences, Inc., (ISI), a time sharing company in Portland, Oregon for use of their DEC PDP-10. By this time Gates, Allen and two other friends Richard Weiland and Kent Evans from Lakeside had formed an organization called the Lakeside Programmers Group. The group managed to obtain a contract from ISI to develop a payroll program in COBOL. Once again this contract provided computer time in exchange for the work required to develop the payroll software. It also provided experience in software development and documentation of a commercial program.

During 1970/71 a Lakeside teacher and former Boeing engineer had the task assigned of creating a program for class scheduling. Unfortunately a flying accident killed the teacher. Gates and his friend Kent Evans worked to refine and complete the program written in FORTRAN. However another tragic accident killed Evans in a mountaineering class. Paul Allen had enrolled at the Washington State University in the fall of 1971. He was now finishing his first year in computer science when Gates enlisted his help to complete the software. They completed the program during the summer of 1972, by which time Gates credit for use of the ISI computer ran out.

During 1971 Gates and Allen contracted to do an analysis program and input data for car traffic counting boxes. These activities resulted in the creation of Gates and Allen's first company called Traf-O-Data. In the fall of 1972 Allen convinced Gates that they could use the Intel 8008 microprocessor in a machine to facilitate the analysis of the traffic tapes. They obtained the services of Paul Gilbert who was an electrical engineering student at the University of Washington to construct this specialized microcomputer. Allen then started developing a simulator program on the university computer that would emulate the Intel 8008 microprocessor.

Gates started sending applications for enrollment to different universities at the end of 1972. Then Gates and Allen got offers from TRW to work on the development of PDP-10 system software. TRW was developing a real-time operating and dispatch system for the Bonneville Power Administration in Vancouver, Washington. TRW was in desperate need of experienced DEC PDP-10 programmers. They both accepted jobs. Allen left university and Gates got permission to finish his final high-school year in an internship with TRW. Access to the DEC PDP-10 allowed Allen to complete his Intel 8008 simulator program. This also allowed Gates to complete the programming for the Traf-O-Data project.

In the fall of 1973 Allen returned to University and Gates entered Harvard University. During his first year, Gates met and became a friend of Steven (Steve) A. Ballmer, who would later become the president of Microsoft. Not completely happy with university life, both Allen and Gates considered taking a year off and applied for jobs at different companies. They were both offered jobs at Honeywell in Massachusetts. Allen accepted a position and moved to Boston. Gates declined the job offer. Paul Allen and Bill Gates would now meet on weekends and some weekday evenings to discuss computers, Traf-O-Data and other possible projects.

In the summer of 1974, Gates obtained a job to computerize a class enrollment project at the University of Washington. Then in the fall of 1974, Gates returned to Harvard

Gates and Allen had looked at developing a BASIC interpreter for the Intel 8008 microprocessor, and concluded it was not powerful enough. Then in April 1974, Intel released the 8080 microprocessor. After reviewing the new microprocessor capabilities, they determined they could develop an effective interpreter for it. Gates and Allen contacted computer manufacturers to determine their interest in a BASIC interpreter for the 8080 microprocessor. However, they were not successful, but this would change with the announcement of the Altair 8800.

6.2 ... Altair/BASIC

In mid-December of 1974, the January 1975 issue of *Popular Electronics* was on the newsstands. It had an exclusive feature article entitled "Altair 8800 -- The most powerful minicomputer project ever presented -- can be built for under $400." The authors of the article were H. Edward Roberts and William Yates of MITS Inc. [301]. Paul Allen read the article and realized this was the opportunity they needed. The Altair 8800 computer used the new Intel 8080 microprocessor that he and Gates had wanted to develop a BASIC interpreter for and now they had a potential market for the software.

Their previous experiences with the Traf-O-Data machine, Allen's simulation program and a modified DEC BASIC interpreter that Gates had written were going to be significant factors. Gates and Allen decided to develop a BASIC interpreter for the Altair 8800 computer. The interpreter with some modification would also be capable of running on other microcomputers using the Intel 8080 microprocessor. The Altair 8800 computer did not have any software. The operator used switches on the front panel to enter instruction codes and data. This process was long, tedious and prone to error.

Gates contacted MITS and advised Roberts that they had a BASIC interpreter. The timing was right. Roberts was just as anxious to have a BASIC interpreter as Gates and Allen were to develop and sell one. In the initial contacts between Roberts and Gates/Allen there were two problems. The first problem was that Gates and Allen had not developed the BASIC interpreter for the Intel 8080 yet. The second problem was that Roberts did not have memory boards at that time capable of storing the BASIC interpreter software. Gates agreed to Roberts suggestion of going to MITS Inc., in a month to demonstrate the program on the Altair.

The next four weeks were intensive. Allen started converting the Intel 8008 simulator software to emulate the Intel 8080 microprocessor. He also adapted assembler and debugger programs for the Intel 8080. Gates started working on the specifications for the BASIC interpreter. To a certain extent some programming languages develop based on earlier experience with similar languages. Gate's BASIC interpreter evolved from DEC's BASIC-Plus. The 4K memory limitation limited the number of features. This memory limitation had to accommodate not just the interpreter but also the user program data.

When the interpreter specification was complete, Gates started writing the assembler instructions for the program. Gates and Allen used the DEC PDP-10 computer at Harvard to create all the programs. Program development consumed any time left after Gates Harvard classes and Allen's job at Honeywell. Gates and Allen also obtained, the assistance of Harvard student Monte Davidoff, to develop the mathematical routines for the interpreter. The interpreter had now grown to require 6K instead of the 4K of memory initially targeted.

Paul Allen demonstrated the BASIC interpreter at MITS Inc., in Albuquerque, New Mexico in late February. The program had not run previously on either an Altair 8800 microcomputer or the Intel 8080 microprocessor. However it was a success. Allen's simulator and Gates interpreter had functioned without a problem. This was an incredible achievement.

Roberts wanted to market the BASIC program as quickly as possible. Allen negotiated some additional time for Gates to refine the final version for release. Paul Allen left Honeywell and joined MITS Inc., to develop software for the Altair 8800 in March. Shortly after joining MITS, Allen became the Director of Software. In April, the headline "Altair BASIC Up and Running" was on the first issue of the Altair users *Computer Notes* newsletter.

MITS began a promotion campaign featuring a mobile van to demonstrate the Altair computer system in April 1975. It traveled all over the country and included a demonstration of a preliminary Version 1.1 of Gates and Allen's BASIC interpreter program. During its stop in California a member of the Homebrew Computer Club appropriated a copy of the paper tape with the BASIC interpreter encoded on it. The club members reproduced the tape and distributed copies to other members.

Gates and Allen signed a contract with MITS in July, just prior to the founding of Micro-Soft. The wording of this contract would be crucial in later developments. It required Gates and Allen to provide BASIC interpreter programs on an exclusive license basis to MITS Inc. MITS agreed to promote and commercialize the program to other companies on a "best efforts" basis. This allowed MITS to sell copies of the BASIC interpreter and pay Gates and Allen a royalty on each sale. Gates and Allen still retained ownership of the software. The agreement specified three versions of the software with varying memory requirements of 4K, 8K and 12K bytes. The compensation to Gates and Allen varied depending on the version. Another arrangement prevailed if the software was sub-licensed to other companies. They also specified that a secrecy agreement be signed by all MITS customers who purchased the software. Gates had concerns about piracy of the software.

Version 2.0 of both the 4K and 8K BASIC interpreter programs had started shipping in July 1975. The software received strong favorable responses in the market. Gates secured the services of Monte Davidoff from Harvard and Chris Larson from Lakeside School to assist in the software development during the Summer.

6.3 ... The Albuquerque Years

In August 1975 Bill Gates and Paul Allen founded a partnership called Micro-Soft in Albuquerque, New Mexico. The company name was an abbreviation of Microcomputer-Software. The company deleted the hyphen in 1976 and capitalized the "S" to form MicroSoft for a period of time. The partnership agreement gave Gates a sixty percent interest and Allen the remaining forty percent. Gates felt he had made a larger contribution.

Gates now started working on refinements for version 3 and an Extended BASIC requiring 12K bytes of memory. However Gates had also received requests to develop a disk version of their program. Gates was having to spend additional time to satisfy these demands and a desire to expand the company.

In the September 1975 issue of the Altair *Computer Notes* newsletter, the editor David Bunnell wrote an article condemning the piracy of the BASIC interpreter. The October issue included an additional article on piracy by Ed Roberts of MITS. Then the February 1976 issue of the newsletter included an open letter by Gates to the hobbyists complaining of the piracy. The subject also received discussion at the first World Altair Computer Convention in March. Then Bill Gates prepared a final letter and appeal that appeared in the April issue of the newsletter with his comments from the March convention. This was the beginning of the major concern by all producers of software for the personal computer market.

In September 1975, Gates enlisted the help of Richard Weiland from the Lakeside Programmers Group. Weiland developed a BASIC interpreter for the new Altair 680B computer that used a Motorola M6800 microprocessor. Allen rewrote the 8080 simulator for the Motorola microprocessor and Weiland had the interpreter completed by January 1976.

Allen and Gates licensed the 6800 BASIC interpreter to MITS for a flat fee. This assured Microsoft of a definite revenue and MITS could charge for the program or provide it free with the hardware.

Copying concerns with the 6800 BASIC interpreter had become a non-issue.

In early 1976, MITS was shipping approximately 1,000 computers a month. However Microsoft was only getting a royalty for its software, at a rate of less than 200 copies per month. Illegal copies were affecting Microsoft revenue.

In February 1976, while still at Harvard, Gates started writing the software for a version of BASIC suitable for a system utilizing disk drive technology. Gates had the software created and operating within a period of two to three weeks. It became known as DISK BASIC.

In April 1976, the company hired its first permanent employee, Marc McDonald. He began work on what became known as Stand-alone Disk BASIC for the National Cash Register (NCR) company. Richard Weiland became the second employee in May with the position of general manager. In mid summer Gates started developing APL (A Programming Language) software. Then Microsoft hired Steve Wood in August, who started working on the development of FORTRAN software. Work also started on the Focal language. DEC initially developed Focal to control scientific instruments. However sales of Microsoft Focal were a failure, that resulted in it being discontinued. Microsoft now started to obtain corporate customers for its BASIC interpreters. Then it opened its first office in Albuquerque in 1976.

Microsoft also started working on a 6502 BASIC interpreter for the MOS Technology microprocessor. Marc McDonald adapted the 6800 simulator to the 6502 and Richard Weiland developed the 6502 BASIC interpreter. The interpreter included a built-in editor for making program changes. The Apple II and other computers used the 6502 microprocessor. In October 1976 Commodore bought MOS Technology and selected the Microsoft 6502 BASIC interpreter for its new PET computer. Commodore placed the BASIC program in ROM that resulted in every Pet computer having Microsoft BASIC.

Microsoft, also offered its simulator, debugger and assembly development software called Develop-80, Develop-68 and Develop-65 for sale without success.

General Electric purchased the Microsoft 8080 BASIC interpreter in late 1976. Then in November 1976 Paul Allen left MITS and joined Microsoft full time. Revenue for the company's first year was over $100,000.

In December 1976, Pertec Computer Corporation signed a letter of intent to purchase MITS Inc. Microsoft had a number of potential customers for sub licensing through MITS during early 1977. However MITS was not promoting the business or using its "best efforts" as agreed to in the licensing agreement with Microsoft. Microsoft advised MITS in April that it was terminating its license agreement with them. In May MITS filed a restraining order that prohibited Microsoft from marketing 8080 BASIC and requested settlement of the contract by arbitration. Then MITS finalized the sale of their company to Pertec in late May. In late 1977 the arbitrator decided in favor of Microsoft. Although Pertec could still sell the BASIC interpreters, they no longer had an exclusive license. Microsoft could now market their products independent of Pertec and also retain the full sales revenue.

Gates left Harvard in January 1977 and was now full time at Microsoft. Then in February, Gates and Allen formalized their partnership agreement. Gates would have a sixty-four percent share and Allen thirty-six percent.

Microsoft had decided to expand its software market from BASIC interpreters to other programming languages such as COBOL, FORTRAN and Pascal. The company also decided to write the software for use with the Digital Research CP/M operating system. The first language to be released was a FORTRAN-80 compiler with a price of $500 in July 1977.

Apple Computer purchased a license for the 6502 BASIC interpreter in August. Previously Apple had used Integer BASIC written by Stephen Wozniak. Apple released the interpreter with modifications by Randy Wigginton of Apple as Applesoft BASIC. Microsoft had also started development of a new version of BASIC for Texas Instruments (TI) that had to be in compliance with a new ANSI BASIC standard. Paul Allen once again created a simulator for the TI TMS9900 microprocessor and Gates hired Bob Greenberg to develop the BASIC software.

Another important contract from NCR required development of a disk version of BASIC for their 8200 terminal. The company assigned Marc McDonald to the project who developed a new disk formatting concept that used a File Allocation Table (FAT). The FAT controlled the sequence of data stored on a disk and improved the performance of disk operations. The company used the concept of a file allocation table in Microsoft Stand-alone Disk BASIC. Also in a Microsoft operating system project called MIDAS and later in QDOS by Tim Paterson of Seattle Computer Products.

Microsoft appointed Steve Wood as general manager to replace Richard Weiland who had left the company in September 1977. With the arbitrator's settlement of the MITS/Microsoft dispute in September, Microsoft now closed a number of contracts that had been pending with other companies. Microsoft was starting to dominate the BASIC language market by the end of 1977. The new major manufacturers of microcomputers such as Apple Computer and Commodore had licensed BASIC interpreters from Microsoft. A large contract was with Radio Shack. Microsoft hired Bob O'Rear in 1977 who then started adapting Microsoft BASIC to the Radio Shack TRS-80. The interpreter would be an alternative to the limited one developed internally by Radio Shack.

By early 1978 the market place had changed. The MITS Altair computer had lost its dominant role. Microsoft's relationship and revenue from Pertec were no longer significant. Microsoft required additional space and started to question Albuquerque as a location for expansion. In March 1978 Microsoft decided to relocate from Albuquerque to Bellevue, a city just east of Seattle in the state of Washington. The move was to take place around the end of the year.

A number of chip makers such as Intel and National Semiconductor were now purchasing Microsoft products. Richard Weiland had returned to Microsoft in January. He worked on the development of COBOL that Microsoft announced as COBOL-80 for the CP/M operating system at a price of $750 in April 1978. The Heath Company also purchased Microsoft BASIC and FORTRAN software in 1978. Then the Heath Company software developer, Gordon Letwin joined Microsoft in late 1978.

In mid 1978 Vern Raburn of GRT (Great Records and Tapes) convinced Gates to supply software to the company for distribution to the retail market. Microsoft released versions of BASIC interpreters for the Processor Technology Sol and Southwest Technical Products computers. Microsoft also released the enhanced Level II BASIC that had been under development for the Radio Shack TRS-80 computer.

In June 1978, Gates negotiated an agreement with Japanese entrepreneur Kazuhiko (Kay) Nishi and his ASCII Corporation. Nishi would be the exclusive agent for Microsoft products in East Asia and receive a lucrative 30 percent commission on all sales. This resulted in the establishment of Microsoft's first international sales office in November, ASCII Microsoft.

Microsoft started developing a simulator program and a BASIC interpreter for the new Intel 8086 microprocessor in 1978. Microsoft hired Jim Lane to develop the simulator software. Bob O'Rear also worked on the simulator software and the 8086 BASIC for the new microprocessor. With Motorola announcing the 68000 and Zilog the Z-8000, 16-bit microprocessors were going to be the future technology

Prior to the move from Albuquerque in December 1978, Gates and Allen had a staff of twelve employees. Microsoft had also finished its first million-dollar sales year.

Figure 6.2: Albuquerque staff prior to move in 1978. Top row, left to right: Steve Wood, Bob Wallace, Jim Lane. Middle row, left to right: Bob O'Rear, Bob Greenberg, Marc McDonald, Gordon Letwin. Bottom row, left to right: Bill Gates, Andrea Lewis, Marla Wood, Paul Allen.

Photograph is courtesy of Microsoft Corporation.

6.4 ... Relocation to Seattle

Microsoft relocated from Albuquerque to new office facilities in Bellevue, Washington in January 1979. The company also obtained its own DEC 2020 minicomputer system.

Microsoft had close to 100 OEM customers for BASIC, FORTRAN-80 and COBOL-80 in March 1979. They also had hired Steve Smith as director of marketing. Work had started on the development of a BASIC compiler and a Pascal programming language.

In the Far East, agent Kay Nishi was generating additional lucrative revenue for Microsoft. Two large Japanese companies, NEC (Nippon Electric Company) and the Ricoh Company signed contracts for Microsoft software.

Vern Raburn of GRT was out of a job due to financial difficulties at the company in 1979. In June he joined Microsoft as president of a new group called the Consumer Products Division. This division would be responsible for selling products directly to the retail market. Also released in 1979 were software products such as TRS-80 Level III BASIC, Typing Tutor and a game adapted by Gordon Letwin for microcomputers called Adventure.

Raburn was also responsible for promoting an interest in developing other software products for the Apple II computers. Apple II computer sales, and more important to Microsoft, software sales for the Apple II, were increasing significantly. This resulted in the concept by Paul Allen to create an interface card for the Apple computer that would allow the CP/M versions of FORTRAN and COBOL to operate on the machine. Microsoft hired Tim Paterson of Seattle Computer Products as a consultant to develop the Apple II card. Microsoft also hired Neil Konzen, an Apple computer programming enthusiast to develop the software. The function of the card evolved from an interface for Microsoft FORTRAN-80 and COBOL-80, to a general interface for the Digital Research CP/M operating system. The card used a Z-80 microprocessor and Microsoft named it the Z-80 SoftCard. Microsoft had to pay Gary Kildall of Digital Research $50,000 for a license to use the CP/M operating system. Subsequently Microsoft hired Don Burtis to improve the card design.

Microsoft completed the 8086 BASIC interpreter developed by Bob O'Rear and introduced it in June 1979. Tim Paterson of Seattle Computer Products had completed a card system for the S-100 bus using the new Intel 8086 microprocessor in May. Paterson had designed the card system and Microsoft tested the new BASIC interpreter using a Cromemco computer. Then Microsoft demonstrated the new BASIC interpreter for the 8086 at the June 1979 National Computer Conference in New York City.

In August 1979 Gates visited EDS (Electronic Data Systems) in Dallas. EDS was a computer service company for mainframes owned by H. Ross Perot. As a result of these discussions, EDS made an offer to purchase Microsoft. However Microsoft rejected the offer due to financial differences.

Microsoft released in 1979, a Macro Assembler language for the Intel 8080 and Zilog Z-80 microprocessors in August, a BASIC Compiler with a price of $395 and the Edit-80 text editor with a price of $120.

Microsoft's revenue had reached $2.4 million with a staff of 28 employees by the end of its 1979 fiscal year. The revenue and number of employees had doubled approximately, during the past year.

Blank page.

Chapter 7 Other Software in the 1970's

7.1 ... Operating Systems

Prior to the introduction of disk drives, software was developed to facilitate the loading or "booting" at startup. An example of this is the Motorola mini operating system for automatic loading called "Mikbug."

IBM developed the floppy disk drive in 1971 and the hard disk drive in 1973. Shugart Associates released their 8-inch floppy disk drive in 1973. With the availability of disk drives, many manufacturers released their own operating systems. However, the dominant system was CP/M from Digital Research.

Digital Research

Gary A. Kildall received a Ph.D., in computer science from the University of Washington, founded Microcomputer Applications Associates (MAA) and became a professor at the Naval Postgraduate School in California in 1972. MAA was the predecessor to the founding of Intergalactic Digital Research by Kildall and his wife Dorothy McEwen, in 1976. The company name changed to Digital Research, Inc., in 1979.

In 1972, Kildall purchased an Intel 4004 microprocessor and developed emulator and assembler programs for it on an IBM System/360 computer at the naval school. While continuing to teach at the naval school, Kildall's MAA provided consulting services to Intel. This resulted in the development of an emulator for the Intel 8008 microprocessor to run on Intel's DEC PDP-10 time sharing system. Kildall also developed a new systems programming language called PL/M (Programming Language for Microcomputers) that was released in 1973. Around this time MAA made a proposal to Intel to develop a disk operating system for the 8000 series microprocessors.

Intel rejected the proposed operating system. However, Kildall continued working on the program and named the new operating system Control Program for

Microprocessors (CP/M). Initially, it would form the basis for the resident programming of PL/M for 8080-based computers with 16K bytes of main memory and Shugart's new disk drive. This required the design of a disk controller and the assistance of a friend John Torode to get it working. The operating system had features such as commands and file naming conventions similar to those used on the DEC PDP-10 system. CP/M included a single-user file system, and used recoverable directory information to determine storage allocation, rather than a traditional linked-list organization. MAA completed CP/M in 1974 and retailed the program for $70. It soon became successful as the dominant 8-bit operating system for microcomputers using the Intel 8000 and Zilog Z-80 series of microprocessors.

In the mid 1970's after several implementations on computer systems with different hardware interfaces, the CP/M software was restructured. CP/M was decomposed into two parts: an invariant part that was written in PL/M and a small variant part was written in assembly language. This small variant module for interfacing to various hardware platforms became known as the Basic Input/Output System (BIOS). Computer suppliers and end users could now create their own physical input/output drivers for CP/M.

In late 1979, Digital released an enhanced Version 2.0 of CP/M that sold for $150. The program had been completely redesigned to support floppy disk drives and high-capacity Winchester disk drives. All disk parameters were moved from the invariant part to a table driven concept in the variant module.

Kildall also developed other programs for use with CP/M. Some of those were an assembly language, text editor and various utilities. Digital Research also developed a multi-terminal operating system called MP/M (Multi-Programming Monitor). It provided real-time processing with multiprogramming and multi-terminal features. The program was compatible with CP/M and sold for $300.

Apple Computer

DOS (Disk Operating System) Version 3.1 was the operating system released with the Apple Disk II drive in June 1978. The disk had 35 tracks with thirteen 256 byte sectors on each track for a total storage capacity of 113K bytes. Earlier versions were not completely functional and therefore not released. Apple released a more stable version, DOS 3.2 in mid-1979.

DOS 3.3 evolved from the release of Apple Pascal programming language in 1979. This release changed the 35 track disk format to sixteen 256 byte sectors with a total disk storage capacity of 143K bytes. Apple developed a utility called Boot 13 to boot the 13-sector-per-track disks.

Other Operating Systems

Bill Levy developed PT-DOS for Process Technology around 1976/77. Radio Shack released TRSDOS for the TRS-80 Mini Disk System in the late 1970's. It was not compatible with CP/M.

7.2 ... Programming Languages

BASIC

A history of the BASIC programming language is provided by Thomas E. Kurtz in *History of Programming Languages* 36], pp. 515-549 and by John G. Kemeny and Thomas E. Kurtz in *Back to BASIC* [115], pp. 1-23. A time chart depicting the evolution of BASIC is provided by Russ Lockwood in a periodical article entitled *The Genealogy of BASIC* [397]. Bill Gates of Microsoft has provided an interesting history of BASIC in a periodical article entitled *The 25th Birthday of BASIC* [394]. Reference Section 2.2 for the initial development of BASIC and Chapter 6 for the Microsoft development of BASIC interpreters for the Altair and other microcomputers.

Dartmouth College made a significant upgrade to their BASIC compiler with the release of version six in September 1971. In 1974, the American National Standards Institute (ANSI) formed a committee to develop standards for the BASIC programming language. This resulted in the release of a standard for Minimal BASIC in 1976 and its official approval in 1978. Work then proceeded on a standard for a "full" BASIC. Dartmouth released version seven of its BASIC compiler in 1979.

Various hardware manufacturers such as Apple Computer, Digital Group, IBM, PolyMorphic Systems and Processor Technology developed BASIC languages for their own computers. However, Tiny BASIC and the following are some of the more significant releases.

Dennis Allison who was a member of the computer science faculty at Stanford University developed Tiny BASIC. The initial version of Tiny BASIC developed by Allison was a simplified BASIC oriented to younger programmers. The program required less than 4K bytes of memory. The *PCC Newsletter* and the initial issue of *Dr. Dobb's Journal of Computer Calisthenics and Orthodontia* in January 1976 provided a detailed description of an extended version of the software. Other programmers such as Tom Pittman and Li-Chen Wang developed and distributed variations of the program for different computers [393].

Robert Uiterwyk developed SwTPC BASIC for the SwTPC 6800 microcomputer in 1975. SwTPC provided low-cost BASIC programs at $1 per kilobyte. A 4K BASIC interpreter cost $4, 8K $8 and 12K $12.

Gordon E. Eubanks developed E-BASIC while working on a masters degree in computer science at the Naval Postgraduate School in California. Eubanks was associated with Gary Kildall and E-BASIC became widely used with the CP/M operating system. Gordon Eubanks, associates Alan Cooper and Keith Parsons developed C-BASIC and founded Compiler Systems, Inc., to market the software. Eubanks subsequently sold the Compiler Systems company to Digital Research and became one of Digital's vice presidents. C-BASIC was a pseudocompiled language developed in 1977 for IMSAI and was included with the CP/M operating system in 1979.

Radio Shack released Level-I BASIC with the TRS-80 Model I computer in August 1977. Steven Leininger developed the interpreter by adapting it from Tiny BASIC. Radio Shack released Level II BASIC developed by

Microsoft for business and advance applications in 1978 and an enhanced Level III version in 1979.

C

Dennis M. Ritchie created the C language at AT&T's Bell Laboratories in 1972. The language was designed to be portable, fast and compact. The UNIX operating system was later reprogrammed using the C language.

FORTRAN

Reference Section 1.4 for the initial development of FORTRAN. Microsoft developed a FORTRAN-80 compiler for the Intel 8080 microprocessor. They announced the program in April 1977 and sold it for $500.

Pascal

Niklaus Wirth developed the Pascal language at the ETH (Eidgenossische Technische Hochschule) in Zurich Switzerland. Pascal evolved from the ALGOL 60 programming language. The main development principals were to provide a language suitable for structured programming and teaching. Wirth drafted a preliminary version in 1968 and the first compiler became operational in 1970.

Kenneth L. Bowles directed the development of UCSD Pascal at the University of California in San Diego. UCSD released the program to users in August 1977 as a complete interactive system for microcomputers and minicomputers. It was initially released for Digital Equipment Corporation (DEC) LSI-11 or other PDP-11 processors, 8080 and Z80 microprocessors. The software system cost $200. Bill Atkinson of Apple Computer, adapted the UCSD Pascal for the Apple II computer in 1979.

Other Languages

Gary Kildall of Digital Research, developed PL/M (Programming Language for Microcomputers) for Intel in 1972. PL/M was a system programming language that developed to provide a simpler alternative to assembly language for the Intel 8000 series of 8-bit microprocessors. It was a refinement of the Stanford University XPL compiler writing language with elements from Burroughs Corporation's ALGOL and IBM's PL/I. Intel marketed the program for the 8000 series of microprocessors in 1973.

BCPL and MESA were systems programming languages developed at the Xerox PARC (Palo Alto Research Center) for the Alto personal computer in the early 1970's.

Alan Kay developed Smalltalk at the Learning Research Group (LRG) of Xerox PARC in 1972. It was the software part of the Dynabook concept and was created for the Alto computer. It is one of the first object-oriented languages and used interactive graphical concepts to create a user friendly environment.

7.3 ... Word Processors

Bravo was a word processing program developed by Butler Lampson and Charles Simonyi for the Xerox Alto personal computer in the early 1970's. It was one of the earliest word processors to feature What-You-See-Is-What-You-Get (WYSIWYG) text display on the terminal screen. Between 1976 and 1978 improvements were incorporated by Simonyi in a new version of the word processor called BravoX. During this time Tim Mott and Larry Tesler developed a text editor called Gypsy that included a new cut-and-paste feature.

The Electric Pencil evolved from a public domain software package called Software Package One (SP-1). The package was distributed by the Southern California Computer Society (SCCS) in the fall of 1975. Michael Shrayer improved the editor portion of the package and called it Extended Software Package 1 (ESP-1). A further upgrade of ESP-1 was called Executer. ESP-1 and Executer were the basis for the first word processor for a microcomputer. Shrayer named it "The Electric Pencil" and it became available in December 1976. The first version was written for the MITS Altair microcomputer. Shrayer founded the Michael Shrayer Software company and other versions of the program were developed for various microcomputers. An improved version, The Electric Pencil II was announced in early 1978.

Seymour Rubinstein was the director of marketing for IMSAI when he left to start his own company, MicroPro International Corporation in late 1978. He hired Bob Barnaby, a programmer who had also worked at IMSAI and created a program called NED (New Editor). Barnaby extended this program into a full-scale word processor for microcomputers. Barnaby developed two programs, a video text editor named Word-Master and a sort/merge program named Super-Sort. MicroPro released Word-Master at a price of $150 in August 1978 and then an improved version named Word-Star at a price of $495 in June 1979. Word-Star was a success and became a dominant word processor used on early CP/M microcomputers.

Apple Writer was created for the Apple II computer by Paul Lutus, at a mountain cabin in the wilderness of Oregon in 1978. Lutus sold the program to Apple Computer for a flat fee of $7,500. Apple Computer released the program that sold for $75 in 1979. It featured automatic search and replacement of words or phrases, justification of text and uppercase and lowercase type.

John Draper developed EasyWriter for the Apple II computer in 1978. Shortly after, Draper met Bill Baker of Information Unlimited Software (IUS) at the third West Coast Computer Faire in the spring of 1979. This meeting resulted in an agreement being reached for IUS to market the program.

Magic Wand was introduced as an easy-to-use word processor in late 1979 by Small Business Applications, Inc.

7.4 ... Spreadsheets

Daniel S. Bricklin conceived the concept for a spreadsheet during his studies for an MBA at the Harvard Business School in the spring of 1978. Bricklin already had a degree in electrical and computer science from MIT. He had also been a software engineer at Digital Equipment Corporation (DEC). The impetus for this concept was the desire to find a way of utilizing a computer to facilitate the financial analysis of varied business situations. A prototype of the program was written in BASIC and called Calculedger, a combination of calculator and ledger.

It was at MIT that Bricklin had become friends with Robert Frankston who would co-develop the software for the spreadsheet program. Frankston had done some programming for Daniel Fylstra of Personal Software, who loaned Bricklin and Frankston an Apple II computer to develop the software.

Bricklin and Frankston formed their own company called Software Arts, Inc., in January 1979 to complete the development of the spreadsheet software. The program was now called VisiCalc, that is an acronym for Visible Calculator. Bricklin developed many of the concepts, data structures, documentation and specifications. Frankston did most of the program coding using assembler language and macros. Assembler was used to improve the speed and to allow the program to run on a 24K byte Apple II computer, 32K bytes with a disk. The limited memory restricted the spreadsheet to 63 columns by 254 rows and reduced the number of features incorporated. Recalculation was limited to across rows or down columns, column widths could be varied, but had to be the same and text could not span columns.

Fylstra offered to sell the program to Apple Computer in January 1979 for $1 million. However Steven Jobs and Mike Markkula rejected the offer. Bill Gates of Microsoft is also reported to have rejected an offer to purchase the program. However, subsequently Arthur Rock and Venrock Associates assisted in financing Personal Software and the new program.

Software Arts signed an agreement with Fylstra of Personal Software, Inc., to market VisiCalc in April 1979. VisiCalc was introduced at the West Coast Computer Faire in May, demonstrated at the National Computer Conference in June, advertised in the September issue of *Byte* (page 51) and released in October for the Apple II computer. VisiCalc was priced at $99.50 then quickly increased to $150 after sales increased dramatically. One limitation was that it could not run on the CP/M operating system, however it was an instant success. The term "killer application" has been credited to the success of VisiCalc. It also became a significant factor in helping to sell Apple II computers.

7.5 ... Databases

Lyall Morill developed a simple database program for microcomputers called WHATSIT? in 1977. WHATSIT? is an acronym for "Wow! How'd All That Stuff get In There?" Morill improved WHATSIT? and Information Unlimited Software (IUS) introduced the program at the second West Coast Computer Faire in the spring of 1978. Bill Baker had previously founded IUS while attending college.

C. Wayne Ratliff was an engineer who adapted a NASA Jet Propulsion Laboratory mainframe database to his IMSAI 8080 microcomputer in his spare time. On completion of the software in August 1979, he named it Vulcan. The software was marketed by Software Consultation Design and Production (SCDP) company and advertised in the *Byte* magazine at a price of $490 without success. The software was subsequently marketed by Ashton-Tate who changed the name to dBASE II in 1981.

Other early database programs were Condor, FMS 80 and Selector.

7.6 ... Miscellaneous

Games

Various games had been developed for use on larger computers in the 1950's and 1960's as described in Sections 1.4 and 2.6. Then in the early 1970's video games that used dedicated processors were introduced and became very popular. Most of these video games used high resolution graphics and sound effects that would subsequently be implemented on more powerful microcomputers.

Nolan K. Bushnell developed the first commercial video game called Computer Space in 1970. It evolved from his interest in games and his previous exposure to the Space Wars game at the University of Utah. Bushnell subsequently founded Atari Corporation (see Section 4.6). This led to Stephen Wozniak developing an Apple Computer version of a game called Breakout that he and Steven Jobs had worked on for Atari. Also released at Apple Computer was a program called Lunar Lander developed by Bob Bishop. Bill Budge also developed a number of game programs, such as Penny Arcade, that he sold to Apple Computer in 1979.

In the early 1970's, Will Crowther developed a non-graphic fantasy game that was set in a cavern world with hidden treasure and challenging features such as dragons, flying horses and trolls. Crowther released the game on the ARPANET. The program was then refined by Don Woods and became known as the Adventure game. It became highly popular and formed the basis for personal computer Adventure games by Adventureland International and Microsoft.

Adventure Land was one of the earliest text adventure games for personal computers. Scott Adams developed the program for the Radio Shack TRS-80 computer in 1978. Adventure Land required a player to search through a magic realm that had wild animals, perils and mysteries to locate treasures. Adams founded Adventure International in 1978 to produce, be a distributor and publisher of other computer games. The game was adapted for the Apple II and other games such as Laser Ball, Fire Copter and Pirate Adventure followed.

Peter R. Jennings developed Microchess initially, for the MOS Technology KIM-1 microcomputer in 1976. Jennings sold the source code for $15. Shortly after, Daniel Fylstra and Jennings founded Personal Software, Inc. to market Microchess and other game programs.

Another chess playing program was SARGON, that was released around 1978. It was developed by Dan and Kathe Spracklen.

Toru Iwatani designed the Pac-Man game at a Japanese company called Namco Limited. It was first introduced in Japan in the late 1970's. Atari licensed the rights for Pac-Man and it became very successful in North America.

Other Software

Radio Shack issued a variety of software in the late 1970's for their TRS-80 systems. Some of the programs were: General Ledger, Inventory Control System, Real Estate, Statistical Analysis and various computer games. The programs were provided on cassettes and 5-inch floppy disks.

Stephen Wozniak developed SWEET16 in 1977 as an interpreter program that was contained in the initial Apple II ROM memory chip. Wozniak called it a 16-bit "metaprocessor" [402]. It was used to manipulate 16-bit pointer data and its arithmetic on the 8-bit Apple II computer.

Mitchell Kapor developed Tiny Troll in 1978/79 with help from Eric Rosenfeld of MIT. The program displayed line charts, multiple regressions, statistical analysis information and had a text editor. The software formed the basis for the later development of VisiPlot and VisiTrend programs.

Conclusion

In the 1970's, software had developed in conjunction with personal computer technology. Initially it had focused on programming languages such as BASIC, and operating systems such as CP/M to support disk drive technology. However in the late 1970's a change in user orientation from the technical enthusiast to the mass market consumer occurred. This was supported by the release of application software such as the VisiCalc spreadsheet.

Part III

1980's -- The IBM/Macintosh Era.

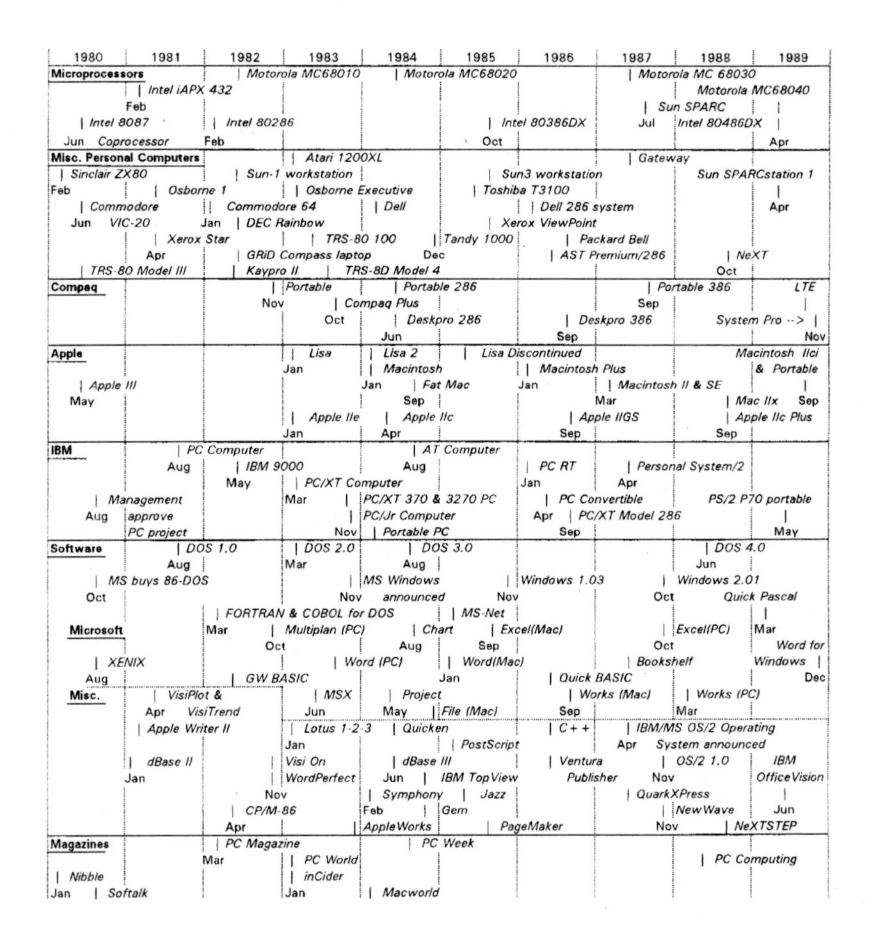

Figure 8.1: A graphical history of personal computers (1980's) - The IBM/Macintosh era.

Chapter 8 Microprocessors in the 1980's

8.1 ... Intel

Microprocessors

The iAPX 432 (Intel Advanced Processor Architecture) which was now a 32-bit microprocessor, was introduced in February 1981. Principals in the development were William Lattin and Justin Rattner. The chip was an advanced design with an innovative architecture. It supported data store using multiple pointer levels, fault tolerance, memory error correction, multiprocessing and object-oriented software. The microprocessor was described as a micromainframe computer in the May 3, 1982 issue of *Fortune* magazine. However, due to performance deficiencies the product was discontinued.

Intel announced the 80186 and 80188 high integration 16-bit internal data path microprocessors in 1982. Both processors were designed for embedded applications in computer peripherals and other electronic products.

The 80286 microprocessor was introduced in February 1982 and was four times more powerful than the 8088. The chip has 134,000 transistors, a 16-bit internal data path and at a clock speed of 8 MHz it has a rating of 1.2 MIPS (million instructions per second). The microprocessor featured on-chip memory management to support multitasking. It also had an on-chip security system for data protection. The memory addressability is 16 megabytes and the microprocessor is available at clock frequencies of 8, 10 and 12.5 MHz. The price at introduction was $360. This microprocessor was selected by IBM for the PC AT computer released in August 1984.

A group of engineers led by John Crawford developed Intel's 80386DX microprocessor. It had full 32-bit capability and preserved software compatibility with the previous 8086 and 80286 architectures. The 80386DX was introduced in October 1985 and was approximately fifteen times more powerful than the 8088.

The chip has 275,000 transistors, a 32-bit internal data path and at a clock speed of 16 MHz a rating of 6 MIPS. The memory addressability is 4 gigabytes and the microprocessor became available at clock frequencies of 16, 20, 25 and 33 MHz. The price at introduction was $299.

The 80386DX was selected by Compaq for the Deskpro computer released in September 1986. This was the first product application of the microprocessor. Intel had tried to get IBM to incorporate the new microprocessor in their product line. However, IBM had concerns regarding the processors power affecting their microcomputer sales and were therefore slow in adopting the 80386. Initially the 80386 microprocessor was not second sourced by Intel to other competitors, other than IBM for a portion of its internal use. The 80386 became a very successful product and started to contribute significantly to the company profits.

The 80386SX microprocessor was introduced in June 1988. The chip has 275,000 transistors, a 32-bit internal bus with a 16-bit external bus and at a clock speed of 16 MHz a rating of 2.5 MIPS. The memory addressability is 16 megabytes and the microprocessor is available at clock frequencies of 16, 20, 25 and 33 MHz.

The 80486DX microprocessor was introduced in April 1989. The chip has 1.2 million transistors, 1.0 micron minimum feature size and a 32-bit bus. This was the first Intel processor to incorporate a Level 1 (L1) cache of 8 KB for faster data access. At a clock speed of 25 MHz it has a rating of 20 MIPS. The microprocessor included an integrated floating-point unit. The memory addressability is 4 gigabytes and the microprocessor became available at clock frequencies of 25, 33, 50, 60, 75, and 100 MHz. The price at introduction was $950.

Coprocessors

The concept for a coprocessor evolved at Intel from the 8086 microprocessor in 1976. This resulted in a floating-point extension to the 8086 instruction set and a systems interface architecture. In 1987 the development of the coprocessor was assigned to the Intel design center in Haifa, Israel.

The 8087 math coprocessor added a set of floating-point instructions to the 8086/88. It was the first implementation of the IEEE standard for floating-point mathematics. Use of the coprocessor resulted in a significant increase in the speed of mathematical computations. The 8087 coprocessor was released in June 1980. Intel announced the 82786 graphics coprocessor in May 1986.

Corporate & Other Activities

Competitive pressures in the microprocessor market from companies such as Motorola and Zilog, resulted in the implementation of a sales campaign called "Operation Crush" in early 1980. Intel had been loosing market share, mainly to Motorola whose microprocessor products were perceived as being superior. Numerous activities were initiated to communicate the overall advantages offered by the company, and the sales personnel were assigned goals to increase the number of design wins for the use of Intel chips in customer products. By the end of 1980, the campaign became very successful. One major design win with far reaching consequences for Intel, was in Boca Raton, Florida for the IBM Personal Computer.

Intel and Advanced Micro Devices (AMD) negotiated a ten year technological exchange agreement, and AMD became a second source for the 8088 microprocessor in February 1982. This was largely the effect of a desire by IBM to have an alternate source for the microprocessor in its new personal computer.

IBM purchased 12 percent of Intel Corporation for $250 million in December 1982. Intel was having financial problems due to intense competition from Japanese manufacturers of memory chips. Then in 1983-84, IBM increased its investment in Intel to 20 percent.

In 1984 Intel approved a project for a line of parallel processing supercomputers. The company also licensed AMD as a second source for the 80286 in 1984.

The semiconductor industry had been enjoying a boom market until mid-1984, when demand slowed dramatically. This resulted in excess capacity across the industry and prices collapsed between 1985 and 1986. At Intel, a reduced demand for microprocessors and termination of the company's DRAM chip business resulted in significant losses in 1986. This resulted in plant closings and the termination of over 8,000 employees. However, in 1987 conditions improved. Intel started to report profitable income and Andrew Grove became the chief executive officer.

During 1986, Intel concluded a technological exchange agreement with IBM. IBM received rights to manufacture up to half of its own requirements for the Intel 80386 microprocessor and to develop an enhanced design for its own use and external sales. Intel received a number of IBM technologies, such as advance chip packaging. IBM started reducing its investment in Intel Corporation in 1986, and completed its divestiture of Intel shares due to its own financial problems in December 1987.

Robert Noyce, who had been moving towards semi-retirement, accepted a position as chief executive officer of SEMATECH, Inc., in mid-1988. SEMATECH is an acronym for SEmiconductor MAnufacturing TECHnology. The company was founded by the U.S. Government and a group of leading U.S. semiconductor manufacturers to conduct research that would help combat competition from Japan.

8.2 ... Motorola

Motorola introduced 10 and 12 MHz versions of the MC68000 microprocessor by the end of 1981. The company then introduced the MC68010 in 1982 and the 32-bit MC68020 in 1984. The MC68020 used 2.5 micron technology, had 200,000 transistors, a 256 byte cache and executed instructions at 2.5 MIPS. Clock speeds are 16-33 MHz.

The MC68030 32-bit unit has all the features of the MC68020 plus a paged memory management unit, separate 256 byte caches for data and instructions and additional enhancements. It executes instructions at 12 MIPS and clock speeds are 16-50 MHz. The MC68030 was introduced in 1987 and was used on the Apple Macintosh IIx computer.

Motorola announced the 88000 family of Reduced Instruction Set Computing (RISC) microprocessors in

1988. They were designed for applications such as multiprocessing and high performance graphics.

The MC68040 is a 32-bit microprocessor that executes instructions at 20 MIPS. It contains 1.2 million transistors, has a 4K byte instruction cache, 4k byte data cache and a floating-point unit. Clock speeds are 25-40 MHz. The MC68040 was announced in April 1989.

8.3 ... Other Microprocessors

The National Semiconductor 16032 was a microprocessor with a 16-bit external data bus and a 32-bit internal bus. It was announced in 1981. The 32032 was the first full 32-bit microprocessor. In 1987, National Semiconductor acquired Fairchild Semiconductor that had been having problems.

The Western Design Center 65802 and 65816 microprocessors were designed by Bill Mensch and announced in 1984. The W65C816 is a 16-bit microprocessor that is used in the Apple IIGS computer.

Gordon Campbell who had previously been with Intel Corporation founded Chips and Technologies, Inc. The company started by producing low cost chip sets for the IBM PC AT computer.

The Zilog Z-80000 is a 32-bit microprocessor and the Z280 is a 16-bit version of the Z-80 that was announced in 1987.

RISC Microprocessors

John Cocke developed the primary concepts for the Reduced Instruction Set Computing (RISC) technology at IBM starting in the 1960's. It evolved from his research on optimizing the interaction between hardware and software. Cocke determined that the overall speed of execution could be increased by reducing the number of complicated instructions to a relatively small set of simple optimized instructions.

The RISC architecture was first implemented on two experimental computers and on the IBM 801 minicomputer in 1978. Then in 1980, IBM's Austin laboratory developed the ROMP (Research Office products MicroProcessor) RISC microprocessor for the office products division. The original intent was to use the microprocessor in a networked office workstation. The first IBM personal computer to use the ROMP microprocessor was the PC RT workstation introduced in January 1986. IBM's mainframe and personal computer divisions did not support the application of RISC technology on their products. This allowed other companies to exploit the technology during the last half of the 1980's.

David Patterson from the University of California at Berkeley, who after evaluating the complexity of the DEC VAX computer instruction set, developed the RISC-I microprocessor in 1980. John L. Hennessy also did research on RISC microprocessors at Stanford University, and with Skip Stritter, also from Stanford and John Moussouris of IBM, they founded MIPS Computer Systems in 1984. MIPS released the 32-bit R2000 RISC microprocessor, that had 185,000 transistors in 1986.

The SPARC RISC microprocessor was developed at Sun Microsystems by Anant Agrawal with assistance from Robert Garner, William Joy and David Patterson from the University of California at Berkeley. SPARC is an acronym for Scalable Processor Architecture. It was introduced by Sun in July 1987 and used on their SPARCstation 1 in 1989.

8.4 ... Other Corporate Developments

The Exxon corporation acquired Zilog Inc., in 1981 then sold it to Zilog executives in 1989.

Texas Instruments introduced its first single-chip digital signal processor (DSP), the TMS320 in 1982. This subsequently became a major product line for the company.

In April 1985, AT&T purchased the assets of Synertek, Inc., a subsidiary of Honeywell for an estimated $25 million. Then in November a French company, Thomson-CSF purchased Mostek, a subsidiary of United Technologies for $70 million.

AMD started litigation to obtain a license as a second source for the 80386 microprocessor in 1987. Intel wanted to restrict the second sourcing of the microprocessor. However, the technological exchange agreement that Intel had agreed to with AMD in 1982 would become a problem for Intel. During the litigation arbitration in 1989, AMD decided to make an independent clone design of the 80386.

Blank page.

Chapter 9 The IBM Corporation

9.1 ... Introduction

Prior to the introduction of the PC computer, IBM was in a state of transition. The US Federal Government Department of Justice was ending a long period of litigation into IBM's monopolistic and anti-competitive practices. Although IBM would be successful in its defense of the governments charges, it had caused problems for the corporation. A moderation in marketing and product line expansion had occurred during this sensitive period of litigation. This resulted in a loss of market share for IBM. This was also the period during which Thomas J. Watson Jr., relinquished his role as the head of IBM. The end of a family dynasty. During this time various organizational changes were evaluated. One of the areas of concern was the size of the organization and the effects of the bureaucracy on new initiatives. Consequently the concept of the Independent Business Units (IBU's) was conceived. Frank T. Cary, the IBM chairman and chief executive officer was quoted as saying that the IBU's "might even teach an elephant (IBM) how to tap dance."

The Entry Level Systems (ELS) unit in Boca Raton, Florida had responsibility for the low cost end of IBM's computer business. It was this division that introduced the unsuccessful IBM 5100 portable computer in September 1975. William C. Lowe was the manager of IBM's Entry Level Systems Unit and was promoted to overall laboratory director in 1978.

It was during the period of 1975 to 1979 that the microcomputer market exploded. It started with the release of the Altair 8800 in January 1975, Commodore Pet in 1976, the Apple II and Tandy Radio Shack TRS-80 microcomputers in 1977. The microcomputers of the 1970's were oriented to the hobbyist type of user. The "hackers" got satisfaction and were indeed fascinated by either the electronic or software complexity of building and operating their own computer. During the late 1970's several microcomputer software releases were laying the foundation for the utilization of microcomputers by business. Those were the word processing program Word-Master in 1978, the Vulcan database in 1979 and of great significance, the financial spreadsheet program VisiCalc in 1979. However at this point in time the business market had not been penetrated to any extent.

The dynamic growth of the microcomputer industry during the late 1970's had not gone unnoticed at IBM. It was recognized by both corporate management and Lowe who initiated a detailed analysis. The leading producers of personal computers in 1979/80 were Apple Computer, Commodore and Tandy Radio Shack. However they had not released products that met the requirements of either the small or corporate business market. This situation was prevalent in not only North America, but also in Europe and Asia. Lowe rationalized that the recognition of IBM as the major international computer manufacturer would be a key factor in the commercial acceptance of a personal computer by their company.

In 1980, IBM released the Model 5120, a desktop version of the unsuccessful 5100 series of portable computers. However, this would not be successful either.

During Lowe's analysis of IBM's possible entry into the personal computer market a number of concerns were identified. The majority of microcomputer developments had been by small entrepreneurial companies. Indeed a number of them such as Apple had been started in a garage. This of course was the opposite extreme from IBM with its extensive bureaucracy. The corporation also tended to engineer all of its components and software in house. This would result in higher than required quality levels that would escalate costs and delay development. Another significant factor was that IBM did not at that time have a microprocessor or the technology. Consequently to compete in the personal computer market would require significant changes at IBM. An organization with greater entrepreneurial type of freedom for development, production and marketing was essential.

Fig. 9.1: William C. Lowe.

Fig. 9.2: Philip D. Estridge.

Figure 9.3: IBM Personal Computer.

Photographs are courtesy of International Business Machines Corporation.

9.2 ... PC Approval and Development

In 1980, John R. Opel was the president of IBM, and all major projects required the approval of the Corporate Management Committee (CMC) at headquarters in Armonk, New York. In July, William Lowe presented to the CMC a personal computer market analysis, his concerns for the product development internally at IBM and two proposals for CMC consideration. One proposal was for IBM to either buy a personal computer company or a personal computer design, such as that from Atari. The second proposal was for IBM to design and build a new personal computer, but to do it outside of the normal corporate structure. To complete this proposal, he requested authorization to assemble a small task force of hand-picked engineers. This group would produce a prototype within thirty days for demonstration to the CMC. Shortly after, the approval to proceed with the new computer proposal was given. Corporate management was anxious to enter the desktop market, the timing had been right.

Funding was granted for twelve engineers to develop the prototype and the detailed commercial proposals. The first person Lowe recruited was Bill Sydnes as manager for the engineering design. Sydnes had been manager of the IBM 5120 desktop computer which had not been a commercial success, but had been developed and produced on time. Lewis Eggebrecht was a principal in the systems engineering design and Joe Bauman was selected to develop the business and manufacturing plans. The rest of the task force was quickly selected. Another member of the founding group was Jack Sams who was in charge of software development. Sydnes and Sams had both been involved in a recent System/23 DataMaster business computer project. This project was delayed by nearly a year due to internal development of the BASIC interpreter. This resulted in the decision to use existing software from outside suppliers. Sams was involved in the initial selection and discussions with Microsoft as a major supplier for software in late July.

The thirty day period to develop and present a prototype to the CMC required that a number of critical decisions be made very quickly by the task force. Some of those were the concept of an open bus architecture, a 16-bit microprocessor, components and peripherals from competitive sources, a software operating system from outside IBM and marketing separate from IBM's sales organization. Maximizing the capabilities of the new computer without affecting the market for IBM's other low end computers required consideration.

In early August Lowe demonstrated the prototype and presented his recommendations for IBM to enter the personal computer market to the Corporate Management Committee. The presentation was a success and the CMC gave approval to form a Product Development Group for the new computer. This group would become one of IBM's Independent Business Units. To maintain confidentiality, code names were assigned to the group and computer. The new group would be known as project "Chess" and the computer as the "Acorn". The timetable required an additional review by the CMC in mid October and the computer to be shipped within one year. The next critical selection for Lowe was a manager for the new group. Lowe had aspirations for higher corporate levels and selected Philip D. Estridge to be the manager of Project Chess in early September.

Don Estridge quickly doubled the engineering staff to twenty six. The final design time frame was extremely short. All of the components had to be state-of-the-art, but be existing and proven in the market. This resulted in a design that was not leading edge, but a conservative product for commercial production and customer acceptance. Some features such as the bus architecture and the keyboard evolved from the IBM System 23 DataMaster computer. David J. Bradley, who had worked on the System/23 DataMaster project, was assigned to develop the control code for the Basic Input/Output System (BIOS). Fully functional prototypes had to be assembled for internal development and outside suppliers of peripherals and software.

By late August IBM was planning for Microsoft Corporation to provide the programming languages.

However the task force was having a problem obtaining a 16-bit version of the popular CP/M operating system from Digital Research, Inc. Then in late September Microsoft made a proposal to supply the operating system (see Section 12.2) and the programming languages. After CMC's final approval of the Chess computer project in October, IBM accepted the Microsoft proposals and a contract was signed in November. IBM also contracted with other software suppliers such as Personal Software to adapt VisiCalc for the new computer.

Other IBM executives were selected to participate in the project and made significant contributions. H. L. Sparks headed marketing and sales, Joseph Sarubbi technical procurement, Dan Wilkie manufacturing and James D'Arezzo communications. D'Arezzo joined the project as manager of communications in January 1981. In March, William Lowe left Boca Raton and became a vice president of the Information Systems Division and general manager of the plant in Rochester, Minnesota. Then in June, Joe Bauman joined Lowe and was replaced by Dan Wilkie as the new director of manufacturing. D'Arezzo in conjunction with Lord, Geller, Federico and Einstein, a New York advertising agency used by IBM, created an advertising campaign based on the Charlie Chaplin tramp characture. The concept provided a friendly and uncomplicated user vision for the new computer introduction that was highly successful. During this time period the name IBM Personal Computer (IBM PC) was selected for the computer. The estimate of the market for the IBM PC was 250,000 units over a five year period. In late July the CMC gave its final approval for the introduction of the IBM PC.

9.3 ... The Original PC

The following are details of the IBM Personal Computer (IBM PC) that was introduced on August 12, 1981 in New York.

The microprocessor selected was the Intel 8088 operating at 4.77 MHz. Internally the microprocessor used the 16-bit instruction set of the Intel 8086 with an 8-bit external data communication bus. An additional socket was provided for the later utilization of the Intel 8087 numeric coprocessor. The memory had 40K bytes of ROM and 16K bytes of RAM, expandable to 64K on the system board and to 256K by adding memory expansion cards. The ROM incorporated the 32K Microsoft Cassette BASIC interpreter and the 8K Basic Input/Output System (BIOS). The BIOS chip provided control of information transfer between elements of the hardware system.

The basic system unit had five 62 pin expansion slots for additional memory, display, printer, communication and game adapter cards. One parallel printer port, one RS-232C serial port and a built in speaker were standard. A separate 83 key adjustable tilt keyboard was connected to the computer by a six foot coiled cable via a serial port. The keyboard incorporated a numeric key pad, ten special function keys and indicator lights to display shift states. The basic system also had an audio cassette recorder connector for mass storage. With a frequency modulator an ordinary television set could be used as a monitor.

Two additional types of display were offered. A monochrome display with a Monochrome Display Adapter (MDA) for business and a color display with a Color Graphics Adapter (CGA) for home use.

The IBM Monochrome Display used an 11.5 inch green-phosphor tube. This monitor required the monochrome adapter card that had 4K bytes of on-board memory. The monitor could display 25 rows of 80 characters. The MDA system provided for 256 characters to support major languages and other items such as business graphics.

The color/graphics monitor adapter enabled connection to a RGB (red-green-blue) monitor, a color television or a black and white monitor. The adapter had 16K bytes of on-board memory and could display two modes of text and three modes of graphics. The first mode of text was 25 rows of 40 characters for color televisions and composite monitors. The second text mode was 25 rows of 80 characters for RGB monitors. The low resolution graphics mode was 100 rows of 160 pixels with 16 colors, but was not supported by IBM. The medium resolution graphics was 200 rows of 320 pixels with 4 colors. The high resolution graphics mode was 200 rows of 640 pixels using a white-on-black image.

The basic unit had provision for two 5.25 inch floppy disk drives manufactured by the Tandon Corporation, The disks were 160K byte single-sided, soft-sectored and double density. The disk operating system was IBM PC-DOS developed by Microsoft.

The printer was an Epson MX-80 with an IBM label. The unit printed bi-directionally at 80 characters per second, with a 9 by 9 dot matrix and a choice of 12 type styles.

Three forms of BASIC were developed by Microsoft and offered by IBM: Cassette BASIC (standard), Disk BASIC and Advanced BASIC, also known as BASICA. Some of the other software available when the PC computer was released were: VisiCalc from Personal Software, three accounting programs from Peachtree Software, EasyWriter word processor from Information Unlimited Software and from Microsoft a Pascal compiler and a fantasy-simulation game called Adventure. IBM also indicated that they would offer Digital Research's CP/M-86 operating system and SofTech Microsystems UCSD p-System which included UCSD Pascal. Communications software was also available to communicate with other computers and for connection to services such as the Dow Jones News/Retrieval Service and The Source.

The basic system unit with 16K bytes of RAM and keyboard sold for $1,565. A system unit with 48K bytes of RAM, keyboard, single floppy disk drive and disk-drive adapter card was $2,235. A monochrome video display was $345 and the printer $755. The combination monochrome display adapter and printer adapter was $335. The 16K, 32K and 64K byte memory expansion cards were $90, $325 and $540 respectively.

A significant marketing decision for IBM was the use of mass merchandising by major retailers such as ComputerLand and Sears, Roebuck and Company to sell the computer. The company also set up a chain of IBM Product Centers in major cities as retail outlets. Large corporate accounts were handled by the Data Processing Division sales force. Another significant decision was the publishing of a *Technical Reference* manual for the IBM PC that provided details of all the system specifications. This was done to facilitate the development of adapter cards and programs by outside suppliers.

The IBM PC was an outstanding success. IBM had orders for 30,000 systems from their own US employees on the announcement day. The only limiting factor on initial sales was the production capacity. Estridge had taken a group of 12 people in 1980 to a work force of 9,500 in 1984. Estridge was named division director of the entry systems business unit in January 1982, and became a vice president of the new Systems Products Division and general manager of entry systems in January 1983. By the end of 1983 IBM had sold 750,000 personal computers.

Figure 9.4: IBM PC/XT Computer.

Figure 9.5: IBM PC AT Computer.

Photographs are courtesy of International Business Machines Corporation.

9.4 ... The Following Models

IBM 9000

The IBM 9000 Instruments System Computer was announced in May 1982 and displayed at the June 1982 COMDEX show in Atlantic City. It was developed by a wholly owned subsidiary IBM Instruments Inc. and marketed as a laboratory instrumentation computer.

The computer used a Motorola MC68000 microprocessor operating at 8 MHz. The memory had 128K bytes of ROM and 128K bytes of RAM expandable to 5.2 megabytes. The unit used a 32-bit Versabus bus standard developed by Motorola. An optional expansion board could accommodate up to five Versabus cards.

The storage system could have up to 4 drives in any combination of 5.25 or 8-inch sizes. The monitor had a 12-inch green-on-black screen capable of displaying 30 lines of 80 characters with a 480 by 768 pixel resolution. The unit had a separate 83-key keyboard, a 57-key user-definable keypad on the main chassis and an optional 200 characters per second in draft mode, four-color dot-matrix printer. IBM developed the real time, multitasking Computer System Operating System (CSOS). The price varied from $5,695 to over $10,000 depending on the configuration.

The PC Series

In February 1982, three projects were initiated that would become the PC AT, PC Junior (PCjr) and the PC/XT. The PCjr was targeted at the low end of the market for home consumers. The PC/XT, with the XT representing extended technology, had a hard disk and was targeted at the professional business market above the PC. The PC AT, with the AT representing Advanced Technology would feature the new Intel 80286 microprocessor. IBM assigned the code name of Circus to the PC AT project.

The Corporate Management Committee (CMC) in Armonk approved all three projects. Product managers for each of the projects were selected to administer the development of the products.

Personal Computer XT (PC/XT)

The product manager selected to develop the PC/XT was Joseph Sarubbi. The PC/XT would be the only project of the three approved in February 1982 to stay on schedule and be released on time.

IBM introduced the PC/XT model in March 1983 in New York City. It included hard disk drive technology but utilized the same microprocessor as the PC. This new model was evolutionary. There had been expectations that IBM would utilize either the Intel 8086 or 80186 microprocessors. However once again IBM had taken a conservative approach to implementation of new technology.

The following are some details of the model. The microprocessor was the Intel 8088, the same as the IBM PC computer. Standard memory was 128K, expandable to 256K on the mother board and to 640K by adding expansion cards. The 40K of ROM contained the Microsoft Basic interpreter and Basic Input Output System (BIOS) software.

A 10 megabyte Winchester hard-disk drive manufactured by Seagate was the significant feature of the standard unit. Also included was a single 5 1/4 inch floppy disk drive, utilizing 360K byte double-sided, double-density disks. An asynchronous communications adapter was standard. The motherboard had eight expansion slots, as compared to five on the PC computer. However three slots were used by the communications adapter, floppy disk drive and hard disk drive adapters. The audio cassette recorder connector that had been on the PC was deleted. IBM also released its first RGB color monitor for both the PC and XT computers.

The cost of the standard unit with 128K of RAM, keyboard, 10 megabyte hard disk drive, 360K floppy disk drive and a asynchronous communications adapter card was $4,995. A monochrome adapter and display was $680. A color graphics adapter and color display monitor was $924. The PC XT model was a huge success and became a workhorse of the business world.

Microsoft made improvements to the operating system software for the PC/XT release. In addition to support for the Winchester hard disk, new features such as a hierarchical file system with sub directories were incorporated into version 2.00 of the operating system. An updated version 2.00 of BASIC was also released that provided advanced support for communications, graphics and music. The generic name of this BASIC interpreter was GWBASIC (Gee Whiz BASIC). The PC-DOS 2.00 operating system and BASIC 2.00 interpreter cost $60 each.

PC Junior (PCjr)

The product manager selected to develop the PC Junior in February 1982 was Bill Sydnes. Sydnes wanted to develop a product for the consumer market that would be sold by mass merchandisers to compete with the Apple II computer at a lower price. Although it would have a somewhat limited capability compared to the PC, its performance capabilities could be improved by the purchase of upgrade features. The code name "Peanut" became associated with the new product. IBM contracted the manufacture of the computer to Teledyne Inc., in Tennessee, a company founded by Arthur Rock and Henry Singleton.

In the summer of 1983, Sydnes resigned from IBM due to differences of opinion with Don Estridge on marketing and other aspects of the PC Junior development. He then joined the Franklin Computer Corporation as vice president for product development. The new manager would be Dave O'Connor. However, O'Connor had inherited design and production problems that delayed the release date from that initially targeted.

The PC Junior (PCjr) was introduced in November 1983 with high expectations as a low-priced home computer. However customer deliveries of the computer did not occur until early 1984. The PCjr had three separate pieces of hardware: the system unit, power transformer and cordless keyboard. Two configurations of the PCjr were released, a standard model and an enhanced model. Both models used the Intel 8088 microprocessor operating at 4.77 MHz with 64K bytes of ROM. The system unit had three expansion slots for 64K bytes of additional memory, a floppy disk drive and a 300-bps (bits per second) internal modem. An expansion bus connector was also provided that could be used to connect a parallel printer. The detached keyboard had 62 keys and used an optical infrared light transmission technology to link between the keyboard and the system unit. The keyboard did not have a numeric key pad or any function keys.

The standard model had 64K bytes of RAM, expandable to 128K and a base price of $669. The enhanced model had 128K bytes of RAM, a capacity to display 80 columns of text and a half-height 5.25-inch, 360K byte double-sided floppy disk drive manufactured by Qume. The enhanced model had a price of $1,269. An IBM Color Display monitor was available at a price of $680.

A new version 2.1 of PC-DOS with a cost of $65 was released for the PCjr. However the memory requirements of the operating system limited the number of application programs that would run on the computer. The system was compatible with the IBM Personal Computer (IBM PC). No other operating systems were offered for the PCjr. The standard model had Cassette BASIC in ROM and an enhanced Cartridge BASIC was available for $75.

In early 1984 sales for the PC Junior were in trouble and production was stopped in June to reduce inventory. The high price, spongy-to-touch "chiclet" style keyboard, limited memory and storage capabilities had resulted in poor customer acceptance.

An advanced version of the PC Junior was introduced in July 1984. Various improvements to enhance the performance were made, such as increased memory and a new typewriter-style keyboard. Then an intensive promotional campaign was launched in the late fall of 1984 to increase lagging sales. However after the holiday season and the end of the promotional campaign sales fell off again. It had been a market failure that resulted in the computer being discontinued in March 1985.

PC/XT 370 and 3270 PC

The Information Systems Division that produced mainframe computers, introduced the PC/XT 370 and 3270 PC computers in October 1983. These products were designed to be a link to IBM mainframe computer systems.

The PC/XT 370 also had a designation of 5160 Model 588. This computer was an enhancement of the PC/XT computer, with three additional boards to emulate IBM System/370 mainframe computers and to function as an IBM 3277 display terminal. In addition to the standard PC/XT Intel 8088 microprocessor, one of the additional boards had three microprocessors. One of the three microprocessors was an Intel 8087 for floating-point arithmetic functions and the other two microprocessors were based on the Motorola MC68000 for emulation of System/370 instructions. The second additional board extended memory by 512K to 768K bytes. The third board provided emulation of the IBM 3277 display terminal. The computer also had one 360K byte floppy disk drive and either a 10 or 20 megabyte hard disk drive. The PC/XT 370 cost $8,995 with a 10 MB hard disk and $11,690 with a 20 MB hard disk. A software package named VM/PC (Virtual Machine/Personal Computer) was required at a cost of $1,000 to interface with a System/370.

The IBM 3270 Personal Computer also had a designation of 5371 with Models 12, 14 and 16 depending on the configuration. The computer combined a standard IBM Personal Computer with an IBM 3270 display terminal. The base computer had 256K bytes of memory, expandable to 640K and a 122-key keyboard with all the keys of a standard PC and a 3270 terminal. A 3270 PC Control Program enabled the computer to concurrently access up to four programs on a host computer, two "notebook" data-storage transfer areas and a PC-DOS application program. The Control Program also allowed a user to define up to seven windows to monitor the programs being accessed. The user could select the color, position and size of any window. A base 3270 Personal Computer with 256K bytes of memory cost $4,130 and the 3270 PC Control Program $300.

Portable PC

The Portable Personal Computer (PC) was introduced in February 1984. The unit measured 8 by 20 by 17 inches and weighed 30 pounds. With this weight, it would become known as a "luggable." The portable used an Intel 8088 microprocessor operating at 4.77 MHz, 40K bytes of ROM and 256K bytes of RAM, expandable to 640K. The unit had one 5.25 inch half-height 360K floppy disk drive with provision for a second drive. A 9 inch amber monitor was built into the unit, seven expansion slots were provided (two used by the floppy disk drive and monitor adapter) and the cost was $2,595. However Compaq had an earlier and better portable which sold at virtually the same price which adversely affected IBM sales and market acceptance.

Personal Computer AT (PC AT)

Two models of the PC AT (Advanced Technology) were introduced in August 1984. This was a significant delay from the release date targeted in February 1982. The variations were a Base model with less memory and no hard disk drive and an Enhanced model.

The microprocessor was the more powerful 16/24-bit Intel 80286 operating at 6 MHz with an optional Intel 80287 Math coprocessor. The permanent memory (ROM) was 64K. The user memory (RAM) was 256K bytes on the Base model and 512K bytes on the Enhanced model. With additional expansion cards, the memory could be expanded to three megabytes on both models.

Both models had one half-height 1.2 megabyte floppy disk drive with provision for a second drive. The Enhanced model had a 20 megabyte hard disk drive. Hard disk drives with up to 40 megabytes capacity could be installed in both models. The models contained eight expansion slots for additional adapter cards. The keyboard was an enhanced version of the PC keyboard. Microsoft released Version 3.00 of PC-DOS and XENIX 286 operating systems for the new models.

The Base model cost $3,995 and the Enhanced model $5,795. The computers were intended as replacements for the XT computer. IBM stated that the computer was designed to be a multitask, multi-user computer.

Problems with the hard disk drive resulted in delayed deliveries of the computer. However the AT computer had good reviews and became a up-market replacement for the XT. The operating speed of the PC AT microprocessor was increased to 8 MHz in 1986.

PC RT

IBM approved development of a workstation computer using the RISC (Reduced Instruction Set Computing) ROMP (Research Office products MicroProcessor) processor in 1983. G. Glenn Henry was the manager of hardware and software system development. The project had the code name of Olympiad and became the PC RT computer.

The PC RT workstation was introduced for work such as CAD (Computer Assisted Design) in January 1986. It utilized a high performance (approximately 2 million instructions per second) IBM ROMP 32-bit RISC processor. An Intel 80286 microprocessor was used as a coprocessor to facilitate program and user interface with the PC family of computers. A Memory Management Unit (MMU) extended the 32 bit processor address to a 40 bit virtual address to provide advanced virtual storage capabilities. The computer had one megabyte of memory, a 1.2 megabyte floppy disk drive and a 40 megabyte hard disk drive. It utilized an AIX (Advanced Interactive Executive) operating system based on AT&T's UNIX System V operating system. The PC RT workstation cost $11,700.

The workstation was not received well due to poor performance as compared to competitive products from Apollo and Sun Microsystems. This resulted in a new project with the code name of RIOS being started in 1986 to develop a new more powerful workstation that would become the RISC System/6000.

PC Convertible

IBM introduced the 5140 PC Convertible (code-named Clamshell) laptop computer in April 1986. The computer name was selected because it could be used as a portable or as a desktop with an expansion box and a larger monitor. The unit weighed 12 pounds and featured an Intel 8088 microprocessor, 256K bytes of memory, two

3.5 inch 720K byte floppy disk drives and a 25-line liquid crystal display (LCD) monitor. The computer cost $2,995. However, the product was not successful due to the use of an older processor, problems with the LCD display and the early use of 3.5 inch floppy disk drives.

PC/XT Model 286

The PC/XT Model 286 was introduced in September 1986. It featured the Intel 80286 microprocessor, 640K bytes of memory, one 1.2 MB floppy disk drive, a 20 megabyte hard disk drive and cost $3,995. However the late introduction and pricing relative to other competitive products resulted in poor sales.

PS/2 Series

The Personal System/2 (PS/2) family of personal computers was introduced in April 1987 (except the Model 25). The "2" in the PS/2 product name, denoted a second generation of personal systems. The Models 50, 60 and 80 had a new architecture with a proprietary Micro Channel Architecture (MCA) bus. MCA was a 32-bit multitasking bus that did not support the previous expansion cards for the PC computer. IBM's intent was to regain control of the open architecture and force clone manufacturers to obtain a MCA license. The preceding models also utilized a new video standard called VGA (Video Graphics Array) that had improved screen resolution. The new OS/2 operating system developed by IBM and Microsoft was also announced for use with the computers.

The following are some details of the various PS/2 models introduced in April. The Model 30 was available in two configurations and featured an Intel 8086 microprocessor operating at 8 MHz, 640K bytes of memory and the PC XT bus. The Model 30-002 had two 720K byte floppy disk drives and cost $1,695. The Model 30-021 had one 720K byte floppy disk drive, a 20 megabyte hard disk and cost $2,295. The Model 50 had an Intel 80286 microprocessor, one megabyte of memory, 1.44 megabyte floppy disk drive, 20 megabyte hard disk drive and cost $3,595. The Model 60 featured an Intel 80286 microprocessor, one megabyte of memory, 1.44 megabyte floppy disk drive, 44 megabyte hard disk drive and cost $5,295. The Model 80 was available in three configurations using the Intel 80386 microprocessor and a 1.44 megabyte floppy disk drive. The Model 80-041 had one megabyte of memory, a 44 megabyte hard disk drive and cost $6,995. The Model 80-071 had two megabytes of memory, a 70 megabyte hard disk drive and cost $8,495. The Model 80-111 had two megabytes of memory, a 115 megabyte hard disk drive and cost $10,995.

The Model 25 was a low cost computer, introduced for business and educational users in August 1987. It featured the Intel 8086 microprocessor, 640K bytes of memory, a 720K byte floppy disk drive and cost $1,395.

A portable version of the PS/2 series, the PS/2 P70 was announced in May 1989. At a weight of 20.8 pounds it would now be called a "luggable." It used an Intel 80386 processor, had 4 MB of RAM (expandable to 8 MB), 120 MB of disk storage, MCA bus and a high-resolution plasma display. The PS/2 P70 received good reviews and had good sales.

The PS/2 series of computers were not well received in the marketplace. IBM had focused on the older Intel 80286 microprocessor rather than the latest 80386 chip. Also the incompatibility of the new MCA bus with old add-on cards, the late release of the new version and general poor acceptance of the OS/2 operating system, all contributed to sales below expectations. The MCA bus was not supported by the industry and became a strategic mistake for IBM.

9.5 ... Software

OS/2 and Microsoft

By the end of 1984, Bob Markell who was a vice president of software and communication products at IBM, had created a task force to determine a suitable operating system for future products. IBM had been working on its own operating system called CP-DOS that would be used for the 286 microprocessor initially and the 386 microprocessor later. IBM also wanted a system that incorporated multitasking, so a user could run more than one application at the same time. Company management had mixed aspirations to develop the new operating system internally independent of Microsoft. However, discussions were held with Microsoft regarding the new system that culminated in the signing of a joint development agreement in June 1985.

The joint development efforts following the agreement led to numerous difficulties between the two different types of corporate styles. IBM was attempting to satisfy many different corporate demands and were adding an increasing number of personnel to the project to maintain the completion schedule. Microsoft was accustomed to software development with a small group of talented programmers. However, Microsoft had conceded final responsibility for the software design to IBM.

Another significant decision that would create subsequent difficulties was the use of assembler language to program the new operating system. This choice and a focus on the Intel 80286 microprocessor for the PS/2 series of computers would add to the complexity and portability of the new system.

In mid 1986 a new concept called Systems Application Architecture (SAA) was approved for implementation. This system provided a common software development environment between the different IBM hardware levels, from personal computers to mainframes. However it also resulted in additional complexity to the software. Also, a new graphical user interface that would be called Presentation Manager, would be developed by the graphics software group in Hursley, England.

The company was also now working on an Extended Edition of the new operating system that Microsoft was excluded from participating in. The Extended Edition included communications and database services. IBM also planned to introduce a set of office applications that would be called OfficeVision for the Extended Edition.

During this period new personal computer hardware was also being developed. The new hardware would have a different bus concept called Micro Channel Architecture (MCA) and an Advanced Basic Input/Output System (ABIOS). With strong enforcement of applicable patents this was going to be IBM's strategy to combat the clones.

IBM announced the new operating system called OS/2 with the Personal System/2 computers in April 1987. A Standard and Extended Version 1.0 were released in November. However it did not include the Presentation Manager software that was now promised for October 1988. Only a few application programs were available. The program cost $325 (twice as much as DOS), required additional memory and storage as compared to DOS, and was not well received.

In May 1988, IBM joined the Open Software Foundation (OSF) that was established to develop a unified UNIX operating system for different platforms. Then IBM purchased a license for the NeXTSTEP operating system from NeXT Computer, Inc. IBM was intent on establishing optional operating systems to OS/2 and PC-DOS.

IBM released Presentation Manager as part of OS/2 Version 1.1 in October 1988. The graphical user interface had been developed by IBM in Boca Raton, Florida, IBM laboratories in Hursley, England and by Microsoft. The graphics were well received. However, the lack of application programs and device drivers, the requirement for additional memory and the pricing adversely affected sales.

During 1989, James A. Cannavino, the new head of the Entry Systems Division began to question the viability of OS/2 and the relationship between IBM and Microsoft. His concerns related to the low acceptance of OS/2, the income Microsoft derived from the PC disk operating system software and the potential impact of a new version of Windows being developed by Microsoft. Software vendors were also expressing concerns regarding the future market share of OS/2 and their significant investments in application programs for the new operating system. Cannavino had even recommended that IBM drop OS/2 in March. However, corporate management rejected his recommendation and instructed him to "build a world class operating system." Cannavino had discussions with Bill Gates and a tenuous agreement was announced at the fall COMDEX show that appeared to support each companies system. However Cannavino had been committed to OS/2, not Microsoft Windows. In late

1989, Version 1.2 of OS/2 was released, however sales of the OS/2 operating system were still well below expectations.

Other Software

Displaywrite was a word processor developed by IBM for the DisplayWriter workstation in 1980. It was one of the few relatively successful application programs written by IBM.

TopView was an IBM character-based user interface that was announced in August 1984. It incorporated windows and multitasking that enabled the running of multiple programs with the ability to switch between them. It was not released until January 1985 and cost $149. However, it was slow, required a lot of memory and did not have a graphical interface. Due to poor acceptance it was withdrawn from the market in June 1987.

IBM released PC Network in conjunction with Microsoft PC-DOS Version 3.1 in March 1985. It was designed to connect the PC-family of computers in a local-area network (LAN).

A software group was formed by Joseph M. Guglielmi in 1987 to develop an office system that would facilitate the communication and sharing of information and software such as databases, desktop publishing, electronic mail, spreadsheets and word processors. The application software used the name OfficeVision for its products. David Liddle who had worked at Xerox PARC, was a principal in the development of the OfficeVision suite of software released in June 1989. However, it was not graphically oriented, priced too high and was not successful. IBM essentially disbanded the software group around 1992.

IBM created a Desktop Software division in 1988. It was established to market personal computer software by IBM and other companies using the IBM logo. However it was not successful either and was terminated around 1992.

9.6 ... Corporate Activities

1980 and 1981 are significant years due to the approval and release of the IBM Personal Computer (see Sections 9.2 and 9.3). John Opel became the chief executive officer of IBM in January 1981.

In January 1982, the Department of Justice withdrew its antitrust suit against IBM. In late 1982, an executive search firm for Apple Computer contacted Don Estridge as a potential candidate for the position of president. However Estridge declined the offer.

Then in December 1982 IBM acquired 12 percent of Intel Corporation stock for $250 million. Intel was having financial problems due in part to Japanese competition in memory chips. IBM made the stock purchase to provide a secure source for its microprocessors and to maintain the viability of domestic chip manufacturing equipment suppliers. This also resulted in Intel licensing the manufacture of the chip to others. IBM now had a second source for its microprocessors. IBM also built a new highly automated factory to mass produce the PC computers.

John Opel became chairman of the board and John F. Akers president of IBM in February 1983. In August, the Personal Computer unit at Boca Raton, Florida became part of a new Entry Systems Division (ESD) and Don Estridge was appointed president of the division. The new divisional organization consolidated major facilities at Boca Raton and Austin, Texas. It also had worldwide responsibilities for product development and management including plants in Greenock, Scotland and Wangaratta, Australia. Joe Bauman became vice president for manufacturing, Joseph Sarubbi director of technologies and Dan Wilkie director of quality assurance and technology for the new division.

Starting in 1983, the company began implementing the traditional bureaucracy at the Entry Systems Division. The freedom enjoyed by the original IBM PC group was coming to an end. H. L. Sparks and James D'Arezzo left IBM and joined Compaq Computer Corporation in 1983. Then in January 1984, Estridge was made a vice president of IBM.

Responsibility for retail dealer sales of all PC products was moved from the Entry Systems Division to the corporate national sales organization in January 1985. John Akers became the chief executive officer in February. In March, Estridge was promoted to vice president of worldwide manufacturing for IBM and William Lowe became president of the Entry Systems Division. Then in a tragic plane crash, Don Estridge was killed at the Dallas-Fort Worth, Texas airport in August. In late 1985, Dan Wilkie resigned from IBM to become president of another company.

Joe Sarubbi retired from IBM and joined the Tandon Corporation as a senior vice president of manufacturing in February 1986. In the spring of 1986, the number of IBM employees peaked at 407,000. During 1986, IBM concluded a technological exchange agreement with the Intel Corporation (See Section 8.4). In June John Akers became chairman of the board, a year that saw significant reductions in IBM's financial performance.

In 1987, it appeared that the financial difficulties encountered in 1986 would continue. Akers initiated the formation of task forces to evaluate the problem. This resulted in the closure of a parts distribution facility and a reduction of 10,000 employees by early retirement and severance package options. IBM also sold the remaining shares of Intel Corporation stock that the company purchased in 1982.

Akers announced a further reorganization to delegate more decision making down to lower levels in the company organization in January 1988. Then in December William Lowe left IBM and joined the Xerox Corporation. He was replaced by James Cannavino who inherited an extremely difficult business situation in the Personal Computer group. The group had lost 1.4 billion dollars in 1998, the PS/2 computer was not selling, the MCA bus and the OS/2 operating system were not accepted by either customers or the industry.

In March 1989, Cannavino made a number of recommendations to the IBM Board to correct the business situation of the PC group. Some of these recommendations were: to significantly reduce the company's focus on the desktop business, increase their participation in the portable and server segment of the business and drop the OS/2 operating system. However, the Board wanted to keep IBM in the PC business, retain the OS/2 operating system and review the relationship with Microsoft. Cannavino was also concerned about reducing IBM's dealer and sales organization costs, and competition from direct sellers such as Dell and Gateway.

In May, Jack D. Kuehler became the president of IBM and Cannavino was promoted to general manager of the Personal Computer group in mid 1989. Around this time Cannavino selected Bob Lawten to analyze IBM's efforts in the portable computer segment of the market. After discussions and agreement with Bill Gates at Microsoft, Cannavino made a recommendation that IBM purchase forty percent of Microsoft. This would motivate both companies to make the relationship work. However, this proposal was rejected by the IBM Board. The reorganization and changes implemented by Cannavino during 1989, resulted in a change from a loss of 1.4 billion dollars in 1988 to a profit of 1.2 billion dollars in 1989.

Figure 9.6: James A. Cannavino.
Photograph is courtesy of IBM Corporation.

Blank page.

Chapter 10 Apple Computer in the 1980's

As a dominant personal computer manufacturer in 1980, Apple Computer had distinct characteristics. It had developed in the Apple II and the related disk drive, technology and ease-of-use features that enabled it to make the transition from hobbyist or technical hacker user to the mass consumer market. These innovations and the release of VisiCalc spreadsheet software resulted in the commercial success of the company.

It had espoused statements such as "Never build a computer you wouldn't want to own" and "One person -- one computer." The visit to the Xerox Palo Alto Research Center (PARC) in 1979 changed the course of product development. The new PARC human interface concepts suited the vision of Apple. It continued the innovative initiatives of Wozniak. The development of the Lisa computer and the Macintosh computer that would be the future of Apple in the 1980's.

10.1 ... Corporate & Other Activities

Public Stock Offering

The initial founders Jobs, Wozniak, Markkula, Holt and the venture capitalists who provided the financing had a tight control of the Apple Computer shares. Markkula also kept a firm control on the later sale and award of Apple shares. Initial shares went through several splits that significantly increased their value. A share distributed before April 1979 was the equivalent of thirty-two shares on the day that Apple went public.

During 1980 the market for new stock issues had improved. In August 1980 the Apple Computer board of directors decided to make a public offering of shares in the company. Apple selected two firms who offered 4.6 million shares of common stock for sale in December 1980. It was a huge success and oversubscribed. On the first day the offered share price of $22 increased to $29. At the end of December 1980 Jobs' ownership in the company was worth about $256 million, Markkula's $239 million, Wozniak's $136 million and Holt's $67 million. The original founders had done extremely well.

1980/82 Activities

The company introduced Apple FORTRAN in January 1980, then they announced the Apple III computer in May (see Section 10.2). In August at a company board meeting the board members decided to implement a new company structure. They changed the organization from a functional one to a product-oriented one. The company created Divisions for the Apple II and Apple III, Lisa, accessories, manufacturing, sales and service.

In February 1981 Wozniak crashed his airplane on take-off from a local airport. Wozniak had serious injuries and for a period of time suffered from amnesia and lapses of consciousness. By early 1981 the employee count had grown rapidly to nearly 2,000. An adjustment required just over 40 employees being terminated in February. The terminations, and the way Mike Scott handled them, had a bad effect on employee morale. Employees described the layoff as "Black Wednesday." A number of activities were converging to undermine the effectiveness of Apple's president Mike Scott. He was experiencing potentially serious health problems with an eye infection. Also he had not impressed management with the poorly handled layoffs. After an executive meeting in March, Markkula requested Scott's resignation and assumed the presidency of the company.

With release of the IBM PC computer in August 1981, Apple made an interesting competitive response. Advertisements in a number of national newspapers published an open letter to IBM. It stated "Welcome, IBM. Seriously. Welcome to the most exciting and important marketplace ..." Jobs received a personal reply from John Opel, the president of IBM thanking him for his comments.

Jobs appeared on the cover of *Time* magazine in February 1982 with the caption "Striking it Rich -- America's Risk Takers." *Life* magazine also featured him a month later. Jobs appeared again in the January 1983 issue of *Time* magazine; this issue named the personal computer as the "Machine of the Year." The article was not complimentary of Jobs, but he was becoming a national personality.

In May 1982, Apple sued Franklin Computer Corporation for patent and copyright infringement. Wozniak, who had been playing a less significant role within the company, decided to return to college and complete his bachelor's degree. Wozniak also arranged and financed a "US Festival" of rock music in September and a second one in 1983. It was during 1982 that Apple, at Jobs' initiative, began the process of giving Apple computers to academic institutions and prisons. In November 1982 Apple held its first AppleFest in San Francisco, California. Then in December, Apple became the first personal computer company to reach a rate of one billion dollars in annual sales.

The company started a project in 1982 called the Apple IIx using an early version of the Western Design Center 65816 microprocessor. The engineering manager was Dan Hillman with some assistance from Wozniak in 1983. However availability and reliability problems with the microprocessor, and potential market conflicts with the other Apple products resulted in the project cancellation in early 1984.

John Sculley

Apple started considering candidates for the presidency in 1982. Markkula had already stated that his role as president was temporary, and the board would not support Jobs' desire for the position. The personal computer market had changed after IBM's entry in 1981. IBM was gaining market share and the other companies engaged in severe price-cutting to maintain sales. The need for a new president at Apple was becoming more important. Apple engaged an executive search firm who approached several candidates, including Don Estridge of IBM. However after an attractive offer from Jobs he decided to stay with IBM. Late in the year they made the initial contacts with John C. Sculley, the president of Pepsi-Cola USA, a subsidiary of PepsiCo.

Figure 10.1: John C. Sculley.
Photograph is courtesy of Apple Computer, Inc.

John Sculley had a bachelor's degree from Brown University and an MBA from the University of Pennsylvania Wharton School. Sculley joined Pepsi-Cola in 1967 and became president of the company in 1977. Markkula offered Sculley $1 million to join Apple, $1 million in annual pay and options for 350,000 shares of Apple stock in the spring of 1983. In addition he received compensation to purchase a house in California and $1 million in severance if he did not work out. Within a few months the stock options would be worth over 9 million dollars. Jobs had made a pointed comment to Sculley when he said "Do you want to spend the rest of your life selling sugared water, or do you want a chance to change the world?" Sculley accepted the offer in April and became the new president and chief executive officer.

Other Activities between 1983 and 1989

The company announced the Apple IIe and Lisa computers in January 1983 (see Sections 10.3 and 10.4 respectively). The publicity described the computers as "evolution and revolution." Bill Atkinson and Rich Page became Apple Fellows in February. Then Apple Computer entered the *Fortune* magazine top 500 companies at number 411 in May and built the one millionth Apple II in June. It was the first computer awarded in a program called "Kids Can't Wait," that provided computers to about 9,000 schools in California. Apple introduced the ProDOS operating system in June and an integrated software package called AppleWorks in November. The company then introduced the Apple III Plus computer and the ImageWriter dot-matrix printer in December. Apple sales had surpassed one billion dollars annual rate by December. However competition from companies such as IBM had reduced profits. An industry wide recession had started that resulted in significant staff reductions. Sculley then decided to reorganize the company into two operational groups; the Apple II and Apple 32 that integrated the Lisa and Macintosh product lines.

The company announced the Lisa 2 (see Section 10.4) and the Macintosh (see Section 10.5) computers in January 1984. The company also released the ProDOS operating system for Apple II computers and reached an out-of-court settlement of their patent infringement suit with Franklin Computer Corporation in January. Then Apple introduced the Apple IIc computer (see Section 10.3) and discontinued development of the Apple III product line in April. Alan Kay who had been a principal at Xerox PARC and chief scientist at Atari, became an Apple Fellow in May. Apple introduced the DuoDisk for the Apple II computer in June for $795, that was essentially two 5.25-inch Disk II drives in a single cabinet. A new version of the word processing program Apple Writer 2.0 was released in September. During the year the company investigated strategic alliances with companies such as AT&T, General Electric, General Motors and Xerox. An alliance would finance major initiatives to increase company penetration in the office systems market. Sales boomed until September, and by November Apple had sold two million Apple II computers. However during the last three months of 1984 an industry wide recession lowered Apple II, Lisa and Macintosh sales significantly below marketing forecasts.

In 1984, Andy Hertzfeld who had worked on the development of software for the Macintosh computer, left Apple Computer and developed a program called Switcher. The program was developed to compete with integrated packages and enabled a user to switch between different applications that could be running simultaneously. Hertzfeld sold the program to Apple Computer who supplied it free with the Macintosh computer.

Apple introduced the Macintosh Office software and an advanced laser printer named LaserWriter priced at $6,999 in January 1985. The printer used a Motorola MC68020 microprocessor and the Adobe PostScript page description language (PDL). The Macintosh Office featured the AppleTalk Personal Network software that allowed a group of Macintosh computers to communicate and be connected to a LaserWriter printer. Unfortunately some key elements of the Macintosh Office software such as the FileServer for sharing information were not ready. Wozniak left Apple in February and started a new company called CL-9 (Cloud-Nine) to develop remote control products for the home. Apple enhanced the Apple IIe computers in March and terminated production of the Lisa computer in April. Different versions of a 3.5-inch UniDisk drive with a capacity of 800K bytes were released in 1985 for the Apple II computers.

As sales deteriorated, the relationship between Jobs and Sculley also deteriorated. In the spring of 1985 they contested in a somewhat acrimonious manner for the leadership of the company. This executive confrontation and a deterioration of sales forced Sculley as CEO of the company to make organizational changes. After securing the support of the board members in April, he persuaded them to relieve Jobs of any operational role in the company. In May the product oriented organization that Mike Scott created in 1980 and Sculley adjusted in 1983, changed back to a functional one. Personnel changes included: Del Yocam, became group executive in charge of all operations; Jean-Louise Gassée, would be in charge of product development; Deborah A. Coleman, world-wide manufacturing; and Michael H. Spindler, all international operations. Then in June, Apple closed three manufacturing plants and laid off 1,200 employees. Finally Apple declared its first quarterly loss.

Another problem for Apple Computer was a lack of application software for the Macintosh computer. This resulted in a campaign to encourage software companies to develop application programs for the Macintosh. Guy Kawasaki who had joined Apple in 1983 and Mike Boich became "software evangelists," that promoted the new campaign. A significant application program for the Macintosh, would be the desktop publishing program called PageMaker.

Jobs resigned in September and started a new company called NeXT Computer, Inc. However when five key personnel joined NeXT, Apple started litigation to stop Jobs and NeXT Computer from using any of its proprietary technology.

Work was started in 1984 to develop an online worldwide network to support the company's dealers. This evolved into a Macintosh system called AppleLink that went online in July 1985. The system was maintained and operated as a joint effort between Apple Computer and General Electric. It featured easy-to-use graphics, icons and windows.

Apple announced the Macintosh Plus and the LaserWriter Plus in January 1986. Sculley became chairman of the company and Apple reached an out-of-court settlement in the litigation with Steve Jobs in January. By February, Jobs had sold his holdings of Apple stock, the separation was now complete and Markkula now became the largest shareholder. Wozniak graduated from the University of California at Berkeley in June. The company announced the Apple IIGS computer and the enhanced Apple IIc in September (see Section 10.3).

Apple introduced the Macintosh SE and Macintosh II (see Section 10.5) in March 1987. Then in April, the company decided to create an independent wholly owned subsidiary called the Claris Corporation, to take over sales and marketing of Apple application software. Apple also introduced the MultiFinder operating system software and the HyperCard software for the Macintosh computer at the MacWorld trade show in Boston in August.

After helping to create the latest versions of the Macintosh computer, Steve Sakoman became head of a new research project in 1987. Sakoman wanted to create a radically different personal information device he named Newton. It would be pen-based, use handwriting recognition and wireless communication. By 1989, a prototype slatelike device had been developed that measured 8.5 by 11 inches. However, the estimated cost had grown from an early target of $2,500 to between $6,000 and $8,000 and Sakoman was having problems with corporate support.

Bill Atkinson developed the HyperCard system software and the HyperTalk programming language in August 1987. This personal software toolkit provided a capability to organize a body of information and then create links and cross-references within it. The data could be text, graphics, video, animation or sound.

In March 1988, Apple started litigation against Microsoft. Apple claimed for infringement of the Macintosh graphics in a new release of MS Windows Version 2.03 software. The suit also named Hewlett-Packard and it's NewWave software in the litigation. Apple introduced the Apple IIc Plus and the Macintosh IIx (see Section 10.5) computers in September. The Macintosh IIx was the first Apple computer to use the Motorola MC68030 microprocessor and 68882 math coprocessor. Claris released AppleWorks GS for the Apple IIGS computer in October.

In July 1989 a judgment in the litigation with Microsoft, significantly reduced Apple Computer's claims. Apple announced the Macintosh Portable and the Macintosh IIci in September 1989 (see Section 10.5). The Macintosh IIci was a high performance version of the Macintosh IIcx. In 1989 Xerox filed a lawsuit against

Apple, stating it had infringed on the PARC copyrights. However the court dismissed most of the lawsuit in early 1990.

Following the company reorganization in 1985 and the introduction of Macintosh products with more memory and hard disk storage, the fiscal condition at Apple started to improve. Although corporate management problems continued. Areas such as executive direction and software development could have been improved. A tug-of-war between profit margin and an open or licensed product architecture to increase the Macintosh market share, resulted in a short term solution with far reaching effects. These years were a turning point for the company. The open architecture of the IBM Personal Computer and all its clones, resulted in a massive market support in both hardware and software development that would be detrimental to Apple's future leadership in the industry. However, sales and profits continued to improve. In the 1989 annual report, the net sales were 5.3 billion dollars and the total number of employees 14,517. An impressive growth compared to 1979. John Sculley as chairman, president and chief executive officer had made significant achievements since his arrival in 1983. However the past was history, as he stated in the annual report "As we look to the new decade, we see a time of enormous opportunity for Apple Computer."

10.2 ... Apple III

The Apple III computer had a difficult gestation from its start in 1978 as the Sara project. A requirement to be able to run Apple II software, limited the microprocessor selection and created difficulties with the design. Wendell Sander was the chief hardware designer. Jobs was involved initially with the product design and affected its completion schedule with numerous demands. Apple set a constricted target date and problems developed around the time Jobs got interested in the Lisa technology. Planning required Apple III development within ten months. It was to be a stopgap product to bridge between an anticipated fall-off in Apple II sales and the introduction of Lisa. The company targeted the Apple III at the small business owner and for professional-managerial users. Apple announced the computer at the National Computer Conference in Anaheim, California in May 1980 and started shipping them in the fall of 1980.

Figure 10.2: Apple III computer.
Photograph is courtesy of Apple Computer, Inc.

The computer used a Synertek 6502A microprocessor with 96K bytes of memory, expandable to 128K. The unit included one built-in 5.25-inch, 143K byte floppy-disk drive and the terminal could display 24 lines of 80-column text.

The computer had a new operating system called SOS (Sophisticated Operating System), built-in Apple Business BASIC and Pascal programming languages. An

Apple II emulation mode enabled some Apple II software to run on the computer. Jeffrey S. Raikes was a principal in the development of the Apple III software.

The company sold Apple III computers as systems. A computer system with a black-and-white monitor, an 80-column thermal dot-matrix printer and the VisiCalc III spreadsheet program sold for $4,500.

The company released a 5 megabyte hard disk called ProFile for use with the system in September 1981. The compressed schedule and the various design changes incorporated had a detrimental effect on the quality of the computer. A number of problems required a redesign of the computer then re-introduction in December. Apple allowed early customers to exchange their old computer for a new one. The company introduced the Apple III Plus with 256K bytes of memory, other improvements and a price of $2,995 in December 1983. However after the introduction of the Macintosh and Lisa 2, the company discontinued the product line in April 1984. The initial poor quality resulted in only sixty-five thousand Apple III's being sold in the three years after its introduction. The resources committed to the release and problems with the Apple III during 1980/81, inhibited enhancements to the Apple II product line.

10.3 ... Apple II's

Apple IIe

During 1981 the Apple II computer group started working on an upgrade that became the Apple IIe. The "e" represented enhanced. Apple intended this upgrade to extend the life of the Apple II. Peter Quinn was the chief engineer and Walt Broedner a principal in the development of the new design. Apple retained the enclosure style of the Apple II computer but completely redesigned the interior. The new design significantly reduced the number of integrated circuit chips. Two custom MOS chips and 64K bit memory chips contributed to this reduction. The company announced the Apple IIe with the Lisa computer in January 1983.

The computer used a MOS 6502A microprocessor with 64K bytes of RAM, expandable to 128K. The storage system supported six 140K byte 5.25-inch floppy disk drives and the terminal could display 24 lines of 40-column text in both uppercase and lowercase characters. The unit had a new 63-key keyboard adapted from the Apple III computer.

The programs in the 16K of ROM were the Applesoft BASIC interpreter, system monitor routine, 80-column display firmware and self-test routines. The company also released the Apple Writer IIe word processor and QuickFile IIe data base application programs. The majority of Apple II and Apple II Plus programs and peripheral cards were compatible with the Apple IIe.

The base list price for a standard unit was $1,395. A typical system with a single Apple Disk II drive and controller, 64K bytes of RAM, 80-column text card, and a monochrome monitor had a price of $1,995. An Apple Disk II drive and controller had a cost of $545. The 80-column text card cost $125 and the extended memory 80-column card that included an additional 64K bytes of memory cost $295.

Apple enhanced the Apple IIe computers with four new high-performance chips in March 1985. The new chips replaced the character-generator, Applesoft and system-monitor ROM chips and the 6502 with a 65C02 microprocessor. The changes provided system improvements, more compatibility with the IIc computer and facilitated the use of mouse-driven software. An additional update of the Apple IIe occurred in January 1987, that incorporated the Apple IIGS keyboard and other improvements.

Apple IIc

Apple introduced the Apple IIc at an "Apple II Forever" conference in San Francisco, California in April 1984. The "c" stood for compact. Peter Quinn who was the chief engineer for the Apple IIe, was also engineering manager of the IIc design team. It was a portable computer that would be in competition with the IBM PCjr. The computer used additional custom integrated circuits as compared to the Apple IIe to reduce the number of chips on the motherboard. Dimensions of the computer were 12 by 11.5 by 2.25 inches and the weight was 7.5 pounds. A German design company styled the attractive case.

The computer used a 65C02 microprocessor and 128K bytes of RAM. The unit had one built-in half-height 5.25-inch 140K byte floppy disk drive and the monitor could display 24 lines of 40 or 80-column text. The computer was a closed-hardware architecture with no expansion slots and the power supply was an external unit. The housing integrated a 63-key keyboard that was functionally a duplicate of the Apple IIe keyboard.

A 9-inch monochrome monitor and a thermal-transfer printer called Scribe were released with the computer introduction. Also announced but not available at the introduction, was a flat-panel display that Apple intended to introduce by the end of 1984.

Apple had updated the IIe ROM software that contained the Applesoft BASIC interpreter and various routines. The computer used the new ProDOS operating system that provided for hierarchical directory structures. Most of the Apple II application programs were capable of running on the IIc computer. The basic computer had a price of $1,295.

Apple introduced an enhanced Apple IIc in September 1986. Then they introduced a less expensive Apple IIc Plus that incorporated a faster 4 MHz version of the 65C02 microprocessor, an internal 3.5-inch disk drive and a built-in power supply in September 1988. The Apple IIc Plus had a price of only $675 or $1,099 with a color monitor.

Apple IIGS

The success of the Apple IIc, the availability of a compatible 16-bit microprocessor and a new Mega II chip were factors in Apple's decision to create the Apple IIGS. The "GS" stands for graphics and sound. The Apple IIGS computer evolved from the 1982/83 Apple IIx project. Principals in the engineering design were Dan Hillman, Harvey Lehtman, Rob Moore and Wozniak. The GS project had several code names during development such as Phoenix, Cortland and Rambo. Hillman and Jay Rickard developed the Mega II as a cost reduction project, that resulted in the integration of most Apple II functions on a single chip. The new 16-bit computer incorporated enhanced graphics, advanced sound capabilities, expanded memory and an Apple II emulation mode. Apple introduced the Apple IIGS in September 1986.

The computer used a Western Design Center W65C816 microprocessor and 256K bytes of RAM, expandable to 8 megabytes. The storage system included support for both 3.5-inch 800K byte and 5.25-inch 140K byte floppy disk drives. The terminal could display 24 lines of 40 or 80-column text. The keyboard was a separate unit with 80 keys and a 14-key numeric keypad. The standard system included a mouse. Apple also introduced an optional 20 megabyte hard disk designated 20SC for use with the computer. A computer system with a monochrome monitor and one 3.5-inch disk drive had a price of about $1,500.

Apple provided a new operating system called ProDOS 16 for the 16-bit native mode and ProDOS 8 for the Apple IIe emulation mode. ROM software included; the Applesoft BASIC interpreter, mouse based system utilities and a desktop environment with similarities to the Macintosh computer. QuickDraw II provided a set of graphic routines. Most of the existing Apple II software was compatible with the Apple IIGS. Apple released a new more powerful and flexible 16-bit operating system called GS/OS in September 1988.

The company released an enhanced Apple IIGS in August 1989. The new computer had 256K bytes of ROM and 1 megabyte of RAM. The ROM software included a number of improvements.

10.4 ... Lisa

William "Trip" Hawkins had developed a marketing plan describing the requirements for the Lisa computer by March 1980. Larry Tesler who had demonstrated the advance Xerox Alto computer systems to Apple in December 1979, joined the Lisa design team in July 1980 and became the manager of software development. Rich Page who was the chief hardware architect, had just completed a prototype of the Lisa computer incorporating a sample Motorola 68000 microprocessor. The Lisa name now denoted Local Integrated Software Architecture.

Once again, as in the Apple III product, Jobs was affecting the design with numerous changes. This resulted in the departure of Ken Rothmueller and the appointment of Wayne Rosing as engineering manager. It also resulted in Mike Scott and Markkula advising Jobs in September that he would no longer be heading the Lisa project. In anticipation of the public stock offering and to placate Jobs they promoted him to chairman of the board. As part of a new corporate reorganization, John Couch became the general manager of the Lisa product group.

Apple had integrated and extended the Xerox PARC (Palo Alto Research Center) graphical concepts in both the hardware and software. Bill Atkinson, Tesler and others developed the operating system and bit-mapped graphics software into an innovative friendly environment for the user. The computer was designed to be intuitive and standard features consistent throughout the system. The desktop user interface utilized the windows concept, icons, a standard user program interface, menu bar, pull-down menus, clipboard and direct manipulation of screen objects by the mouse. Apple had achieved a new "state of the art" for personal computer software.

However Apple was having problems with the Twiggy floppy disk drive for Lisa. They decided to redesign the drive and have it produced by the Alps Electric Company, a Japanese manufacturer. Apple provided a preview of the Lisa computer to the Manhattan East Coast media then officially announced its introduction in January 1983. However the late delivery of disk drives from Alps Electric delayed shipment of computers until May.

The computer used a Motorola MC68000 microprocessor with 1 megabyte of RAM. The storage system had two Twiggy 5.25-inch 860K byte floppy disk drives and a separate 5 megabyte Winchester-type hard disk named ProFile. The computer housing enclosed the 12-inch monitor and two floppy disk drives. The keyboard was a separate unit and included a numeric key pad. The system utilized a one button mouse for control of the screen cursor.

Figure 10.3: Lisa computer.
Photograph is courtesy of Apple Computer, Inc.

The software group developed the operating system, Window Manager, QuickDraw graphics, Desktop Manager for file and program manipulation and LisaGuide instruction package. Apple used software and the MC68000 microprocessor to generate the video display. The company also developed a suite of seven application programs called the Lisa Office System. LisaDraw provided drawing capability for lines, boxes, circles and other features with mouse control. LisaWrite was a what-you-see-is-what-you-get word processor developed by Tom Malloy, another Xerox PARC recruit. LisaCalc was a sophisticated spreadsheet program. LisaGraph was a graphing program for creating bar, line, mixed bar and line, scatter and pie charts. LisaList was a database program with searching, sorting and reporting capabilities. LisaProject was a PERT (Program Evaluation and Review Technique) program with capabilities for displaying Gantt and task charts. LisaTerminal was a communications program with emulation capabilities for the DEC VT52, DEC VT100 and Teletype ASR-33 terminals. AppleNet software was also available for connecting multiple Lisa installations.

A Lisa system with one megabyte of RAM, two floppy disk drives, a ProFile hard disk and seven application programs sold for $9,995. A C.Itoh dot-matrix printer cost about $700 and a Qume letter quality printer was about $2,100. The company targeted the Lisa computers as office systems with pricing that excluded Apple's traditional personal user. Unfortunately at the Lisa announcement, Jobs told reporters of the new Macintosh computer. He stated that the Macintosh would cost $2,000 compared to $10,000 for the Lisa. This and the lack of compatibility between the two computers would affect future sales of the Lisa computer. Apple unbundled the suite of software and reduced the Lisa price to $6,995 in September.

Lisa 2

Apple announced the Lisa 2 family of computers at the annual shareholders meeting in January 1984. The company changed the design and pricing to counteract marketing concerns related to the new Macintosh computer.

Apple released three models for what they called the Apple 32 SuperMicro product line. Those three models were the Lisa 2, Lisa 2/5 and Lisa 2/10. Each of the models used the same Motorola MC68000 microprocessor as the Lisa and had 512K bytes of user memory. The number of floppy disk drives on each model changed from two 5.25-inch Twiggy's to a single 3.5-inch Sony drive as used on the Macintosh. The Lisa 2 had no hard drive and sold for $3,495. The Lisa 2/5 had an external 5-megabyte ProFile hard drive and sold for $4,495. The Lisa 2/10 had an internal 10-megabyte hard drive and sold for $5,495.

An operating system software package called MacWorks was available that enabled all three models to run Macintosh application programs. However one potential problem was that the Macintosh pixel display was square, whereas the Lisa pixel display was rectangular. The Lisa Office application programs required a model with a hard drive and a memory card to extend the memory to one megabyte. Apple also announced a new AppleBus for a small-scale local-area network to connect peripherals. It also facilitated the transfer of files between the Lisa and Macintosh computer systems.

Conclusion

Corporate America did not accept the Lisa computer as an office system. The Lisa software and files were not compatible with either IBM or the Macintosh and the price was too high. Sales were significantly below marketing forecasts in 1984. Then in January 1985 Apple renamed the Lisa 2/10 computer Macintosh XL and reduced the price to $3,995. The XL denoted extra-large or ex-Lisa. Apple discontinued the other two Lisa models. This however did not result in any significant sales increase and Apple discontinued the computer in April 1985.

10.5 ... Macintoshes

Development and Release

Under Jef Raskin's direction, Brian Howard and Burrell Smith had completed prototypes using the Motorola M6809E microprocessor. Raskin had also hired Guy "Bud" L. Tribble to develop the Macintosh software. In September 1980 the board considered cancellation of the project due to problems with the Apple III and Lisa computers. After Jobs' separation from the Lisa product development in September, he started looking at the low-cost Macintosh project. Jobs now questioned Raskin's selection in 1979 of the Motorola M6809E microprocessor with its limited capabilities. Jobs supported a proposal by Burrell Smith and Bud Tribble to change the microprocessor to the Motorola MC68000 and had a new prototype constructed by December. The new design had many capabilities comparable to the Lisa, at significantly lower cost. Jobs perceived the Macintosh as being a suitable successor to the Apple II.

Jobs became the Macintosh manager and was given authorization to change the Macintosh development from "project" to "product" status. The time frame for the new product development would be twelve months. In January 1981 Jobs increased the staff by moving key Apple II people to the Macintosh development group. Principals in the hardware development were Burrell Smith and Rod Holt. In the software development the principals were Bud Tribble the software manager, Bill Atkinson, Andy Hertzfeld, Bruce Horn and Randy Wigginton. Industrial designers Jerry Manock and Terry Oyama designed the computer enclosure. Differences in management and technical issues developed between Jobs and Raskin that resulted in Raskin leaving the development group in 1981, and resigning from Apple Computer in March 1982. Jobs had also supported the removal of Mike Scott as president in March 1981 and was now in a stronger position to control the destiny of the Macintosh computer within the company. Apple established a target date of early 1982 to ship the new Macintosh.

Jobs made a number of significant design decisions during 1981. The "footprint" of the new computer would be no larger than a telephone directory to minimize the space occupied by the computer on a desktop. After various mockups of the case, Manock finalized the enclosure design by early summer. The enclosure integrated the monitor and floppy disk drive within the case. The unit had a detached keyboard. There would be no expansion slots; Apple decided that software would be the means to expand the capabilities of the computer. A set of software tools within ROM would facilitate program development and provide a consistency in the user interface. Apple had outside suppliers write most of the application programs.

In the spring of 1981, Jobs visited Paul Allen and Bill Gates at Microsoft to discuss the requirements for software on the Macintosh computer. Jobs wanted

Microsoft to supply a spreadsheet, a chart program and a BASIC interpreter for shipment with the Macintosh. At this time Microsoft was busy developing software for the new IBM Personal Computer. However after visiting Apple and seeing a presentation of the Macintosh, they reached an agreement in January 1982 to provide the software requested by Jobs. Microsoft adapted the spreadsheet program from Multiplan, the charting program became MacGraph and work on the adaptation of a BASIC interpreter began.

The executive management approved production of the Macintosh computer in December 1981 with shipment date targeted for October 1982. However in early 1982 the introduction date of the Macintosh changed to May 1983. Jobs decided to assemble the Macintosh computer using advanced robotic techniques in a highly automated factory. They would also utilize a cost-effective Japanese concept of "just-in-time" for delivery of production parts.

The software manager, Bud Tribble left Apple in 1982. Robert L. Belleville replaced him and became director of Macintosh engineering. Bill Atkinson, Steve Capps, Andy Hertzfeld, Bruce Horn and Larry Kenyon designed the operating system. Bill Atkinson had the experience of being a principal in the development of the Lisa operating system. Horn and Capps developed the Finder program for file and program control in the desktop environment. A drawing program named MacSketch that became MacPaint, and an interface program called Toolbox by Hertzfeld were also under development internally. Apple assigned Donn Denman to develop a BASIC programming language called MacBASIC. A potential marketing problem for the Macintosh and Lisa computers was developing. It was their lack of compatibility in operating systems, programs and data files.

Jobs had arranged for Randy Wigginton to write a word processing program for the Macintosh, when he decided to leave Apple. However Jobs was determined to have Wigginton write the word processing program. In December 1981 Jobs offered Wigginton royalties up to $1 million if he developed the software on-time for the Macintosh delivery date. Wigginton demonstrated his first pass at his word processing program in early 1982. It was obvious that the Macintosh screen had trouble displaying 80 columns of text. The screen display resolution had changed from the original 256 by 256 pixels to 384 by 256 pixels and now for the new requirements to 512 horizontal by 342 vertical pixels. This allowed the lines to break on the screen at the same place they break on the printer, a What-You-See-Is-What-You-Get (WYSIWYG) system.

Jobs was also having problems with the Macintosh name. Although Raskin had changed the spelling, McIntosh Laboratories claimed it was a phonetic infringement of their trademark. Apple subsequently made a payment to McIntosh Laboratories to license the rights to use the Macintosh name.

In January 1983, problems with the Twiggy floppy disk drive resulted in the computer introduction date being moved to August. By the summer, Alps Electric was still having problems, Apple therefore decided to use a new 3.5-inch disk drive developed by Sony. The disk had a greater capacity and a more protective plastic case. However this resulted in the computer introduction date being delayed again. Apple announced the introduction to the public in a dramatic Orwellian commercial during the Super Bowl football game in January 1984. The official introduction was two days later on January 24th, at the annual shareholder's meeting.

The computer used a Motorola MC68000 microprocessor with 128K bytes of RAM. The storage system had one integral 3.5-inch 400K byte floppy disk drive from Sony. The disk controller used a single large-scale integrated chip called IWM (Integrated Woz Machine). It was a one-chip integration of the disk controller functions as developed by Wozniak for the Apple Disk II drive. The display was a 9-inch monochrome monitor. The computer had a small 9.75 by 10.9 inch footprint and was 13.5 inches high with no expansion slots and a separate keyboard. The computer system utilized a single-button mouse similar to the Lisa mouse.

Figure 10.4: Macintosh computer.
Photograph is courtesy of Apple Computer, Inc.

The 64K of ROM contained; the operating system, a Finder program and a set of routines called the User-Interface Toolbox. The Toolbox routines controlled the mouse, windows, menus, text editing and other features of the user interface. A QuickDraw graphics program developed by Bill Atkinson was also in ROM. Two application programs were available from Apple. The first was MacPaint, also created by Atkinson and the second was the word processing program called MacWrite, written by Randy Wigginton. Microsoft had also adapted the spreadsheet program Multiplan and a Microsoft BASIC interpreter for the Macintosh. Other programs announced for release in 1984 included; Pascal, Assembler/Debugger, Logo, MacDraw, MacProject, MacTerminal and the Word processing program.

The price of the Macintosh was $2,495, an ImageWriter printer $595 and a second 3.5-inch disk drive $495. Prior to introduction Apple devised a unique marketing strategy to increase initial sales. They formed an Apple University Consortium (AUC) that offered the Macintosh to students and faculty for a flat $1,000. This program was a huge success that enabled the Macintosh to penetrate the educational market.

Shortly after the enthusiastic introduction users noted two limitations that impeded the utilization of the Macintosh. One limitation was the storage system and the other was memory. Apple released a second 3.5-inch external drive in the early summer. The other storage limitation was the lack of a hard drive. However the significant limitation was the 128K bytes of user memory. The bit-mapped video system and extensive use of graphics utilized a portion of the RAM memory. This reduced the memory available for application programs and files. It also resulted in a slowdown of program execution. Users were saying that the computer was a weakling, slow and under-powered. Macintosh sales were below market forecasts by summer. In response to this and a lack of application software, Apple lowered the price to $1,995. Then Apple started development of the TurboMac and released the "Fat Mac" with 512K bytes of RAM memory in September.

In November 1985, Gates and Sculley signed a controversial agreement related to the similarities between Microsoft Windows software and the Apple Lisa/Macintosh graphics. The agreement recognized that Microsoft had used derivatives of the Apple graphics and allowed their use in Microsoft Windows and other application programs. Also in return for an extension of the license for Microsoft BASIC (Applesoft BASIC) on the Apple II computer, Apple agreed to terminate the completion of MacBASIC that Donn Denman had been developing. Microsoft also agreed to upgrade the Word application program and delay the Excel spreadsheet for the IBM Personal Computer.

Macintosh Plus

Apple announced the Macintosh Plus with one megabyte of memory in January 1986. The computer had an improved disk drive with greater capacity, cursor keys and a numeric key pad. The Macintosh Plus now included provision for the connection of a hard disk drive. The computer had a price of $2,599. Apple also released the LaserWriter Plus printer that had a price of $6,798.

Raskin and Jobs concept of a closed system had limited the power and expansion capabilities of the Macintosh. However Apple changed this with the introduction of the Macintosh SE and II series of computers.

Macintosh SE

Apple introduced the Macintosh SE (System Expansion) that was an upgrade of the Macintosh Plus at the AppleWorld conference in March 1987. The SE model targeted the business market.

The computer had one expansion slot for a plug-in board, two internal disk drives and a heavy duty power supply with a cooling fan. One of the internal drives could be a 20 MB hard disk. A rewritten ROM provided a speed improvement. The Macintosh SE had a price of $2,769.

Macintosh II

Apple also introduced the Macintosh II at the AppleWorld conference in Los Angeles, California in March 1987. It was a second generation of the Macintosh family for advanced users. Mike Dhuey was a principal in the design of the computer. It featured an open architecture, a more powerful 32-bit microprocessor, built-in hard disk storage, a separate color or monochrome display and network capabilities that would enable connection to the IBM world. This release preceded the IBM announcement of the PS/2 (Personal System/2) series of computers by one month.

The computer used a Motorola MC68020 microprocessor with a floating-point coprocessor and one megabyte of RAM, expandable to 8 MB. The Mac II had about four times the speed of a Macintosh SE. The unit had an open NuBus architecture developed at MIT. The 13-inch color or 12-inch monochrome monitor could display 640 by 400 pixels as compared to the 512 by 342 pixel display on the previous Macintosh computers. The computer included custom sound chips to digitize audio input or output.

The basic computer with one megabyte of memory had a cost of $3,898. A 40 MB hard disk drive cost $1,599 and a color monitor with a display board had a cost of $1,547. A complete system had a total price of $7,044. Apple released an implementation of the AT&T UNIX operating system called A/UX, for the Macintosh in 1988.

Other Macintosh Developments

In March 1988, a software project called Pink was started to develop a next-generation operating system. The new system would incorporate advance features, including object-based technology and preemptive multitasking. Erich Ringewald was the initial project leader. Another development that became the Blue project was started to improve the current operating system. This became the System 7 operating system released in 1991.

Apple introduced the Macintosh IIx in September 1988. It used the Motorola MC68030 microprocessor and 68882 math coprocessor. The computer had a price of $7,769. Early in 1989, Apple released the Macintosh SE/30 that used the Motorola MC68030 microprocessor. Then shortly after, Apple released the powerful modular Macintosh IIcx. Apple introduced the Macintosh IIci in September 1989. It was a high-performance version of the Mac IIx operating at 25 MHz. The computer had a price of $6,269.

Macintosh Portable

Apple introduced the Macintosh Portable in September 1989. The computer was available in two models: a model with a single floppy-disk drive and a model with both a floppy-disk drive and an internal 40 megabyte hard disk drive.

The computer used a Motorola CMOS 68000 microprocessor with 1 megabyte of RAM expandable to 2 megabytes. The unit included a built-in 3.5-inch 1.4 megabyte floppy-disk drive and an active matrix liquid crystal display with a screen resolution of 640 by 400 pixels.

The computer was 15.25 inches wide by 14.83 inches deep and the height varied from 2 to 4 inches, front to back. The weight without a hard disk drive was 13.75 pounds. The keyboard had 63 keys and a unique arrangement for locating either a trackball pointing device or an 18-key numeric keypad on the left or right hand side of the keyboard. The computer used lead acid batteries with a power management system controlled by a 6502 microprocessor. This provided 8 to 10 hours of operation on a single battery charge.

Chapter 11 Competitive Computers

As 1980 began, Tandy/Radio Shack had about 40 percent of the personal computer market, Apple Computer was in second place and Commodore in third. The balance of the market was shared by a number of small entrepreneurial companies.

In 1981, the market started to change significantly with the release of the IBM Personal Computer. This computer had instant acceptability, especially in the business segment of the market. In the home consumer segment of the market, a price war evolved between 1982 an 1983 that had a significant impact on companies such as Atari, Commodore, Radio Shack and Texas Instruments. The combination of a requirement for compatibility with the IBM PC and aggressive competition resulted in major financial problems for companies such as Atari, Osborne Computer and Texas Instruments. A dominant company that survived these competitive pressures was Tandy Radio/Shack.

11.1 ... Tandy Radio/Shack

Reference Section 4.5 for the founding of the Tandy Corporation and computers released in the 1970's.

TRS-80 Model III

The TRS-80 Model III computer was introduced in July 1980. The physical configuration was similar to the TRS-80 Model II. The computer, monitor, keyboard and space for two optional 5.25 inch floppy disk drives were enclosed in a one-piece molded housing. The computer used a Z-80 microprocessor with 4K bytes of RAM which was expandable. The non-detachable keyboard had 65 keys and included a numeric keypad. The unit included a high-resolution 12 inch monitor. The computer was sold in three configurations with a price ranging from $699 to $2,495. A single floppy disk drive cost $849.

TRS-80 Color Computer

The TRS-80 Color Computer, also known as the CoCo was a low cost unit that was announced in the summer of 1980. The computer and a 53 key keyboard were an integrated unit. The computer used the Motorola 6899E microprocessor with 4K bytes of RAM expandable to 16K. A television modulator was built-in for a separate color television display unit. The graphics resolution was 128 by 192 using 4 colors and 256 by 192 using only one color. The basic computer sold for $399, a 13 inch color video receiver was also $399 and a cassette recorder was available for $59.95. The major price war that started in 1982, resulted in the price of the computer dropping to $199 by June 1983.

TRS -80 Pocket Computer

The TRS-80 Pocket Computer measured 7 by 2.75 by 0.5 inches and used four mercury batteries. It included a 57 key keyboard and a 24 character liquid crystal display. The computer used two 4-bit CMOS microprocessors with 7K bytes of ROM and 1.9K bytes of RAM. The unit sold for $249 and was introduced in mid 1980.

TRS-80 Model 16

The TRS-80 Model 16 was announced in January 1982. It utilized the Motorola 68000 and Z80 microprocessors, with 128K bytes of RAM and one eight inch disk drive. Microsoft provided a XENIX operating system for the computer.

TRS-80 Model 100

This was the first "notebook" computer and was conceived by Kay Nishi and supported by Bill Gates of Microsoft in 1981/82. Kyoto Ceramics (Kyocera) of Japan manufactured the computer. Radio Shack paid Microsoft a royalty and announced the TRS-80 Model 100 Portable Computer in March 1983.

The basic configuration had a Intel 80C85 operating at 2.4 MHz, 32K bytes of ROM and 8 to 32K bytes of RAM. The unit had a cassette interface and a display that provided 8 lines of 40 characters with a graphics resolution of 64 by 240 pixels.

This portable unit measured 11.6-inches wide by 2-inches high by 8.25-inches deep and weighed 3 pounds, 14 ounces. It had an integrated liquid crystal display (LCD) and a full-size keyboard with 12 function keys. The unit operated on four AA batteries for about 20 hours or on an optional AC adapter. A 300-baud modem was also built-in..

ROM had a collection of software developed by Microsoft and included: operating system, BASIC interpreter, text editor, communications program, address-book program and appointment-calendar program. The unit varied in price from $799 for a computer with 8K bytes of RAM to $1,134 for a unit with 32K bytes of RAM.

TRS-80 Model 4

This model was released in 1983. It used the Z80 microprocessor and had 64K bytes of memory. The display could show 24 lines of 80 columns.

Tandy 1000

The Tandy 1000 computer was the companies first IBM compatible product and was introduced in December 1984.

11.2 ... Commodore

Reference Section 4.4 for the founding of Commodore International and computers released in the 1970's.

CBM 8032

The CBM 8032 was introduced as a business computer at the National Computer Conference in May 1980. CBM denoted Commodore Business Machine. The basic configuration had a MOS 6502 microprocessor, 18K bytes of ROM and 32K bytes of RAM expandable to 96K. The storage system had dual 5.25-inch floppy-disk drives with 500K bytes of storage per drive. The display had a

12-inch screen that could display 25 lines of 80 characters.

The unit included a keyboard with a numeric keypad, a cassette interface and a BASIC programming language. Separate single and dual floppy disk drive units were available in various storage capacities. The computer was priced at $2,829.95.

SuperPET

The SuperPET 9000 series were an advance PET series based on the CBM 8032 system. The computer used a MOS 6502/6809 microprocessor with 36K bytes of ROM, 96K bytes of RAM and 2K bytes of screen RAM. A green phosphor screen could display 25 lines of 40 characters.

VIC-20

In 1979/1980 Commodore started to focus its new product development at the low price end of the market. The company wanted to produce a low cost, friendly consumer product for the mass market.

Prototypes of the VIC-20 were demonstrated at the Consumer Electronics Show (CES) in June 1980 and production models at the CES in January 1981. The VIC prefix stands for Video Interface Computer and was derived from the name of the Video Interface Chip developed in 1978.

The basic configuration had a MOS 6502A microprocessor, 16k bytes of ROM and 5K bytes of RAM expandable to 32K. The display had 23 lines of 22 characters.

The unit could be connected to a color television set or a video monitor. The computer included an integrated keyboard with four programmable function keys. External connectors were provided for a cassette recorder, floppy disk drive, program cartridge, games port, TV/monitor and printer. The VIC BASIC was a Commodore modified version of Microsoft BASIC.. The computer was priced at $299.95 initially and was an immediate success. By January of 1983 Commodore had sold one million VIC-20's. Starting in 1982, a major price war ensued that resulted in the price of the VIC-20 dropping to $89 by June 1983.

Commodore 64

The Commodore 64 computer was announced in January 1982. The basic configuration had a MOS 6510 microprocessor, 20K bytes of ROM and 64K bytes of RAM. The unit had a 40 column display with 16 colors.

The unit had Sprite graphics and a music synthesizer chip. It was released in September 1982 at a price of $595. In 1983 it was released to mass merchandisers and the price dropped below $400.

MAX Machine

Initially developed with the name of Ultimax as a low cost computer with a target price of $149. It had only 2K bytes of memory, a 25 line by 40 column display and contained a music synthesizer chip. The MAX Machine was announced at the Consumer Electronics Show in June 1982. It was not successful and was withdrawn shortly after introduction.

Commodore SX-64

The Commodore SX-64 is a portable version of the Commodore 64 computer which was announced at the Consumers Electronics Show in January 1983.

Commodore PLUS/4

The Commodore PLUS/4 was initially called the 264 series. The computer was introduced at the Consumer Electronics Show in January 1984. The computer featured a built-in integrated software program called 3-Plus-1. The program included a word processor, spreadsheet, business graphics and a database.

In January 1984 Jack Tramiel resigned from Commodore and was replaced by Marshall Smith as president. Following this a significant number of Commodore executives departed from the company. Shortly after leaving Commodore Tramiel founded a new company called Tramel Technology Limited (TTL). In July, TTL acquired Atari from Warner Communications Inc., in a shared ownership agreement. Then in August, Commodore acquired the Amiga Computer Corporation. Commodore released the Amiga 1000 computer in 1985 that sold for $1,295. It featured outstanding graphics, a multitasking windowing operating system and sound capabilities.

11.3 ... Osborne

Adam Osborne graduated with a doctorate degree in chemical engineering from the University of Delaware in 1968. Osborne then worked for Intel and wrote technical manuals for the early microprocessors. In 1975 he wrote a book entitled *An Introduction to Microcomputers* [44]. This was one of the first popular books to describe the microprocessor in an introductory manner. It was the start of Osborne as an author and publisher. He also wrote a computer opinion column for the trade press called "From the Fountainhead." In 1979, Osborne sold his publishing company to McGraw-Hill. Osborne decided to enter the personal computer market and founded Osborne Computer Corporation in 1980.

Osborne 1

Osborne hired Lee Felsenstein to design a low cost portable computer. The Osborne 1 computer was introduced in March 1981 at the West Coast Computer Faire.

The basic configuration consisted of a Z-80A microprocessor operating at 4 MHz with 4K bytes of ROM and 60K bytes of RAM. The storage system had two 5.25-inch floppy-disk drives with a capacity of 100K bytes on each disk. The unit had a built-in 5-inch video monitor. The display was organized as 32 lines of 128 characters with 24 lines of 52 characters visible and scrolling to view the remainder of screen.

This portable unit measured approximately 20.5 inches wide by 9 inches high by 13 inches deep and weighed about 24 pounds. The integrated unit included the video monitor, two 5.25-inch floppy disk drives and detachable keyboard. The keyboard had 69 keys with a numeric keypad. The numeric keys could be programmed to operate as function keys.

Osborne arranged to obtain CP/M from Digital Research, Word-Star from MicroPro and SuperCalc from Sorcim at very reasonable prices. This allowed him to incorporate them as a bundled software package with the machine. C-BASIC and Microsoft MBASIC languages were also included for programming. The entire system, computer and software was priced at $1,795 and was extremely successful.

Osborne Executive

An improved portable model called the Osborne Executive with a larger display was announced in early 1983. The basic configuration had a Z-80A microprocessor operating at 4 MHz and 10K bytes of ROM and 128K bytes of RAM. The storage system had two half-height 5.25-inch floppy-disk drives with a capacity of 204K bytes on each disk. The display was a 7-inch amber monitor capable of displaying 24 lines of 80 characters.

The Osborne Executive computer had the same case as the Osborne 1 computer but had a slightly heavier weight of 28 pounds. The entire keyboard was software programmable.

A complete package of software was provided with the system and included: CP/M Plus and UCSD p-System operating systems, MBASIC, C-BASIC, Word-Star, SuperCalc and a database management program called Personal Pearl. The complete system including software cost $2,495.

An upgrade model called the Executive II was also announced. This unit included an additional board with an Intel 8088 microprocessor, 128K bytes of memory, 640 by 200-dot high-resolution monochrome graphics and MS-DOS operating system. This additional board and software provided a degree of IBM compatibility. The complete system cost $3,195.

Rapid expansion of the company during 1982 accompanied by new management personnel resulted in financial difficulties. Then the early announcement of the new Osborne Executive computer in early 1983 ahead of availability, significantly reduced sales and cash flow. The result was Osborne Computer Corporation went bankrupt in September 1983. This era is chronicled by Osborne in his book *Hyper-Growth: The Rise and Fall of Osborne Computer Corporation* [161]. In the spring of 1984 Osborne founded a new company Paperback Software

International. The company markets low cost software for small companies.

11.4 ... Kaypro

The company was initially founded as Non-Linear Systems in 1953 by Andrew Kay to produce digital voltmeters. Due to market conditions Kay decided in 1980 to enter the personal computer market. The company name was subsequently changed to Kaypro Corporation.

Kaycomp II and Kaypro II

The computer developed was a portable unit with disk drives, keyboard, nine-inch monitor, bundled software and sold for $1,795. The computer was named Kaycomp II and was introduced in April 1982. The computer name was later changed to Kaypro II.

The basic configuration had a Zilog Z80 operating at 2.5 MHz with 64K bytes of RAM. The storage system had two double-density, dual-sided 193K byte 5.25-inch floppy-disk drives built-in. The unit had a 9-inch green-phosphor monitor that displayed 24 lines of 80 columns.

The portable unit measured 18 by 15.5 by 8 inches and weighed 26 pounds. The detachable keyboard had 86 keys which included a 14 key numeric keypad.

The operating system was CP/M and both a MBASIC interpreter and SBASIC compiler languages were provided. Other programs included with the system were: PerfectCalc, PerfectFiler, PerfectSpeller, PerfectWriter, Profit Plan and Word Plus.

The Kaypro 4 and 10, computers were subsequently released. The Kaypro 4 had diskette capacity of 392K bytes and the Kaypro 10 had both increased diskette capacity and a 10 megabyte hard disk.

11.5 ... Compaq

Rod Canion, James Harris and William Murto founded the Compaq Computer Corporation in February 1982. The name Compaq was selected to provide a unique identity associated with the initial focus of producing a "compact" computer. This focus resulted in a proposal to produce a portable computer that would avoid direct competition with IBM. Each of the founders had been senior managers at Texas Instruments (TI). Canion had been manager of three TI Product Customer Centers, Harris a vice-president of TI engineering and Murto a former vice-president of TI marketing and sales. The company received initial financing from the Sevin-Rosen Partners, a high-tech venture capital firm. Benjamin M. Rosen, a partner in the venture capital firm subsequently became chairman of the board for Compaq. H.L. Sparks and other marketing executives were recruited from IBM in 1982, to manage the marketing and sales organization. The company went public in December 1983.

Figure 11.1: Founders of Compaq Computer.
Photograph is courtesy of Compaq Computer Corporation.

Compaq was the first company to introduce a successful IBM compatible computer. The Canadian Hyperion computer preceded it, but it was not successful. Compaq achieved the compatibility by reverse engineering the IBM ROM BIOS chip. They also obtained the cooperation of Microsoft, to adapt a BASIC interpreter and MS-DOS for the new portable computer that provided identical functions and compatibility with the IBM PC.

Compaq Portable

The Compaq Portable was announced in November 1982. The basic configuration had an Intel 8088 microprocessor with 128K bytes of RAM expandable to 256K. The storage system had a 320K byte double-sided 5.25-inch floppy-disk built-in. The unit had a 9-inch green monochrome monitor capable of displaying 25 lines of 80 characters.

Figure 11.2: Compaq Portable computer.
Photograph is courtesy of Compaq Computer Corporation.

The portable unit measured 20-inches wide by 8.5-inches high by 15.3-inches deep and weighed 28 pounds. At this weight, a more correct term would have been "luggable." The integrated unit included the video display, a 5.25-inch floppy-disk drive with provision for a second drive and a detachable keyboard. The IBM style of keyboard had 83 keys, 10-key numeric keypad and 10-key function pad. A socket was provided for the addition of the Intel 8087 coprocessor. MS-DOS was the operating system

The Compaq Portable was priced lower than the IBM PC. A basic unit with 128K bytes of RAM and one 320K byte disk drive cost $2,995. A two-disk-drive system cost $3,590.

Other Models

The Compaq Plus with a 10MB hard disk drive was introduced in October 1983.

The Deskpro 286 was Compaq's first desktop computer and was introduced in June 1984. The Deskpro 386 computer was released in September 1986. It was the first IBM compatible computer to use the Intel 32-bit 80386 microprocessor. A special memory management software system was developed by Microsoft. It was released eight months ahead of a similar IBM product.

The Compaq Portable 286 was released in June 1984 and the Compaq Portable 386 in September 1987.

The Compaq LTE was the companies first notebook PC and was introduced in October 1989.

The SystemPro was Compaq's first server computer, it used EISA (Extended Industry Standard Architecture) and was introduced in November 1989.

11.6 ... NeXT

Steven P. Jobs founded NeXT Computer Inc., in September 1985 following his departure from Apple Computer, Inc. Five key Apple personnel joined Jobs in the formation of NeXT. They were Susan Barnes the Macintosh controller, George Crow an engineering manager, Dean Lewin the higher education marketing manager, Rich Page who was an Apple Fellow and Bud

Tribble the Macintosh software manager. The NeXT computer was developed with an initial market target of the higher-educational users. Input from various schools and universities influenced the design. In early 1987 Ross H. Perot invested twenty million dollars in the company. The NeXT computer was released in October 1988.

The basic configuration had a Motorola 68030 microprocessor operating at 25 MHz with a 68882 math coprocessor and 8 megabytes of RAM expandable to 16 megabytes. The storage system used the first commercially available erasable optical drive with a removable 256 megabyte capacity cartridge. A fast NuBus architecture with four 32-bit expansion slots was incorporated. The unit had a 17-inch monochrome monitor with a resolution of 1,120 by 832 pixels.

The computer had advanced VLSI (Very Large Scale Integration) technology and a built-in Motorola digital signal processor. NeXT had developed its own UNIX based operating system named NeXTSTEP. The operating system was included free with the computer that had an impressive graphical interface. On introduction it was termed the machine for the 90's. However the price of $6,500 for the computer and $1,995 for the NeXT laser printer, would be a limiting factor on initial sales.

11.7 ... Miscellaneous

Access Portable Computer

The basic configuration had a Zilog Z-80A microprocessor operating at 4 MHz with 8K bytes of EPROM and 64K bytes of RAM. the storage system had two double-density 5.25 inch 184K byte floppy-disk drives. The unit had a 7-inch amber monitor that could display 25 line of 80 characters.

The portable unit measured 16.1 by 10 by 10.8 inches and weighed 33 pounds. The computer had a detachable keyboard, a built-in Epson MX-80 dot-matrix printer, two 5.25-inch floppy-disk drives, an acoustic and direct connect modems. Included with the computer was a set of software: CP/M operating system, Microsoft's MBASIC, C-BASIC, PerfectWriter,

PerfectSpeller, PerfectCalc, PerfectFiler and TELCOMU communications program. The complete system was priced at $2,495.

Acer

Stan Shih was a principal in the June 1976 founding of Multitech International Corporation in Taiwan. Initially the company produced consumer electronic products. The company name was changed to Acer Incorporated in 1987. North American operations are controlled by Acer America Corporation. Acer is a highly decentralized company that manufactures most of its own components and peripherals. The company is now a major worldwide supplier of personal computers.

Acorn

The Acorn computer was an important British microcomputer that was developed by Acorn Computers Ltd., in conjunction with the BBC (British Broadcasting Corporation) in the early 1980's. Two models were produced, the BBC Models A and B.

The Model B had a basic configuration that used a MOS 6502 microprocessor operating at 2 MHz, 16K bytes of ROM and 32K bytes of RAM. A disk drive was available. The unit could display 24 lines of 40 columns or 25 lines of 80 columns. High-resolution graphics was 256 by 640 pixels in 2 colors.

The unit had a full-size typewriter style keyboard with function keys. The ROM memory included a BBC BASIC interpreter and a 6502 assembler. The Model B computer was priced at 399 pounds sterling and the Model A with fewer features at 299 pounds sterling.

Apple II Clones

The first Apple II clone was announced by Franklin Computer Corporation in the spring of 1982. It was named the Ace 100. Apple Computer successfully sued Franklin for copying Apple's system software at a Federal Court in January 1983. This resulted in the bankruptcy of Franklin Computer by the end of 1984.

The Laser 128 was a clone of the Apple II and was available by the end of 1986. It was a low cost computer produced by Video Technology. An improved version named the 128EX/2 was subsequently released.

Other manufacturers in the Far East and Europe produced clones with names such as the Apollo II, Orange Computer and Pineapple.

AST Research

Safi Qureshey, Albert Wong and Thomas C. Yuen started a high-tech consulting firm that incorporated as AST Research, Inc. in 1980. The company name prefix is derived from the initial of their first names. AST Research started by producing peripheral cards. The company went public in 1984 and introduced its first personal computer, the Premium/286 in 1986. Albert Wong left the company in 1988.

Atari

For the founding of the Atari Corporation and its first computers see Section 4.6. By 1982 Atari was one of the largest manufacturers of computer-based video games. However, in late 1982 sales of video games started to drop. During this period Atari decided to shift a significant portion of its production to overseas manufacturing. This resulted in massive layoffs of its personnel beginning in February 1983 and continued into the summer. Deteriorating conditions resulted in the resignation of Atari director Raymond Kassar in July 1983 and the recruitment of James J. Morgan to replace him. By the end of 1983, Atari had lost more than $500 million. Atari had announced the 1200XL computer in January 1983.

Then in July 1984 Atari was acquired by Jack Tramiel of Tramel Technology Limited. Tramiel had left Commodore International in January and negotiated a shared ownership agreement with Warner Communications Inc. Within a month Tramiel significantly reduced Atari's personnel again and its physical facilities.

Staff were immediately assigned to developing a new computer for introduction at the January 1985 Consumer Electronics Show. The new computer would use a 16-bit microprocessor.

Corvus Concept

Advertised as "a complete system for automating the modern office" in early 1982. It featured a rotatable monitor that could be turned 90 degrees for display in portrait or landscape modes.

The basic configuration had a Motorola 68000 microprocessor operating at 8 MHz and 256K bytes of RAM, expandable to 512K. The storage system had a separate 8-inch floppy-disk drive and a separate hard-disk drive with 6, 10 or 20 megabyte capacities. The unit had a 14-inch monochrome monitor with a 8.5-inch by 11.5-inch viewing surface. It could display 63 lines of 90 columns in the portrait position or 47 lines of 117 columns in the landscape position.

The unit had a standard 91-key keyboard, numeric keypad and 10 software-definable function keys. A package of software was available from Corvus: operating system, logical spreadsheet, EdWord word processor, Pascal, FORTRAN and a CP/M emulator. The basic unit cost $4,995, floppy-disk $1,500 and $2,495 for a 6 megabyte hard-drive.

DEC

Digital Equipment Corporation (DEC) entered the microcomputer marketplace in April 1982. The computers were of a higher quality, and higher price. Cofounder Kenneth Olsen stated "DEC is not in the consumer business." Three products were introduced: the DECmate II -- a word processing system with limited computing capabilities, the Professional 300 series and the Rainbow 100 series. Barry James Folsom was a principal in the engineering design.

The Rainbow series were general purpose computers that consisted of the 100 and 100+ models. The Rainbow 100 had a basic configuration that used a Z80A microprocessor and a 16-bit Intel 8088 microprocessor. The memory had 64K bytes of RAM expandable to 256K. The storage system had two 5.25-inch 400K byte floppy-disk drives built-in. The display was a separate 12-inch unit that could display 24 lines of either 80 or 132 columns. The unit had a separate keyboard with 103 keys, a numeric pad and 20 function keys. A significant feature of the computer was the two microprocessors that allowed application software to run on either MS-DOS or CP/M operating systems. The price was $4,190. The main differences in the Rainbow 100+ was that main memory was increased to 128K bytes, expandable to 896K and a 10 MB hard disk drive was included.

The Professional series consisted of the Models 325 and 350. These were personal minicomputers based on the 16-bit PDP-11 architecture. Both models used the DEC three-chip F-11 CPU. Both models had 512K bytes of RAM, expandable to 768K on the Model 325 and to 1,280K on the Model 350. Both models had two 5.25-inch 400K byte floppy-disk drives and the Model 350 included a 10 MB hard disk drive. Two operating systems -- P/OS and RT-11 based on PDP-11 systems were available.

Sales were significantly below expectations. DEC had a bureaucratic organization and poor marketing capabilities that were not suited to the fast changing personal computer market. Another factor was problems with the initial Rainbow computer running IBM software. This resulted in DEC terminating production of the computers in February 1985

Dell

Michael S. Dell, purchased his first computer, an Apple II, at the age fifteen in 1980. He then took it apart to improve his understanding of how it was designed and made. Shortly after the introduction of IBM Personal Computer he started buying, upgrading and selling IBM compatible computers. This modest beginning was the start of Dell's method of direct sell to the end user.

While attending the University of Texas, Dell created a company called PC's Limited to market his computers in January 1984. PC's Limited soon attained sales of between $50,000 and $80,000 per month. Dell left university and incorporated the company as Dell Computer Corporation in May. Products were still sold under the PC's Limited brand name. The strategic orientation of the company was to market more sophisticated, build-to-order, PC compatible computer systems directly to end customers. This eliminated the markup to dealers and resellers and provided a significant cost advantage in product pricing.

Figure 11.3: Michael S. Dell.
Photograph is courtesy of Dell Computer Corporation.

To improve profitability Dell decided to produce his own computer. This was simplified by using a chip set for the Intel 286 microprocessor supplied by Chips and Technology, Inc. The design was contracted to an engineer name Jay Bell for $2,000. In 1986, Dell introduced at the spring COMDEX show the fastest performing 286-based computer system. The 12 megahertz machine was priced at $1,995 compared to IBM's 6 megahertz machine priced at $3,995.

By the end of 1986, Dell Computer had sales of about $60 million. The company was growing, and opened Dell UK in June 1987. To finance an aggressive growth of the company, Dell arranged with Goldman Sacks & Company for a private capital offering in October. The company made its first public offering of shares in June 1988.

In mid 1987, the company began to change its advertising from the use of PC's Limited to the use of the Dell Computer trademark. The change to the use of the new trademark on all product lines was completed in April 1988.

Epson

For Epson corporate history and printer development see Section 17.5.

The HX-20 was the first "notebook" computer and was announced in November 1981. The basic configuration had two Motorola 6301 microprocessors (a CMOS version of the 6801) operating at 614 kHz, 40K bytes of ROM and 16K bytes of RAM expandable to 32K. A microcassette was used for storage. The unit could display 4 lines of 20 characters(32 by 120 dots). The unit measured 11.3-inches wide by 8.5-inches deep by 1.7-inches high and weighed 3 pounds, 13 ounces. It had an integrated liquid-crystal-display (LCD), standard-size keyboard, 24-character-per-line printer and a built-in cassette interface. A Microsoft BASIC interpreter and a word processor called Skiwriter were also included with the unit which was priced at $795.

The HX-40 computer was a 3.5 pound notebook computer with a CP/M operating system, 32K bytes of ROM and 64K bytes of RAM. It had a liquid crystal display with 8 lines of 40 characters.

The Geneva computer was a 4 pound notebook computer with a CP/M operating system, 32K bytes of ROM and 64 K bytes of RAM. It had a 8 line by 80 character liquid crystal display (LCD).

The QX Series was Epson's first line of personal computers and were introduced in 1983. The QX-10 had a 8-bit CP/M-80 compatible operating system. an optional 16-bit MS-DOS board was available. The QX-11 used an Intel 8088 microprocessor with 256K bytes of RAM expandable to 512K. It also had two 3.5 inch floppy disk drives and a 12 inch green monochrome monitor with a resolution of 640 by 400 pixels. The QX-16 had a 16-bit microprocessor with an MS-DOS operating system and a 8-bit microprocessor with a CP/M-80 operating system. The unit had 512K bytes of RAM and dual 5.25 inch floppy disk drives.

Epson's first IBM PC compatible computers were the Equity Series which were introduced in 1985. The Equity I was compatible with the IBM PC computer. It used a Intel 8088 microprocessor with 256K bytes of RAM. It had a single 5.25 inch floppy disk drive. The Equity II was compatible with the IBM PC/XT computer. It used a NEC V-30 microprocessor which was compatible with the Intel 8086 microprocessor. It had 640K bytes of RAM and was available with a single 5.25 inch and a 20 megabyte hard disk drive. The Equity III was compatible with the IBM PC AT computer. It used a Intel 80286 microprocessor with 640K bytes of RAM. It was available with a 5.25 inch floppy disk drive, 20 megabyte and 40 megabyte hard disk drives.

Gateway 2000

Theodore W. Waitt started assembling computer systems in 1987 and founded Gateway 2000, Inc., with his brother Norm Waitt and Mike Hammond in 1988. Waitt selected the company name to suggest that their computers were the gateway to the 21st century. Gateway is a major direct marketer of personal computers.

GRiD

John Ellenby founded GRiD Systems Corporation in 1979. Prior to the founding of GRiD Systems, Ellenby worked for Xerox at PARC (Palo Alto Research Center) and was responsible for the development of the Alto II computer. The company has focused on the high quality segment of the portable computer market, rather than the low cost segment. The company was acquired by the Tandy Corporation in 1988.

A portable IBM compatible laptop computer named the Compass I was introduced in April 1982. The computer weighed less than ten pounds, had a flat electroluminescent screen and a built-in modem. To achieve greater durability and reliability the models utilized bubble memory for secondary storage instead of floppy-disk drives. The Compass II model was introduced in June 1984.

GRiDCase was a portable IBM compatible laptop computer that weighed 12 pounds and included a built-in 3.5 inch floppy-disk drive. The GRiDCase was introduced in April 1985 and the GRiDCase Plus in September 1986.

The GRiDPad is a tablet styled portable introduced in 1989. The "Pad" in GRiDPad stands for "pen and display." The unit did not have a keyboard, but used a stylus for input. Physically it measured 12.5 by 9.25 by 1.5 inches and weighed 4.5 pounds. It used an Intel 80C86 processor, had 1 MB of RAM with MS-DOS in ROM. The LCD screen was 8 by 5 inches and the CGA resolution was 640 by 400 pixels. The GRiDPad sold for $2,370.

Hewlett-Packard Company

See Section 2.4 for the founding of Hewlett-Packard and Chapter 4 for some of the company's early products. In the early 1980's a number of computer product lines were developed: Series 70, portable computers., Series 80, desktop computers for technical users., Series 100, desktop computers for business users and Series 200, high performance computers for engineering applications. The following are examples of products from each of these series.

The HP-75 Portable computer had 48K bytes of ROM and 16K bytes of RAM expandable to 24K. The storage system had a built-in magnetic card reader with a storage capacity of 1,300 bytes. The unit had a 1-line, 32 character liquid-crystal-display (LCD). The portable unit measured 11.1 by 6 by 1.1 inches and weighed 1 pound, 10 ounces. The integrated keyboard had 65 keys. A BASIC interpreter, text editor, file manager, clock/calendar and appointment scheduler programs were included. The portable unit was priced at $995.

The physical configuration of the HP-85 computer included a 5-inch video display, thermal printer, data cartridge drive and a keyboard. The microprocessor was a 8-bit custom HP design with 32K bytes of ROM and 16K bytes of RAM expandable to 32K. The 5-inch monitor screen could display 265 by 192 dots in graphics mode or 16 lines of text. An Enhanced BASIC interpreter was included in ROM memory. The computer was priced at $3,250 and was available in January 1980.

The HP-150 computer with a touch screen was introduced in September 1983. The innovative "touch screen" allowed a user to point to commands on the screen as an alternative to either keyboard entry or mouse activated selection. The basic configuration had an Intel 16-bit 8088 Microprocessor and 256K bytes of RAM expandable to 640K. The storage system had two 3.5-inch diskette drives with 246K bytes storage capacity. The unit had a 9-inch green monochrome screen that could display 27 lines of 80 characters. A 5 or 14.7 megabyte hard disk drive could be substituted for one of the 3.5-inch diskette drives. The separate keyboard had 107 keys, including a numeric keypad and cursor control pad. The computer used the MS-DOS operating system.

The Series 200 models used the Motorola 68000 microprocessor. The first product released was the Model 226, the Model 216 was the lowest cost product and the Model 236 the most expensive.

The HP Vectra PC was introduced in 1985.

In April 1989, HP purchased the Apollo Computer company for $476 million and merged it into its own workstation product line.

Hyperion

The Hyperion was a portable computer developed by the Dynalogic Corporation that was founded by Murray Bell in Ottawa, Canada. In 1981, Dynalogic encountered financial difficulties. This resulted in 80 percent of the company being acquired by the Bytec Management Corporation, a venture capital company that was co-founded by Michael Cowpland. Dynalogic introduced the portable computer at the 1982 spring COMDEX show in Atlantic City, New Jersey. The unit was compatible with the IBM Personal Computer. To obtain compatibility, The IBM BIOS was reverse engineered and MS-DOS was used. The unit used an Intel 8088 microprocessor, had two 5.25-inch floppy disk drives, a 7-inch amber monitor, a detachable keyboard, weighed 28 pounds and was priced at $4,950 (Canadian).

The Hyperion portable computer was acclaimed at the COMDEX show. It had preceded the Compaq portable computer by about six months. However, starting with the production release in January 1983, things did not proceed smoothly. A company reorganization resulted in Bell leaving in March. In October, Cowpland merged several of his companies into Bytec-Comterm to produce the computer. Then starting in late 1983, a large number of customer complaints were received about disk drive failures due to overheating. This resulted in a significant drop in sales and by the fall of 1984 production was terminated.

Lobo Max-80

The Lobo Max-80 computer was produced by Lobo International in Goleta, California around 1982-83. The basic configuration consisted of a Z-80B microprocessor operating at 5 MHz, 128K bytes of RAM, storage capacity of four 3.5 or 5.25-inch drives and a hard drive.

NEC

The Japanese Nippon Electric Company (NEC) introduced the PC-100 personal computer in October 1983. The computer used the Intel 8086 microprocessor with 128K bytes of RAM, expandable to 768K. The display used a full-bit mapped screen with a resolution of 512 by 720 pixels.

NEC was an early developer of portable computers. It introduced the MultiSpeed portable around 1986 that featured a multiple speed processor. In 1989, NEC introduced the UltraLite notebook type of computer. The UltraLite measured 11.75 by 8.33 by 1.4 inches, weighed 4.4 pounds and with a 1 MB silicon hard disk drive it was priced at $2,999. The unit used the NEC V-30 microprocessor, had 640 KB of memory and a 4.25 by 8.25 inch backlit LCD could display 25 rows by 80 columns.

Packard Bell

Packard Bell Electronics, Inc. was founded in 1926 as a producer of radios. It expanded its consumer electronic products to include television during the 1970's. In the 1960's, Packard Bell produced hybrid mainframe computer systems and was purchased by Teledyne in 1968. However, it was not successful. In late 1985, Beny Alagem purchased the Packard Bell name from

Teledyne. Then in 1986, Beny Alagem, Jason Barzilay and Alex Sandel founded a personal computer company using the Packard Bell Electronics name. Packard Bell became one of the dominant companies supplying personal computers through innovative marketing and aggressive pricing. In 1988 it had attained a position of number 6 in the USA market share.

Silicon Graphics

James H. Clark, an associate professor of electrical engineering at Stanford University and six other graduate students founded Silicon Graphics, Inc. (SGI) in 1982. Clark had designed an integrated circuit chip in 1981, called the Geometry Engine, that enabled the rapid display of three-dimensional graphics. This formed the basis for SGI's introduction of the IRIS 1000 3-D terminal and the IRIS Graphics Library in November 1983 and the IRIS 1400 3-D workstation in late 1984. The workstations would range in price from $40,000 to $50,000. One notable use of SGI workstations was in the creation of the dinosaur graphics for the movie *Jurassic Park*.

Former Hewlett-Packard executive Edward R. McCracken joined Silicon Graphics as chief executive officer in 1984 and the company went public in 1986. A dissatisfaction with corporate management and the strategic direction of the company, resulted in Clark leaving SGI in January 1994.

Sinclair

Clive M. Sinclair began his career as a writer of periodical articles, technical manuals and books in England. He started Sinclair Radionics to produce electronic kits for amplifiers, radios and other products in July 1961. In the 1970's the company developed and released consumer products such as electronic calculators, miniature televisions and digital watches. However by the late 1970's the company encountered financial difficulties due to various product problems and an extremely competitive calculator market. Sinclair obtained government financial assistance starting in 1976 that led to him losing control of the company.

Sinclair Research Ltd., designed and produced the ZX80, ZX81 and QL "Quantum Leap" computers.

The ZX80 computer was widely advertised as a low cost computer at a price of $199.95 and was announced in North America in February 1980. The computer used a Zilog 8-bit Z-80A microprocessor operating at 3.25 MHz. Memory consisted of 4K bytes of ROM that contained a BASIC interpreter and 1K bytes of RAM expandable to 16K bytes. The unit had a built-in RF modulator for connection to a monochrome television set which could display 24 lines of 32 characters. An integrated plastic membrane keyboard had 40 pressure sensitive keys. A connector for connection of a cassette recorder was provided.

The ZX81 computer was built by the Timex Corporation and announced in March 1981. This computer subsequently became the Timex/Sinclair 1000 computer.

Southwest Technical Products

Southwest Technical Products released the SwTPC 6809 in 1980. It used the Motorola MC 6809 microprocessor with 8K bytes of RAM. The unit motherboard had eight 50-pin slots and eight 30-pin slots. Three boards were provided: a processor board, a programmable memory board and a serial-interface card. The unit sold for $495 as a kit, or $595 assembled and tested.

Sun Microsystems

Sun Microsystems, Inc. has not been known as a supplier of personal computers, it is however a dominant manufacturer of engineering and scientific workstations. Andreas Bechtolsheim, William N. Joy, Vinod Khosla and Scott G. McNealy founded the company in January 1982.

Khosla had previously helped to found a company called Daisy Systems in 1980 that provided computer-aided engineering systems (CAE). However, after a year-and-a-half with the company, Khosla left to fulfill a desire of producing a low-cost general purpose computer for engineers and scientists that could be connected to a network. His investigation for this new computer led him to Bechtolsheim, who had developed hardware for a network at Stanford University. Khosla and Bechtolsheim then contacted financier Robert Sackman of U.S. Venture Partners who provided part of the initial financing for the founding of the new company. Khosla offered a friend Scott McNealy, who was a manufacturing manager at Onyx Computer the position of director of manufacturing operations and the status as a founder. The founders now required a specialist in the UNIX operating system used by the engineering and scientific community. This resulted in a meeting with William Joy who had been a principal in the design of Berkeley UNIX, a version of UNIX developed at the Berkeley campus of the University of California. Joy accepted an offer to join Sun and was also given founder status. Khosla became the chief executive officer.

A prototype of the first graphics workstation based on Bechtolsheim's previous design was prepared by May 1982. It was a low cost open system with networking capabilities suitable for the CAD/CAM market place. The workstation was named Sun-1, used a Motorola MC68000 microprocessor, had one megabyte of memory and incorporated the UNIX operating system. A production version called the Sun-2 was released in late 1982 that used a refinement of the Berkeley UNIX operating system called SunOS.

Khosla's management style resulted in a conflict with the Sun board and his resignation in the fall of 1984. He was succeeded by Scott McNealy as interim CEO and as permanent CEO in 1985. The Sun-3 series of workstations that used the Motorola MC68020 microprocessor, were introduced in September. This new workstation became highly successful.

During 1985, Sun introduced the slogan "The Network is the Computer." This was an extension of the concept whereby the power of a personal computer is significantly increased by a connection to a network. It also led to the concept of the low cost network computer.

In the mid 1980's, Sun started finding new sales for its workstations in the automobile manufacturing and financial trading markets. The company was doing very well, and went public in March 1986.

Sun had decided that it required a more powerful microprocessor than Motorola could provide for future networking capabilities. It therefore decided to develop its own RISC microprocessor and recruited Anant Agrawal in April 1984 to lead the design team. The new design was given the name Scalable Processor Architecture (SPARC), and was introduced in July 1987.

The success of the Sun-3 workstation resulted in Sun displacing the Apollo Computer company as the leading producer of technical workstations by 1987. In October, Sun recruited Ed Zander from Apollo Computer as a vice president of marketing. Zander became the chief operating officer and is now the second in command at Sun.

In January 1988, AT&T formed an alliance with Sun by purchasing 7.5 percent of the company, estimated to be worth $320 million. The intent of the alliance was to merge the AT&T UNIX and Sun UNIX into one unified operating system. It was also hoped that the other major companies with UNIX systems would participate in the unification. However, they decided not to and formed the Open Software Foundation (OSF) in May 1988 to create their own unified UNIX system. The alliance was not successful and resulted in AT&T selling its investment in Sun in 1991. However, it did lead to the development of Sun's UNIX based Solaris operating system.

In 1988, Sun introduced the 386i personal computer. The *i* stood for "integrated" operating system, that contained a blend of UNIX and DOS features. However, sales were extremely poor and it was discontinued shortly after.

Sun introduced the SPARCstation 1, its first RISC technology workstation using the new SPARC microprocessor in April 1989. However, the company also reported its first quarterly loss in June. This resulted in organizational changes and layoffs.

Texas Instruments (TI)

For earlier microcomputer releases by Texas Instruments in the late 1970's see Section 4.7. In January 1983, Texas Instruments introduced the TI Professional computer (TIPC) for professionals and small business users. The TIPC used an Intel 8088 microprocessor with 8K bytes of ROM, 64K bytes of dynamic RAM, a single 320K byte, 5.25 inch floppy disk drive and an optional internal floppy disk drive or 10 MB hard disk drive. The TIPC architecture supported an optional speech recognition system.

Texas Instruments encountered problems with the TI-99/4 computer released in 1979 and released the TI-99/4A in the summer of 1983. This was an improved computer with more memory and a lower price.

The TI CC40 (Compact Computer 40) was a small inexpensive portable computer introduced in the early 1980's. It measured 9.2 by 5.7 by 1 inches and had a 1-line, 31 character liquid crystal display (LCD). The memory was 6.2K bytes of RAM and a TI BASIC interpreter was included with the unit which was priced at $250.

The TI products were not successful and the company suffered significant financial losses in the highly competitive market between 1982 and 1983. This resulted in the company announcing its withdrawal from the home computer market in October 1983.

Timex/Sinclair 1000

The Timex/Sinclair 1000 is essentially the same as the Sinclair ZX81 computer. It was manufactured by Timex Corporation and was available in 1982 at a price of $99.95.

Toshiba

The Tokyo Shibaura Electric Company, or Toshiba Corporation was formed by the merger of two large Japanese electrical equipment manufacturers in 1939. The company made significant investments in information, communication and semiconductor technology in the early 1980's. Toshiba introduced the first 1-megabit DRAM chip in 1985 and the T3100 laptop computer in 1986. Toshiba is now a leading producer of laptop computers.

Victor 9000 and Sirius 1

In 1980 Chuck Peddle left Commodore to form his own company Sirius Systems Technology to produce a 16 bit microcomputer for the business market. Prior to this he had been a principal in the design of the 6800 microprocessor at Motorola Inc., designer of the 6502 microprocessor at MOS Technology and the PET microcomputer at Commodore International Inc.

To improve the marketing capabilities for the new computer he became associated with Victor Business Products, an established producer of calculators and cash registers to form a new company Victor Technologies Inc. Victor would concentrate on the United States market and Sirius on the international market.

The design was started in December 1980 and the first prototype was shown in April 1981. Two computer products evolved from the same design, the Victor 9000 and the Sirius 1. They were essentially identical except for the industrial design.

The basic configuration used an Intel 8088 microprocessor, 16K bytes of ROM and 128K bytes of RAM expandable to 896K. The storage system had two 612K byte 5.25-inch single-sided floppy-disk drives built-in. The unit had a high-resolution (800 by 400) green-phosphor monitor.

The unit came with a choice of three keyboards which were detached and had up to 103 keys. The operating systems were MS-DOS and CP/M-86. The basic configuration was priced at $4,495. Initially the computer was a success in both North America and Europe. However by summer of 1983 Victor Technologies was having financial difficulties and filed for bankruptcy in February 1984. Later in the year a Swedish company Datronic purchased the company.

Xerox

Xerox had acquired Shugart Associates, a disk drive manufacturer in 1977, and appointed one of its cofounders Donald J. Massaro as president of the Dallas, Texas Office Products Division in 1979. Massaro supported the Systems Development Division (SDD) and the

"Star" office automation system project. In February 1980, Massaro appointed David Liddle to head a group responsible for production of the Star computer. The computer was introduced as the 8010 "Star" Information System at the National Computer Conference (NCC) in April 1981.

The basic configuration had a Xerox developed MSI (Medium Scale Integration) processing unit that was about three times as fast as an Alto computer. The computer had 512K bytes of RAM and a 10 or 29 megabyte hard disk storage system. The display terminal had a bit-mapped 10.5 by 13.5 inch screen with a resolution of 72 pixels per inch, horizontally and vertically. Extensive research extended the graphical desktop user interface developed at PARC. New developments included overlapping windows, a menu bar and enhanced icons. A two-button mouse was used to position the cursor and for other desktop control activities. Other features incorporated were items such as a What-You-See-Is-What-You-Get (WYSIWYG) document editor that used a 16-bit character set to accommodate foreign languages and capability for distributed personal computing via an Ethernet connection. However, the cost of $16,595, a closed system with the software only available from Xerox and a lack of a financial spreadsheet resulted in its low market acceptance.

To expand the product line, a crash project for the Xerox 820 personal computer was initiated. However the computer costs were excessive and the technology dated which resulted in its failure.

To combat the low acceptance of the Star system, Xerox initiated an improved design of the Star computer. The new design was released as the 6085 "ViewPoint" workstation in 1985. The ViewPoint workstation incorporated an improved performance MESA processor, optional IBM Personal Computer compatible processor, one megabyte of memory (expandable to 4 megabytes), 5.25 inch floppy disk drive, a 10 to 80 megabytes hard disk and an Ethernet connection. The base system initially cost $6,340.

PARC also developed a powerful processor called Dorado, that was at least ten times as powerful as an Alto computer.

Zenith

Zenith Data Systems (ZDS) was another manufacturer that cloned the IBM Personal Computer in the early 1980's, ZDS did this by reverse engineering the BIOS software chip similar to that done by Compaq.

ZDS was also an early manufacturer of portable computers for the U.S.A. government. The Zenith MiniSport was an early lightweight that measured 12.5 by 9.8 by 1.33 inches and weighed 5.9 pounds. It used an Intel 80C88 processor, had 640 KB of RAM, a 2.5-inch 720 KB floppy disk drive and an 8.5 by 3.25 inch backlit transfective display.

The Bull company purchased ZDS in 1989.

Other Companies

Excaliber Technologies was founded in 1980 to produce a computer called Powerstation that was designed for executives. Dennis Barnhart founded Eagle Computer in November 1981 to produce personal computers. The Aquarius computer was announced by Mattel in January 1983. Coleco Industries, a toy and video-game manufacturer introduced the Adam home computer that included a letter-quality printer with a low price of $599 in June 1983. Convergent Technologies announced a three pound portable computer called Workslate with a built-in spreadsheet in late 1983. Many other companies entered the personal computer market during the 1980's such as Morrow's Micro Decisions.

Chapter 12 Microsoft in the 1980's

As Microsoft entered the 1980's, it derived most of its revenue from the sale of BASIC interpreters. In the next few years this would change significantly.

12.1 ... Corporate & Other Activities

1980 Activities

In early 1980 Microsoft decided to get into operating systems and acquired a license for UNIX from AT&T (see Sections 2.6 and 12.3). Then in March, Microsoft introduced their new Z-80 SoftCard (see Sections 6.4 and 17.6) at the West Coast Computer Faire. This CP/M interface card for the Apple II computer was an immediate success.

In 1980, Steve Wood who was the general manager, decided to leave and go to Datapoint. Gates replaced him with an old friend Steven A. Ballmer, as assistant to the president in June 1980.

A phone call from IBM in July 1980 was to have a major impact on Microsoft. Initially an inquiry to obtain programming languages for the proposed IBM PC computer, it evolved into a requirement that included the Disk Operating System and application software (see Section 12.2).

Starting in November 1980, David F. Marquardt of Technology Venture Investors had discussions with Microsoft regarding a plan to change the partnership into a corporation. This resulted in Microsoft becoming Microsoft, Inc., in June 1981. The ownership and percentage of shares was divided between the principals as follows: Bill Gates 53, Paul Allen 31, Steve Ballmer 8 and Vern Raburn 4. Then in September, Technology Venture Investors purchased 5 percent of the company for $1,000,000.

Apple Macintosh

Steve Jobs and other members of the Apple Macintosh development team had discussions with Microsoft between spring and August of 1981. Jobs wanted Microsoft to supply application software for the new Macintosh computer. Then in October 1981, Gates and members of the Microsoft application group visited Apple for a demonstration of the Macintosh computer. This demonstration impressed Microsoft with the future potential of the Macintosh.

Radio Shack Model 100

During 1981, Kay Nishi through a Japanese associate evaluated an 8-line by 40-character liquid-crystal-display(LCD) from Hitachi. Nishi and Gates decided that this would form the basis for a general-purpose portable computer. They developed a specification for the machine that would include a BASIC interpreter, word processor, communication program and address book in ROM. Then in 1982, they decided to have Kyoto Ceramics (Kyocera) of Japan manufacture the computer. The planning determined a scheme for the world-wide marketing of the computer. In the Far East it would be by NEC as the PC-8200, in Europe by Olivetti as the M-10 and in the Americas by Tandy Radio Shack as the Model 100. Microsoft would receive a royalty on each unit sold. Radio Shack released the Model 100 in March 1983. It became the first laptop computer.

Relocation and Administrative Changes

In November 1981 the number of employees at Microsoft had increased to 100. Two new employees in 1981 were Jeffrey (Jeff) S. Raikes, a Stanford MBA who had been with Apple Computer and Chris Peters. Both would subsequently become vice presidents of Microsoft. The company had also moved to a new office building in Bellevue. Then in July 1982 Microsoft hired its first President, James C. Towne. Towne was an executive at Tektronix, a manufacturer of oscilloscopes and test equipment. Gates became executive vice president, with responsibility for all product related activities, and remained chairman of the board and CEO. However after a short period of time Towne did not satisfy Gates who started looking for a replacement.

Other 1982 Activities

In January 1982 Microsoft signed an agreement to supply Apple with a spreadsheet, a business graphics program and a database. In the agreement Jobs had a clause added restraining Microsoft from releasing similar graphics application software to other customers for a twelve month period after the introduction of the Macintosh. Gates amended it to be no later than January 1983. Apple then provided prototypes of the Macintosh to Microsoft for software development.

By early 1982 Tim Paterson had completed an update to the PC operating system that increased the disk capacity from 160 to 320K bytes. At the end of March he left Microsoft and returned to Seattle Computer Products. Subsequently he started his own company called Falcon Technology.

Scott D. Oki, an MBA graduate from the University of Colorado, joined Microsoft in early 1982. Shortly after he presented a business plan for an international group to handle marketing and sales outside the USA. Gates approved the proposal and Oki became director of international operations in September. Microsoft International was a success, and became a significant source of revenue and profits for Microsoft.

During 1982, it became apparent to Microsoft that they required a graphical user interface for their disk operating system. Apple Computer had demonstrated the Macintosh system to Microsoft in late 1981 and VisiCorp had displayed their new VisiOn system in the fall of 1982. This started the development of a new graphics user interface for DOS that would become Windows (see Section 12.4).

In early 1982, Microsoft reached an agreement with Compaq to supply MS-DOS and BASIC software. Compaq required the software for a portable computer, that would be the first IBM compatible computer. The software compatibility, had a potential for conflict with IBM. Microsoft could now sell the BASIC interpreter for use on other IBM clones. Microsoft also reached agreements with Hewlett-Packard and Digital Equipment Corporation to supply software during 1982.

Figure 12.1: William H. Gates and Paul G. Allen. Photograph is courtesy of Microsoft Corporation.

In September 1982, Paul Allen was on a European trip with Gates when he developed some lumps on his neck. The diagnosis determined that he had Hodgkin's disease. After treatment Allen decided to resign from Microsoft in February 1983. Allen started his own software company called the Asymetrix Corporation in 1985.

Microsoft signed a number of contracts for its software products during 1982. A BASIC interpreter for Hitachi in Japan and MS-DOS for Victor Technologies are examples of programs that generated significant revenue for Microsoft.

In late 1982, Vern Raburn left Microsoft and joined Lotus Development Corporation as general manager. By the end of 1982, Microsoft employment had doubled to 200 employees and sales were $34 million.

During 1982/83, Microsoft did extensive development of the Macintosh software. Microsoft had numerous difficulties. The operating system and graphics user interface were in transitional development. The hardware changes such as screen resolution, disk drive configuration and the amount of memory delayed software development.

1983-85 Activities

Initial proposals for the formation Microsoft Press evolved in March 1983. The concept was to establish a publishing facility that would provide high quality computer texts and enhance Microsoft marketing efforts. The first manager was Nahum Stiskin, then Min S. Yee in May 1985.

Microsoft hired Raleigh Roark in 1982, to be in charge of hardware development. He became a principal in the development of the Microsoft mouse. David Strong, a Seattle designer, styled the mouse. Then a Japanese company called Alps Electric developed it into a product that Microsoft introduced in May 1983. Two versions were available: a mouse for the IBM Personal Computer powered from the computer bus through an add-on board and a mouse that obtained its power from the computer serial port. The mouse and interface software had a price of $195.

By early 1983, differences in the management of the company had developed between Gates and James Towne. This resulted in Towne leaving Microsoft in June. Microsoft offered Jon A. Shirley the presidency of Microsoft at the May 1983 National Computer Conference in Anaheim. He was the vice-president of computer merchandising at Tandy Corporation and had been with the company for twenty-four years. He accepted the offer in June and became the president and chief operating officer of Microsoft in August.

Microsoft also hired Rowland Hanson in early 1983 as vice president of corporate communications. Hanson implemented changes to enhance the corporate image of Microsoft. One change was a new orientation in the product naming policy to emphasize the company name. For example the word processor name changed from Multi-Tool

Word to Microsoft Word and the Interface Manager became Microsoft Windows. Frank M. (Pete) Higgins also joined the company in 1983 and subsequently became a vice president of Microsoft.

MSX was an 8-bit software/hardware system initiated by Kay Nishi for the Japanese market. The MSX computer system used a Z-80 microprocessor with graphics, sound, color-TV output and included a BASIC interpreter. Tim Paterson developed an 8-bit version of DOS called MSX-DOS for the computer system. Microsoft announced the MSX system in June 1983.

Microsoft concluded a new agreement with Apple Computer just prior to the release of the Macintosh computer in January 1984. It canceled the previous agreement of January 1982 and allowed Microsoft to market its own programs for the Macintosh computer. They announced Multiplan and Microsoft BASIC for the Macintosh at the computer release in January. However due to a lack of testing, the software had problems. Then in December, Microsoft released the Chart and File programs for the Macintosh in December. The company also started to adapt Microsoft Word to the Macintosh computer.

Between 1982 and 1985, Bill Gates featured in several magazines. It started with the cover of *Money* magazine in November 1982. In 1984 there was an article in the January issue of *Fortune* magazine. Then early in the year he had a profile in the *People* magazine and in April he was on the cover of *Time* magazine. Then in February 1985, he featured in the *Good Housekeeping* magazine. He was becoming a national figure. The marketing group were fostering this publicity of Gates. They also promoted concepts such as: "Microsoft Aims to be the IBM of software" and "a computer on every desk and in every home."

Gates had been directing most of the software development. However the scope of this responsibility was affecting his effectiveness. In August 1984, Jon Shirley implemented a reorganization to place Steve Ballmer in charge of the Systems Division. Also, Microsoft recruited Ida Cole from Apple Computer to be in charge of the Applications Division. Shirley also hired Francis J. Gaudette as vice president of finance and administration in September, to organize the financial activities of the company.

In February 1985, Ida Cole became a vice president of Microsoft in charge of application software. Then in 1986, her responsibilities changed from application software to international products.

In the fall of 1985, Apple Computer started expressing legal concerns regarding the similarity between Windows and the Macintosh user interface. Microsoft had also expressed concern regarding the development of MacBASIC by Apple. This resulted in meetings between Gates and Sculley. Shortly after the release of Windows, they signed an agreement in November that permitted Microsoft to use certain visual features of the Macintosh and Apple Computer stopped development of MacBASIC.

During 1985 Microsoft initiated actions to evaluate and adapt CD-ROM and multimedia technology. One of the significant problems was the use of the Philips CD-I (Compact Disk - Interactive) disk format and its interface with MS-DOS. This resulted in Raleigh Roark being assigned to head a CD-ROM group that developed a disk format named MS-CD for MS-DOS and Macintosh computers. Then in November, Microsoft was a participant in the adaptation of the High Sierra Proposal for a standardized disk format. Microsoft also contracted with the Cytation company to develop a multimedia encyclopedia for a planned CD-ROM conference.

1986-89 Activities

In January 1986, Microsoft bought Cytation and appointed its founder Tom Lopez as head of a new CD-ROM division. Cytation's CD-ROM reference disk called CD-Write was renamed Bookshelf. Then in March, Microsoft sponsored the first CD-ROM conference in Seattle, Washington. At the March 1987 CD-ROM conference, Art Kaiman from the RCA company demonstrated Digital Video Interactive (DVI) technology recorded on a CD-ROM disk. It was an impressive multimedia display running on an IBM computer using the Microsoft disk operating system.

Figure 12.2: Microsoft headquarters in Redmond, Washington.
Photograph is courtesy of Microsoft Corporation.

Construction of new headquarters had began in 1985. The new location was a 400-acre wooded site known as Sherwood Forest in Redmond, Washington. The initial buildings were completed and Microsoft moved to the new corporate campus in February 1986.

During 1984/85 Microsoft was under increasing pressure to make a public offering of the company shares. The employees stock incentives required a market to realize the true value for their shares. Also at a certain number of stock holders the Securities Exchange Commission would be requiring Microsoft to register the stock. Microsoft selected Goldman Sachs & Company and Alex. Brown & Sons to underwrite the public offering in December 1985. The prospectus showed that the largest shareholders were Gates, Paul Allen, Steve Ballmer and Technology Venture Investors. Gates owned forty-nine percent of the shares and Paul Allen twenty-eight percent. Other major stockholders included Gordon

Letwin, Jon Shirley, Charles Simonyi, and Gates parents. An offering price of $21 a share was established and the stock first traded to the public in March 1986. On the first day of trading the shares opened at $25.75 a share, peaked at $29.25 and closed at $27.75. Gates share of the company was worth over $300 million on the first day of trading. The stock continued to rise and by 1987 Gates was a billionaire.

About this time Gates has stated "...I proposed to IBM that they buy up to 30 percent of Microsoft -- at a bargain price -- so it would share in our good fortune, good or bad." Gates hoped this would help resolve some of the difficulties Microsoft was having, with the IBM joint development for the new OS/2 operating system. However IBM declined the offer.

Shortly after the public stock offering, Gates terminated the East Asia marketing agreement with Kay Nishi. Gates then recruited Susumu Furukawa from Nishi's ASCII Corporation to be the head of a new Microsoft Japanese subsidiary. Gates then recruited Chris Larson. He was an old friend from Lakeside School and helped establish the new subsidiary in Japan. Another important addition to Microsoft staff in 1986, was Paul A. Maritz who would subsequently become a vice president of Microsoft.

Microsoft encountered additional legal problems with the MS-DOS licenses held by Seattle Computer Products and Falcon Technology in 1986. Tim Paterson had started Falcon Technology after leaving Microsoft and Seattle Computer Products. Both were having financial problems and were considering selling the rights to their licenses. Microsoft obtained the Falcon license by purchasing the company in early 1986. Microsoft then obtained the license held by Seattle Computer Products, in an out-of-court settlement during litigation in December. The purchase of the license rights cost Microsoft about $1 million each.

Microsoft acquired Dynamical Systems Research, Inc. and its personnel in June 1986, for $1.5 million in Microsoft stock. Microsoft bought the company to obtain a clone of the IBM TopView software. Two principals in the company were Nathan P. Myhrvold, who would subsequently become a vice president of Microsoft, and Dave Weise. Microsoft also acquired Forethought, Inc., from its founder Rob Campbell for $12 million in July 1987. The reason for the acquisition was to obtain the PowerPoint graphics presentation program for the Macintosh computer.

In March 1988, Apple Computer filed a lawsuit claiming that Microsoft Windows Version 2.03 and Hewlett-Packard's NewWave programs copied the "look and feel" of the Macintosh. Speculation stated that Apple was really trying to inhibit or counteract development of the IBM Presentation Manager graphical user interface. Microsoft filed a countersuit against Apple for slander with intent to inhibit Windows development. In March 1989 the judgment on the lawsuit favored Apple and sent the Microsoft stock into a steep decline. Additional litigation resulted in the number of items in contention being significantly reduced. Then in July 1989, the judge threw out 179 items of alleged similarity. This left only 10 items in dispute.

Microsoft reorganized the CD-ROM division in mid 1988 and changed the name to Multimedia Systems division. Then Rob Glaser replaced Tom Lopez who had started the CD-ROM division. In October, Intel purchased the Digital Video Interactive (DVI) system demonstrated at the 1987 Microsoft CD-ROM conference.

By 1988 application software such as the Windows word processor project Cashmere and the Windows database project Omega were not meeting schedules. Consequently Microsoft hired Michael J. Maples who had responsibilities for software strategy at IBM, as vice president of the applications division in June. Then in 1989, Maples reorganized the applications division into smaller business units with a narrower market focus.

Between 1988 and 1989, Gates acquired between four to five acres of lakefront property at a cost of about $5 million. Located on Lake Washington, it would be the site of his future luxurious home.

In March 1989 Microsoft purchased close to 20 percent of Santa Cruz Operation (SCO), Inc., for a reported $25 million. Santa Cruz had previously ported the XENIX operating system software to other computers for Microsoft. However more important strategically, Santa Cruz was a member of the Open Software Foundation (OSF) that Microsoft had chosen not to join. Then another significant event with a potential effect on Microsoft occurred. IBM purchased a license for the UNIX based NeXTSTEP operating system.

Microsoft also acquired a California company called Bauer in July 1989. The company specialized in printer technology that included TrueImage fonts and printer driver software.

In September 1989 Gates incorporated his own separate company called Home Computer Systems. Gates founded the company to analyze the potential market for a mixture of electronic media and a photo data base of still art. The company subsequently arranged contracts with institutions such as the National Gallery of London and the Seattle Art Museum. The company name changed subsequently to Interactive Home Systems, which became Continuum and then to the Corbis Corporation.

At the end of Microsoft's 1989 fiscal year in June, net revenue was $803.5 million and the number of employees 4,037. Systems and languages accounted for 44 percent of revenue, applications 42 percent and hardware and books the remaining 14 percent. Geographically domestic revenue was 43 percent and international revenue 55 percent. Then in December 1989, it was announced that Jon Shirley wanted to retire in June 1990. A search began for a suitable replacement.

12.2 ... IBM PC Software

Initial Discussions

Bill Gates received a phone call from Jack Sams of IBM in late July 1980. Sams was a member of the small project team doing the initial concept analysis for a proposed personal computer. He was in charge of software development and arranged a meeting for the next day at Microsoft. IBM had Microsoft sign a nondisclosure agreement. This first meeting with Gates and Ballmer was of a general exploratory nature by IBM, to evaluate

Microsoft capabilities. IBM did not disclose any requirements specific to the proposed computer at this meeting. However, Sams did recommend to his manager William Lowe, that they use Microsoft software.

In mid August, after the IBM Corporate Management Committee (CMC) approved the personal computer project, IBM requested a second meeting. IBM had Microsoft sign a more detailed nondisclosure agreement and both companies had legal representatives at the meeting. IBM now revealed details of project Chess and the proposed personal computer with the code name of Acorn. IBM also advised that they wanted Microsoft to supply a series of programming languages for the new computer: BASIC, COBOL, FORTRAN and Pascal. IBM required the BASIC software by April 1981.

Operating System

During the mid-August West Coast trip, IBM attempted to negotiate with Digital Research to obtain the CP/M operating system for the Acorn computer project. However Digital Research would not sign the IBM nondisclosure agreement. Another factor affecting Digital Research's involvement was that they were not committing company resources to releasing a 16-bit version of CP/M till sometime next year.

Then in late August, IBM and Microsoft discussed alternative operating systems that could replace CP/M. The XENIX operating system was available from Microsoft, but the Acorn computer would not have the resources required by the software. However Paul Allen had been aware by early August of the 16-bit operating system developed by Tim Paterson at Seattle Computer Products. Paterson had developed the software called QDOS (Quick and Dirty Operating System) for use with the company's 8086 card system. QDOS had many similarities to CP/M but used a file allocation table (FAT) developed by Microsoft for controlling the disk file format and space allocation. IBM was now encouraging Microsoft to supply the operating system. In September, Microsoft made an agreement with Seattle Computer Products to license the 16-bit operating system. Seattle Computer Products now called the software 86-DOS (Disk Operating System). The license fees cost Microsoft a total of $25,000.

The Contract

In late September, Gates gave a presentation to IBM in Boca Raton, Florida formalizing their proposals to supply the languages requested and an operating system. Following the presentation legal negotiations proceeded to specify the terms for price, delivery and licensing. Microsoft and IBM signed the contract for the software in early November 1980. Microsoft would supply BASIC, COBOL, FORTRAN, Pascal, an Assembler and a Disk Operating System (DOS). A significant item in the contract provided Microsoft with marketing rights to sell the operating system to other companies. IBM wanted the disk operating system by January 1981. Microsoft was already behind schedule.

Software Development

Microsoft now started using the Intel 8086 simulator program, that they started developing in 1978, for software development on the Acorn prototype. IBM did not deliver a prototype of the Acorn computer until December. The simulator program allowed Microsoft to start software development and continue when they had reliability/availability problems with the Acorn prototype.

A top priority at Microsoft was the development of the operating system. Final testing of the programming languages and other application programs required a functional operating system. Microsoft assigned Bob O'Rear to adapt the 86-DOS to the specific requirements of the Acorn computer and the Basic Input/Output System (BIOS) software. O'Rear had started at Microsoft in 1977. He was now working with Tim Paterson at Seattle Computer Products and David Bradley at IBM who was developing the BIOS. Paterson developed a simple text editor called EDLIN that Microsoft included in the DOS program. A preliminary version of 86-DOS was operating in February 1981. Paterson left Seattle Computer Products and joined Microsoft in May.

Digital Research became aware of the adaptation of 86-DOS to the IBM computer project. They were now expressing their concern regarding the similarity of 86-DOS to CP/M. To avoid potential litigation IBM agreed with Gary Kildall of Digital Research to offer his CP/M-86 operating system for the new computer.

The preceding developed concerns at Microsoft regarding the control of 86-DOS. In June 1981, Microsoft made an offer to Rod Brock, the owner of Seattle Computer Products to purchase the 86-DOS software. In July, Microsoft signed an agreement that purchased all rights to 86-DOS for an additional $50,000. The total cost, after including the initial license fee of $25,000 was $75,000.

Microsoft agreed to provide Seattle Computer Products with unlimited rights to the operating system and future improvements for use in their products. The agreement also provided beneficial terms for other Microsoft programming languages.

IBM released the operating system as PC-DOS in August 1981 and as MS-DOS by Microsoft. Microsoft subsequently either licensed or released other versions of the operating system, with different names such as SB-86 and ZDOS.

Other Software

The other urgent requirement from Microsoft was for the ROM BASIC. Mike Courtney, a previous developer of APL at Microsoft, worked on the BASIC interpreter. Paul Allen worked on the advanced versions of BASIC that included DISK BASIC. Gates also got involved in certain aspects of the software. Microsoft finished the ROM BASIC in March 1981.

Microsoft hired Richard Leeds in June 1981 and assigned him to develop a 16-bit version of COBOL for the IBM PC. Microsoft offered other programs being sold by their Consumer Products Division to IBM. Those were Adventure, Olympic Decathlon, Time Manager and Typing Tutor. Also offered was a spreadsheet program just being developed at Microsoft called Electronic Paper.

12.3 ... Operating Systems

XENIX

Gates obtained a license for a standard version of the AT&T UNIX operating system in February 1980. Microsoft then adapted the operating system for 16-bit microcomputers and announced it as XENIX in August. Microsoft hired a company called Santa Cruz Operation (SCO), Inc., to port the software to various computers. One of the first customers for XENIX was the 3Com Corporation that Bob Metcalfe had co-founded in 1979.

Microsoft introduced Version 3.0 of XENIX in April 1983. Then Microsoft released XENIX 286 in August 1984 for the IBM PC AT computer.

Subsequent IBM Activities

After the introduction of the IBM PC computer in August 1981 Microsoft licensed the 16-bit operating system to Lifeboat Associates. They were a major software vendor and sold the operating system under the name of Software Bus-86 (SB-86).

Microsoft now sold the operating system to many OEM customers who were developing Intel 16-bit computers. The operating system sold initially with different names depending on the source. Microsoft used MS-DOS, IBM PC-DOS, Lifeboat Associates SB-86, Zenith ZDOS and so on. Later Microsoft would restrict this proliferation of names and insist on MS-DOS for all implementations other than IBM.

See Appendix B for a description of the different versions of DOS and the release dates.

MSX-DOS

Tim Paterson developed an 8-bit version of MS-DOS called MSX-DOS. Microsoft developed the software for the MSX hardware system and released it in June 1983.

OS/2 and IBM

Microsoft participated in meetings of an IBM task force formed to evaluate operating systems for their personal computers. In June 1985 Microsoft and IBM signed an agreement to jointly develop a new operating system for future products. Initially the new operating system had the name of Advanced DOS. This joint development agreement resulted in many difficulties for the two organizations. Then the accidental death of Don Estridge of IBM in August and his replacement by William Lowe did not help. IBM's bureaucratic type of organization for software development was in contrast to Microsoft's use of a small group of talented programmers. Gordon Letwin was in charge of the Microsoft development group.

In April 1986 Microsoft agreed to modify the Windows software to accommodate IBM's requirements for the new operating system. They also agreed to provide Windows compatibility for IBM's TopView applications. This resulted in Microsoft acquiring a company called Dynamical Systems Research, Inc., that had developed a TopView clone called Mondrian.

During 1986/87 a number of developments at IBM resulted in additional difficulties at Microsoft. In mid 1986, IBM advised Microsoft of a new concept being implemented, called Systems Application Architecture (SAA). This software would enable the linking of various hardware levels, from personal computers to mainframes. Following the SAA advisement, Microsoft became aware of the new IBM personal computer hardware incorporating the Micro Channel Architecture (MCA) bus and an Advanced Input/Output System (ABIOS) chip. IBM also advised Microsoft that they were developing an Extended Edition of the new operating system. However they also stated that IBM would develop the software without the participation of Microsoft. These changes had a significant impact on the joint software development activities for the new operating system

In April 1987, IBM announced that the new operating system with the name of OS/2 (Operating System/2). Also, IBM had selected the name Presentation Manager for the graphical user interface and would subsequently incorporate it as a part of OS/2. These activities had effectively negated Microsoft's development efforts under the joint development agreement. It also was having a serious impact on the future potential of Windows. IBM released OS/2 Version 1.0 in November 1987, for the IBM PS/2 computer and other computers using the Intel 80286 and 80386 microprocessors. Presentation Manager was finally released in October 1988.

During 1989, the relationship between IBM and Microsoft did not improve. James Cannavino, who had replaced William Lowe as head of the IBM Entry Systems Division was also concerned about the relationship. However, Microsoft was now concentrating on a new version of Windows.

NT (New Technology)

During the deterioration in the relationship with IBM between 1987 and 1988, Microsoft initiated a new project called Psycho to develop a future operating system that would replace OS/2. Nathan Myhrvold headed the project that would incorporate portability with capabilities to accommodate reduced instruction set computing (RISC) technology. Myhrvold then licensed a UNIX based operating system technology called Mach. The project would now incorporate features of Mach, an ability to run on different microprocessors and systems with multiprocessors.

Then in October 1988, David N. Cutler who had been a principal in the design of the DEC Virtual Memory operating System (VMS) for VAX computers joined Microsoft. Cutler and his design team started working on the new operating system that Microsoft named NT, representing New Technology.

12.4 ... Windows

The advance demonstrations of the Apple Macintosh computer graphics system during 1981, and the VisiCorp demonstration of VisiOn at COMDEX in November 1982, added impetus to the development of a graphical user interface at Microsoft. VisiCorp had established a lead in the development of a graphical multi-window operating environment using the mouse.

Microsoft had started a project called Interface Manager in late 1981. However with the new competitive developments, Microsoft reviewed and extended the specifications of the interface in late 1982. Rao Remala became responsible for the window manager and Dan McCabe did the graphics. An intensive marketing effort began to advise OEM customers that Microsoft also had a graphical windows software system under development. However Gates could not obtain the support of IBM, who decided to develop their own interface that would become TopView. Related events in early 1983, were the introduction of the Apple Lisa computer with its innovative graphics in January, and the release by Microsoft of DOS Version 2 for the IBM PC XT computer in March. A primitive demonstration of the Interface Manager program was developed by McCabe and Remala in April. By summer, a change in corporate marketing strategy resulted in the program being renamed Microsoft Windows.

In the fall, Charles Simonyi recruited Scott MacGregor whom he had known at Xerox PARC to be the head of the Windows development team. The team now included Marlin Eller, a mathematician, and Steve Wood. Eller would develop the graphical device interface, Remala the user interface and Wood the system kernel. In November, Microsoft announced Windows in New York. That same month VisiCorp released VisiOn and Quarterdeck announced a graphical system called DESQ.

At the 1983 fall COMDEX convention Microsoft did intensive marketing of Windows, although the product was far from being complete. MacGregor's team had developed a new demonstration program that could display Multiplan, Word and Chart running at the same time. Then due to a lack of IBM support for Windows, Microsoft stated that it would retail the program for less than $100 and promised to release the software in April 1984. However, they had completely underestimated the magnitude of the programming effort required. This resulted in the release date being changed to the fall. By June, Microsoft had firmly committed itself to establishing Windows as a standard graphics user interface. However, IBM was still not supporting Windows and announced their character-based interface called

TopView in August. A related event in August, was Microsoft's release of DOS Version 3 for the IBM PC AT computer. That same month, a company reorganization resulted in MacGregor reporting to Steve Ballmer instead of Bill Gates.

Gates wanted the Windows program to be more like the Macintosh. This resulted in Neil Konzen who had worked on application programs for the Macintosh, being assigned to the Windows team in August. Gates and Konzen were very critical of the Windows software. The result was a redesign of the Windows software to make it more like the Macintosh. This was also intended to simplify the adaptation of Microsoft's application software to either the IBM Personal Computer or the Macintosh. A number of Macintosh features were added such as: calendar, clock, control panel, games and an elementary word processor. Another late change requested by Gates was keyboard equivalents for all mouse operations. Other potential problems were the use of a less intuitive tiled window display and compatibility problems with DOS. The redesign and increase in program size resulted in a new target release date of June 1985. The delays and conflicts with Ballmer resulted in Scott MacGregor leaving Microsoft in the spring of 1985.

In May, Microsoft demonstrated an advanced version of Windows at COMDEX. In June, they released a test version of Windows to software developers and computer manufacturers. Finally in November 1985, at the fall COMDEX show, Microsoft released Windows Version 1.03 as a retail package listing for $99. The release date was a significant change from the original promise of April 1984, and the probable cause for the use of the term "vaporware."

The program featured multitasking that enabled users to work with several programs at the same time, and to easily switch between them. However, the program operated in real mode, not the safer protected mode and had a maximum memory limitation of one megabyte. Microsoft received mixed reviews due to its slow speed, the windows could not overlap and lack of application programs utilizing windows technology. The issue of slow speed was related to the users available memory. The release package stated that a minimum of 256K was required, but 512K was recommended. However, even with the recommended memory users were not happy with the speed. It was not a successful conclusion.

Between 1986 and 1987 Microsoft assigned a lot of human resources to the joint development of the IBM operating system OS/2. This resulted in the Windows team being reduced significantly. However, it did include Rao Remala and Dave Weise. Microsoft released Version 2.01 of Windows in October 1987, with the Excel spreadsheet program. Then they released Version 2.03 of Windows and Windows 386 in January 1988. Version 2 featured overlapping windows, access to EMS memory and movable icons. Windows 386 was a Version 2 optimized for the more powerful Intel 80386 microprocessor. Version 2.1 of Windows was released in June and was renamed Windows 286. However it was still not a commercial success.

Early in the summer of 1988, Weise started to incorporate protected mode features in Windows that Murray Sargent had developed for a program debugger. Sargent was a physics professor at the University of Arizona who was working for Microsoft during the summer. Weise also utilized EMS (Expanded Memory Specification) capabilities that overcame some of the memory limitations of Windows. The program with these two significant improvements was successfully demonstrated by Weise in August. However, a related event of some concern, was the release of the Presentation Manager program by IBM in October.

The Windows team incorporated enhancements to the graphics, such as three dimensional buttons. Then during 1989 TrueType font technology was obtained from Apple Computer, in exchange for TrueImage font technology that Microsoft obtained when it acquired a company called Bauer. Gates was determined to have a successful Window product. Contributing to this momentum was an increasing number of application programs being released by other companies for Windows. Another significant factor was the availability of personal computers, with more powerful microprocessors to handle the graphics user interface. Windows would become a successful product in 1990.

12.5 ... Languages

In 1981, Microsoft developed a version of BASIC for the Epson HX-20 laptop computer. Then in the spring of 1982, Microsoft released GWBASIC (acronym for Gee Whiz BASIC), that included support for advance graphics. Microsoft also developed an IBM compatible BASIC for the Compaq portable computer in 1982.

Microsoft designed the BASIC interpreter released with the IBM PC computer on an 8-bit computer architecture. Microsoft had been getting reports that the performance of the PC BASIC was no better than the 8-bit Apple II computer. Mike Courtney programmed a new BASIC interpreter and optimized it for 16-bit computers and version 2.0 of the disk operating system. Microsoft released the new BASIC in March 1983, at the same time as PC-DOS 2.0 for the IBM PC/XT computer.

Microsoft released COBOL and FORTRAN for MS-DOS in March 1982. Then they released the programming languages C and Pascal for MS-DOS in April 1983.

In late 1983, Microsoft quickly developed a Macintosh version of BASIC to compete with the Apple MacBASIC that was having delays in completion. Microsoft released the BASIC interpreter for the Macintosh at the same time as the Macintosh computer in January 1984.

Microsoft released QuickBASIC in mid 1986, with a structure and programming environment similar to Borland's Turbo Pascal. Microsoft subsequently released an improved version of QuickBASIC to compete with Borland's Turbo BASIC.

Microsoft released Quick Pascal in March 1989, to compete with Borland's Turbo Pascal. Then they developed Quick C to compete with Borland's Turbo C.

Languages became less significant to Microsoft as the 1980's progressed. Users were not programming. Application programs, operating systems and the Windows graphic user interface had become the dominant consumer software.

12.6 ... Application Programs

Microsoft hired a consultant Paul Heckel, who had been at Xerox PARC, to evaluate the requirements for a new spreadsheet in May 1980. Heckel suggested that Microsoft develop a spreadsheet similar to VisiCalc with menus and an improved user interface. In late 1980, Microsoft assigned programmer Mark Mathews to develop the software. The spreadsheet program became known as Electronic Paper.

Then in November 1980 Charles Simonyi made his initial contacts with Microsoft. As a Hungarian teenager Simonyi had developed his programming skills on a Russian Ural II vacuum tube computer. After working in Denmark and studying at Berkeley in California, he started working for Xerox PARC (Palo Alto Research Center). Simonyi co-developed an innovative word processing program called Bravo for the Alto computer. In November, Simonyi submitted a far reaching plan for application software development to Gates and Steve Ballmer. The proposal included plans to use the latest graphical concepts pioneered by Xerox in new word processors, spreadsheets, databases and other programs. In February 1981, Microsoft appointed Simonyi director of applications development at Microsoft.

One of his early concepts was for the development of core software with a consistent graphical user interface for each application program. The core software facilitated the development of programs for different computer platforms. The graphical interface became known as the Multi-Tool Interface and utilized mouse control.

Another concept developed by Simonyi became known as "Hungarian" notation. This concept applied a naming convention to variables, functions and macros. Simonyi has stated that its use will "improve the precision and speed of thinking and communicating." The Hungarian notation convention simplifies the reading of source code by other developers. Microsoft uses it primarily in the development of application software.

Spreadsheets

Multiplan

After Simonyi's arrival at Microsoft in 1981, he modified the spreadsheet program Electronic Paper, to incorporate the core software concept, the Multi-Tool Interface, windows and other improvements. Principal programmers for the software were Doug Klunder, Bob Mathews and Dave Moore. Microsoft renamed the program Multiplan, then released it for the Apple II and Osborne computers in August 1982, and for the IBM PC in October. Multiplan had unique features. Some of those were: Windows that enabled display of separate areas of the spreadsheet, menus, named cells, help screens and automatic recalculation. However initially the program was slow in operation. This was quite obvious when compared to Lotus 1-2-3 released in November 1982. Contributing to the poor performance was an IBM requirement that the program operate on a PC with only 64K bytes of memory. However Microsoft could readily adapt the software adapted to many different computers, that resulted in significant sales of the program.

Microsoft released an enhanced update, Version 1.1 in February 1984. An increase in the memory requirements improved the performance. It was also available for numerous computer platforms and other languages including Japanese. Microsoft then adapted the program for the Apple Macintosh computer and announced it in August 1984. However, Neil Konzen had to rewrite the program due to a number of problems. Multiplan had difficulties competing with Lotus 1-2-3 in North America. However it had highly successful sales overseas.

Microsoft released Version 2.0 of Multiplan for the PC computer in October 1985. Then they released a faster Version 3.0 in January 1987.

Excel

During 1983, Lotus 1-2-3 had replaced VisiCalc and Multiplan as the dominant spreadsheet program. Microsoft felt that improvements to Multiplan would not be enough to compete with Lotus 1-2-3. Microsoft required a new innovative product. After extensive review of competing products, Jabe Blumenthal who had worked in marketing, Doug Klunder and others defined the design of a new spreadsheet with advanced capabilities. Microsoft assigned Klunder, who had worked on Multiplan to develop the software. The project now had the code name of Odyssey and a completion target of mid 1984.

In early 1984, Lotus Development Corporation was concentrating their efforts for the Apple Macintosh, on a new integrated program called Jazz. Lotus 1-2-3 was also in a very dominant position in the IBM PC market. Gates therefore decided in March, to change the initial introduction of Odyssey, from the IBM PC to the Macintosh computer with 512K bytes of memory. This change in computer platforms resulted in a delay. Microsoft now targeted the program for completion in nine months. At the end of the nine months Klunder left Microsoft for a short period. Philip Florence who had come from Wang Laboratories, replaced him.

During 1984, Microsoft considered a number of different names for the advanced Odyssey spreadsheet project. Then Microsoft selected the name Excel that a branch manager had submitted. Microsoft announced Excel for the Macintosh in May 1985, but it was not available for release until September at a price of $395. The program was highly successful and reported to be even better than Lotus 1-2-3 on a PC computer. Microsoft released Version 1.5 of Excel in May 1988 and Version 2.2 in May 1989.

Microsoft started work on Excel for Windows on an IBM Personal Computer after the Macintosh release. The software design had a separate layer of code that isolated the program from the Macintosh and DOS/Windows operating system. This facilitated the adaptation of the software for Windows. Jeff Harbers was a principal in the new software development. The program incorporated capabilities for Lotus 1-2-3 file interchange and use on the OS/2 operating system. Microsoft released Windows Excel in October 1987. This release also included Version 2.01 of Microsoft Windows and featured overlapping windows and movable icons. In October of

1989, Microsoft released Excel for Presentation Manager and the OS/2 operating system.

Word Processors

Word

Development of a word processor started in mid 1982. Simonyi and Richard Brodie who had worked with Simonyi at Xerox PARC developed the program. The program initially had the name Multi-Tool Word.

The word processor utilized advance graphical concepts similar to the Xerox Bravo program and mouse control for selection, changing and deletion of text. The program incorporated the concept of "What-You-See-Is-What-You-Get" (WYSIWYG). It was the first word processing program to display boldface, italics, underlining, sub and superscripts. It also featured multiple windows, ability to work on multiple documents, temporary storage of deleted text and style sheets for automatic formatting of a document. Microsoft also incorporated the capability to use laser printers.

Microsoft introduced Multi-Tool Word in the spring of 1983. Then in the summer, Microsoft reoriented its product naming policy to emphasize the company name. This resulted in the word processor being renamed Microsoft Word. Microsoft introduced the word processor in September. Microsoft Word had a price of $475 with a mouse, or $375 without. Microsoft provided free demonstration copies of the program to subscribers of the *PC World* magazine in the November 1983 issue.

The program received mixed reviews and initial sales were below expectations. Microsoft released improvements to the program during 1984, and Version 2.0 that included a spelling checker and word counter in February 1985. Then Microsoft released Version 3.0, that included a sophisticated on-line tutorial in April 1986 and Version 4.0 in November 1987. The new versions improved the popularity of the program and the market share.

Macintosh Word

Microsoft released Version 1.0 of Macintosh Word in January 1985 and a revised version in June. The company then announced major improvements to Macintosh Word in October 1986 that they incorporated into Version 3.0 (Microsoft did not release a Version 2) released in February 1987. However the program had a number of problems that resulted in a free upgrade. Microsoft released Version 4.0 in March 1989.

Word for Windows

The company initially developed Word for Windows under the code name of Cashmere. Richard Brodie was a principal in the early development of the software. However, Brodie left Microsoft and the project name changed to Opus in 1986. After many delays, Microsoft released Version 1.0 in December 1989. However it had a number of problems that resulted in improvements being incorporated in a subsequent release.

At the end of the 1980's the two leading word processor programs were WordPerfect and the different versions of Microsoft Word. WordPerfect was in the number one position. However with the Macintosh application included, Microsoft Word was closing the gap.

Databases

Omega is the code name of a database project started in the early 1980's for use with Windows. However the project had problems and Microsoft terminated the development in 1990. Following the termination of Omega, the company started a new entry level database project with the code name of Cirrus.

Integrated Packages

Works for the Macintosh

The initial development was a program called Mouseworks for the Macintosh computer by Don Williams who had worked for Apple Computer. The program included a word processor, spreadsheet, database and communications module. Microsoft obtained the rights to market the program and it in September 1986.

Works for the PC

Development of Works for the PC computer began in late 1985. It was to be an easy-to-use integrated program for the low-end of the IBM Personal Computer market. The program integrated a word processor, spreadsheet, database and a communications module. Richard Weiland was a principal in the programming. Microsoft changed the software, from an initial Windows design, to text mode with a "Windows look" due to performance considerations. Microsoft then decided to incorporate a tutorial with the program in June 1986. Barry Linnett headed the tutorial development. Microsoft released Works for the IBM Personal Computer in March 1988. It was a highly successful product.

Other Programs

Microsoft introduced Project for Windows that facilitated the planning and management of projects in May 1984. The Project program incorporated capabilities for critical path planning, cost analysis and scheduling. Then Microsoft released Microsoft Chart for the IBM PC and Macintosh in August.

PowerPoint is a presentation graphics program that Microsoft obtained when it acquired Forethought, Inc., in July 1987. The program can create overheads, slides and on-line presentations.

Mail is an e-mail program that originated from two products acquired by Microsoft called MacMail and PCMail. Publisher is a desktop publishing program introduced in 1988.

Microsoft released MS-Net with MS-DOS Version 3.1 in March 1985. This provided user network access to a shared hard disk and files. Then after difficulties with a joint development agreement with 3Com Corporation, Microsoft announced LAN Manager for networked OS/2 systems in October 1989.

Multimedia

Microsoft established a CD-ROM division in 1985. The division then started development of an encyclopedia CD-ROM disk. Microsoft based the text on the *Funk and Wagnalls Encyclopedia*. However the project would have many delays.

The division then released its first CD-ROM multimedia disk that included graphics and sound, called Bookshelf in September 1987. The disk now contains a collection of reference works. It includes *The American Heritage Dictionary*, *The Original Roget's Thesaurus*, *The Columbia Dictionary of Quotations*, *The Concise Columbia Encyclopedia*, *Hammond Intermediate World Atlas*, *The People's Chronology* and *The World Almanac and Book of Facts*.

Then in mid 1988, work started on the development of a new technical standard called Multimedia PC (MPC) for CD-ROM disks. A consortium of companies supported the standard. They include AT&T, NEC, Olivetti, Philips, Tandy and Zenith.

Chapter 13 Other Software in the 1980's

The 1980's began with a continuation in the shift from the technical programming language enthusiast, to the increased use of application software. Business productivity programs such as databases, spreadsheets and word processors became a major segment of the software market. The release of more powerful microprocessors, less expensive memory and storage devices accelerated these changes. Ease-of-use and user friendly were the terms used to describe the new focus for software in the 1980's. Incorporation of innovative graphics helped to establish this new focus. These market demands and the introduction of the IBM Personal Computer in 1981, resulted in major changes and a rapid expansion of the software industry. However, between 1983 and 1984, a severe downturn in the personal computer industry, caused significant financial problems for a number of software companies. The operating system segment of the software market also experienced significant change.

13.1 ... Operating Systems

Seattle Computer Products

Seattle Computer Products, Inc., had developed an Intel 8086 microprocessor card for the S-100 bus in May 1979. They required a 16-bit disk operating system and proposed to use Digital Research CP/M-86 which had been promised for the end of 1979. In April 1980 CP/M-86 was not available and Seattle Computer Products decided to develop its own operating system. It was written by Tim Paterson and called QDOS (Quick and Dirty Operating System) because it was created so quickly (in two man-months). QDOS 0.1 was released in August 1980.

QDOS was similar to CP/M. Paterson obtained compatibility with CP/M by incorporating a translator that converted 8080 instructions into 8086 instructions. He then provided equivalent CP/M functions to operate on the 8086 microprocessor. Paterson also improved the data storage capabilities and file organization of QDOS as compared to CP/M by using the Microsoft concept of a file allocation table (FAT). This concept controlled the disk format and space allocation.

Seattle Computer Products contacted Microsoft in early August regarding adapting 8086 BASIC to QDOS and a possible cross-licensing agreement. In September 1980 Microsoft purchased non-exclusive rights to the Seattle Computer Products operating system. Then in November 1980 Microsoft signed a contract with IBM to provide a variety of software including an operating system for their new PC Computer. A new version of the operating system called 86-DOS 0.3 was released in late 1980. Version 1.0 was released in April 1981 which was very similar to Microsoft MS-DOS. Then Tim Paterson left Seattle Computer Products and joined Microsoft in May 1981. Microsoft purchased all rights to the Seattle Computer Products disk operating system. in July 1981

Digital Research

In late August 1980, IBM visited Digital Research to negotiate the possible use or adaptation of CP/M for its new PC computer. However Digital Research would not sign an IBM non-disclosure agreement that resulted in a termination of the meeting. Then in September it became apparent that Digital Research would not assign the resources required to provide a 16-bit version of CP/M in the time schedule required by IBM. This resulted in IBM selecting Microsoft to provide the PC computer operating system.

The negotiations with IBM were indicative of company organizational problems. This resulted in John Rowley being hired as president in November 1981.

CP/M-86 was a 16-bit version of the CP/M operating system developed by Digital Research for the Intel 8086 microprocessor. Some early copies were available in 1981, however it was not released for use on the IBM Personal Computer until April 1982. The software was priced from $175 to $240, which was considerably higher than the $60 IBM charged for the Microsoft operating system. The price was subsequently reduced, but the late release, higher price and a lack of support from IBM limited sales.

Concurrent CP/M-86 is a program developed by Digital Research during 1982/83 to enable multitasking. This provided the capability of running up to four processes or programs at the same time. The company also developed another operating system called Concurrent DOS for the DOS environment in 1984.

Then in May 1988, Digital Research released Version 3.3 of DR-DOS (Digital Research - Disk Operating System), an operating system compatible with MS-DOS. This first release of DR-DOS was followed by Version 3.40 in January 1989 and Version 3.41 in June 1989.

Microsoft

Microsoft entered the operating system segment of the software market when it developed XENIX in August 1980. Then with the release of the IBM PC computer in August 1981, Microsoft became a significant provider of operating system software (see Sections 12.2 and 12.3). See Appendix B for a description of the different versions and corresponding release dates of DOS.

IBM

IBM contracted with Microsoft to provide the operating system for its PC computer released in August 1981. IBM added some utilities to the Microsoft operating system and called it PC-DOS. Then in 1984 IBM entered into a joint development agreement with Microsoft to develop a new operating system that became OS/2 (see Section 9.5).

Apple Computer

Apple Computer released the Sophisticated Operating System (SOS) for the Apple III computer in May 1980. It was one of the earliest operating systems to have installable device drivers.

The Professional Disk Operating System (ProDOS) was developed for the Apple II computer and evolved from the Apple III computer SOS operating system. It also resulted from a requirement to provide an interface with peripheral devices other than the Disk II drive, a hierarchical directory structure and peripheral device drivers. ProDOS was released in January 1984 and was subsequently renamed ProDOS 8.

The Professional Disk Operating System 16 (ProDOS 16) was a 16-bit operating system released with the Apple IIGS computer in September 1986.

The GS/OS native operating system was released for the Apple IIGS computer in September 1988. It incorporated the concept of File System Translators (FST's) to determine disk format for the selection of device drivers. Version 5 of the Apple IIGS System software was released in May 1989. It was a significant upgrade of the system software with improvements in performance, graphics and file operations.

Other Operating Systems

Douglas L. Michels founded The Santa Cruz Operation (SCO), Inc. in 1979. The company started developing UNIX-based operating systems in the early 1980's.

UCSD p-System is an operating system that included UCSD Pascal. It was provided by SofTech Microsystems for the IBM Personal Computer released in 1981. It had a price of about $450.

Mach is a UNIX based micro-kernel operating system developed by Richard (Rick) Rashid at Carnegie-Mellon University in the mid 1980's. It was designed to be portable to many types of hardware and multiprocessor computers. It formed the basis for the NeXTSTEP operating system and influenced the design of the Microsoft Windows NT operating system.

NeXTSTEP is a UNIX-based operating system released with the NeXT computer in October 1988. Avidis Tevanian who had been one of the lead designers of the UNIX Mach system at Carnegie-Mellon University, was the chief software designer at NeXT. IBM subsequently licensed NeXTSTEP with the intent of adapting the system for its new workstation.

Operating System User Interfaces

Operating system user interfaces received a lot of attention from hardware and software suppliers between 1980 and 1983. Apple Computer had developed a graphical user interface for the Lisa and Macintosh computers. VisiCorp demonstrated VisiOn in late 1982 and Microsoft had initiated an extensive development of a system that would become Windows (see Section 12.4) in 1983.

VisiOn is a multiwindow graphical environment program that was demonstrated by VisiCorp (formerly Personal Software) at the November 1982 COMDEX show in Las Vegas. VisiCorp released the program in November 1983. It had taken between two to three years to develop under the code name of Quasar. William T. Coleman was the group manager responsible for the development of VisiOn and related application programs. The software interfaced between the IBM PC operating system and user application programs. It utilized high resolution graphics, a mouse and had its own application programs. It was machine and device independent. The program cost $495 at introduction, then after poor sales the price was reduced to $95. However VisiOn could only run programs written for the interface. The only programs available at the release were VisiCalc, VisiGraph and VisiWord from VisiCorp. This limitation was detrimental to its widespread acceptance. VisiOn created financial difficulties for VisiCorp. Control Data Corporation subsequently purchased the VisiOn software.

DESQ is a windowing system developed by Quarterdeck Office Systems in May 1984. It could multitask DOS programs but did not have a graphical interface. It was not successful. Quarterdeck subsequently developed DESQview for use with IBM TopView in July 1985.

Digital Research developed GEM (Graphics Environment Manager). Lee Lorenzen was a principal in the software development. It had a graphical user interface with a "look and feel" similar to the Macintosh computer, but could not multitask DOS programs. It was demonstrated at the COMDEX show in October 1984. The company also developed a number of application programs for use with the system. During

1985 changes were made to the user interface, due to threatened litigation by Apple because of the similarities to Macintosh.

Hewlett-Packard developed NewWave PC that is an interface for use with Microsoft Windows. HP announced NewWave in November 1987. It included additional features to make Windows easier to use that had similarities to the Apple Macintosh system. This resulted in litigation by Apple Computer in March 1988.

Berkeley Softworks released GEOS (Graphic Environment Operating System) that was ProDOS-compatible in March 1988. It had a Mac-style desktop, word processor and spelling checker.

Other operating system interfaces are Microsoft Windows (see Section 12.4), IBM Presentation Manager and TopView (see Section 9.5) and a TopView clone called Mondrian. Metaphor Computer Systems is a company founded by David Liddle and Donald Massaro in 1983, that developed a windows-like graphical user interface for the PC computer.

The Massachusetts Institute of Technology (MIT) developed the X Window System for the UNIX operating system. The X Window System is a graphical communications interface that provides a standard way of controlling graphic displays from one X Window System to another. Graphical user interfaces that have been developed for the X Window System are DECwindows, Open Look by UNIX International and OSF/Motif by the Open Software Foundation.

13.2 ... Programming Languages

BASIC

As a result of demands for structured programming concepts, Dartmouth College developed a powerful BASIC compiler in 1983, called Dartmouth Structured BASIC, also known as SBASIC. After many conflicting requirements and technological changes since the formation of a standards committee in 1974, the American National Standards Institute (ANSI) completed a draft for a "full" BASIC standard in 1983. This led Dartmouth College to develop a compiler based on the new standard that would be portable to most personal computers. Dartmouth released this new compiler named True BASIC in 1984.

During the 1980's, the availability of more powerful computers and increased memory, resulted in a move from interpreters to compilers for new BASIC programming languages. Examples of these are the Digital Research CBASIC-86 compiler, Microsoft QuickBASIC and the Tandy Radio Shack RSBASIC compiler

Borland announced Turbo BASIC at the fall COMDEX show in November 1986.

C

Borland released Turbo C at a price of $99.95 in the late 1980's.

C++

Bjarne Stroustrup developed the C++ programming language at Bell Telephone Laboratories. C++ is an object-oriented extension of the C language. C++ became available in 1986.

Logo

In the summer of 1982 Logo became available for the Apple II and TI-99/4A computers.

Modula-2

Modula-2 is a general purpose systems implementation language based on the use of modules. It was developed by Niklaus Wirth after a number of years research on the capabilities of Pascal. The initial Modula language was described by Wirth in 1977 and an improved version Modula-2 was available in 1982.

The first commercial implementation of Modula-2 was announced by Volition Systems in December 1982. It was released for a number of computers including the Apple II, Apple III and 8080/Z80 based systems.

Oberon

Oberon is an object-oriented systems programming language developed by professors Niklaus Wirth and Jurg Gutknecht at the ETH (Eidgenossische Technische Hochschule) in Zurich, Switzerland. The language is a distillation of the best features from Pascal and Modula-2. Oberon is smaller and simpler than its predecessors. The software was developed in conjunction with CERES, which was a single-board graphics workstation. Work on the language began in 1985. Wirth named the language in tribute to the precision of the Voyager spacecraft as it flew past Uranus's moon of the same name (Oberon) in 1988. Implementations are available for Apple Mac II's, DEC DECstations, IBM RISC System/6000 workstations, Intel based PC's and Sun SPARCstations.

Pascal

Microsoft developed a Pascal compiler for the IBM PC in April 1983 that had a price of $300. Then in March 1989 Microsoft released Quick Pascal.

UCSD Pascal was originally developed at the University of California, San Diego (UCSD). The UCSD system included the programming language and the UCSD operating system and was supplied by SofTech Microsystems, Inc.

Philippe Kahn moved from France to the USA and founded Borland International, Inc. in 1983. Turbo Pascal was the first product developed by Kahn. The compiler was introduced by an advertisement in the

November 1983 issue of *Byte* magazine at a price of only $49.95. It was an innovative program with an integrated programming environment. It facilitated program development by having a built-in text editor from which one could compile, correct errors and run the program. The program was an immediate success. Some other programs released by Borland are the Paradox database, Quattro Pro spreadsheet and Sidekick.

PL/I

Digital Research developed a three-pass compiler for PL/I in 1980. The compiler was written in PL/M and was based on the G Subset of PL/I which was an adaptation for minicomputers. The first two passes produced symbol tables and intermediate language suitable for various hardware systems. The third pass, optimized the code and developed the final machine code for a specific system.

13.3 ... Word Processors

The 1980's was a period of transition for word processing. The market changed from companies such as IBM, Lanier and Wang Laboratories supplying dedicated word processors, to the use of general purpose personal computers and word processing software from other independent companies. The dominant companies changed in 1983 from Wang to MicroPro providing Word-Star, then to the WordPerfect Corporation in 1986. However by 1989, Microsoft Word and WordPerfect had close to equal shares of the market.

EasyWriter

In 1981, IBM negotiated with Bill Baker of Information Unlimited Software (IUS) to adapt the Apple II EasyWriter word processor for their new personal computer. An agreement was reached and John Draper with assistance from Larry Weiss of IUS developed the program for the IBM PC. It had a price of $175 and was released with the support of IBM in August 1981. The program did not receive good reviews. However, it had good initial sales, because it was the only word processor available for the IBM PC at its introduction. Improvements were made to the program, but it did not compete successfully with later word processors. In 1983, Computer Associates International purchased IUS from Bill Baker for over $10 million.

Word-Star

For the founding of MicroPro International and the introduction of Word-Star in 1979, see Section 7.3. MicroPro adapted Word-Star to the IBM PC in mid 1982 and it quickly gained a dominant share of the market. Other company products were CalcStar, DataStar and InfoStar, MailMerge and SpellStar. The company became a public corporation in March 1984. A new version of the word processor named Word-Star 2000 with a new interface was introduced at the fall COMDEX show in 1984. However it functioned slower and received a poor reception. This was a turning point in the dominance of MicroPro as the leading supplier of word processors. The company changed its name to WordStar International in 1989.

WordPerfect

Alan Ashton who was a Ph.D. graduate in computer science from the University of Utah, started developing the specification for a word processing program in the summer of 1977. The specification defined innovative features for word processors at that time. It included text scrolling, use of function keys and automatic on-screen editing. In 1978, Bruce Bastian started working with Ashton to develop the software for the word processor.

A simplified version of the software called P-Edit for program editing was released and sold by an associate Don Owens. Ashton, Bastian and Owens then formed a company called Satellite Software International (SSI) in September 1979 to market P-Edit and the new word processor. The word processor was completed in March 1980 and called SSI*WP. The software only worked on a Data General computer system and the retail price of the program was $5,500. However the program was easy to use and fast.

In October 1980, W. E. Pete Peterson who was a brother-in-law of Bastian joined the company. At the end of 1981 Owens was removed as an officer of the company and Peterson became manager of sales and marketing and subsequently an executive vice president.

In early 1982 the company started adapting P-Edit for the IBM Personal Computer and completed the conversion in August. Work on the conversion of SSI*WP to the IBM PC was completed in the fall of 1982. The name WordPerfect was selected for the program, then it was announced to the press in October. WordPerfect was released as version 2.20 and shipped in November. The program had innovative features and the company provided excellent customer support.

In early 1983, SSI purchased Don Owens share of the company for $139,000. Ashton and Bastian now owned 50 percent each of the SSI stock. Peterson subsequently received a small percentage of the shares. Versions 3.0 and 4.0 were released at the 1983 and 1984 fall COMDEX shows respectively. Also in 1984, the company name was changed to SSI Software. With the release of version 4.0, reviews were very favorable and sales set new records in 1985.

Other company products were MathPlan (later named PlanPerfect), SSI*Data (later named DataPerfect), a legal-time-and-billing system named SSI*Legal and a version of the Forth programming language named SSI*Forth. In 1986, the company name was changed to WordPerfect Corporation and WordPerfect became the leading word processing program. Executive WordPerfect, a "junior" version of WordPerfect was released for portable computers in May 1987. In 1988 the company released WordPerfect Office incorporating electronic mail for networks, a version of WordPerfect for the Apple Macintosh in April and version 5.0 of WordPerfect for the IBM PC in May

In 1989, the tenth anniversary of the company founding, WordPerfect had achieved significant success. However with the release of graphical user interfaces by IBM in OS/2 and by Microsoft with Windows, increasing demands were occurring in the marketplace for WordPerfect to provide a graphical version of its word processor. Due to competitive concerns with Microsoft Word and higher expectations for the success of the IBM OS/2 operating system, the company decided to emphasize the development of a version of WordPerfect for OS/2 first.

Other Word Processors

Paul Lutus developed a new word processor for the Apple II computer in 1980, with a number of improvements as compared to his previous Apple Writer program. Lutus negotiated a royalty agreement for the program with Apple Computer instead of the flat fee received previously. The program called Apple Writer II was released by Apple Computer in 1981. It became a very popular word processor for the Apple II computer and made Lutus quite wealthy.

Timothy E. Gill founded Quark, Inc., in 1981. Gill developed an early word processor called Word Juggler for the Apple III computer. Farhad Fred Ebrahimi purchased half of the company in 1986.

MultiMate is the name of a series of word processing programs initially developed by MultiMate International Corporation. The company was established in 1982 and was acquired by Ashton-Tate, in December 1985. The first program developed was MultiMate Professional Word Processor. An improved version with additional advanced features was called MultiMate Advantage Professional Word Processor. An easy-to-use version of the program called MultiMate Executive Word Processor is also available.

Camila Wilson developed Volkswriter and founded Lifetree Software Inc., in 1982. Volkswriter was introduced at the West Coast Computer Faire in March 1982. The program was written in Microsoft Pascal. It was one of the earliest effective word processors for the IBM PC computer. It was priced at $195 and a deluxe version at $295.

Bank Street Writer is a word processing program that was developed by a group of Boston programmers for Apple II computers. Brøderbund Software obtained the publication rights and released it in December 1982. Brøderbund subsequently released it for Atari computers.

pfs:Write was developed by Software Publishing Corporation in the early 1980's. It became a popular word processing program with significant market penetration. See Section 13.5 for initial developments at Software Publishing Corporation.

Other word processing programs such as the Apple Computer MacWrite, IBM Displaywrite, Lotus Ami Pro (developed by the Samna company and subsequently became WordPro), Microsoft Word and Sierra On-Line Homeword shared the market. See Section 12.6 for details of Microsoft Multi-Tool Word, Word and Word for Windows word processing programs.

13.4 ... Spreadsheets

Lotus 1-2-3

Lotus 1-2-3 was co-developed by Mitchell D. Kapor and Jonathan M. Sachs in 1981. Sachs was a graduate in mathematics from MIT and spent fourteen years studying and working at various positions at MIT. In the mid seventies Sachs left MIT and supervised the development of an operating system at Data General. Following this he co-founded Concentric Data Systems where he designed a spreadsheet to run on Data General hardware. Kapor is a graduate in psychology from Yale University and partially completed a masters degree at MIT. In 1978/79 Kapor co-developed a program called Tiny Troll and in 1981 two programs called VisiPlot and VisiTrend.

In 1981 Sachs and Kapor reached an agreement to adapt Sachs spreadsheet program for the new IBM Personal Computer. Sachs had strong technical experience, especially in assembly language and Kapor had successful commercial experience in program development, with special skills in the design of the user interface. Kapor and Sachs assigned the code name of TR10 to the software development project.

To finance the initial development and company startup they contacted Sevin-Rosen Partners. Benjamin M. Rosen was a venture capitalist who had purchased Tiny Troll from Kapor. Rosen and his partner L. J. Sevin and other investors entered into an agreement to help finance with the development and introduction of the spreadsheet software. Lotus Development Corporation was founded in April 1982. The name Lotus was selected by Kapor, and comes from India where it is associated with the concept of perfect enlightenment.

The program featured natural-order recalculation, integrated graphics capability for charting, had a limited database capability and provided a computer based user tutorial. However the program required 128K bytes of memory. The software was optimized for a PC computer with the increased memory capacity and took advantage of the more powerful capabilities of the Intel 8088 microprocessor. The software was very fast in operation and achieved extra speed by going around DOS.

Lotus 1-2-3, was announced in October 1982, demonstrated at the November 1982 COMDEX show in Las Vegas and shipped in January 1983. It was an immediate success and soon replaced VisiCalc as the dominant spreadsheet program.

The company became a public corporation in October 1983. An improved version 2.0 of Lotus 1-2-3 was released in November 1985. Jim P. Manzi became president of Lotus in 1984 and chairman after the departure of Mitchell Kapor in July 1986. Lotus 1-2-3/3 was a new improved version of Lotus 1-2-3 that was announced in April 1987 for IBM OS/2 systems. In September 1987 Lotus announced it would be delayed, it was finally released in June 1989.

VisiCalc

Reference Section 7.4 for the founding of Software Arts and the initial development of VisiCalc in 1979.

In February 1980, Robert Frankston of Software Arts developed the DIF format for VisiCalc to facilitate data transfer. During 1980/81 Software Arts adapted VisiCalc to other computers such as the Atari, Commodore Pet, IBM PC and Radio Shack TRS-80. VisiCalc Advanced Version was released with additional features at the National Computer Conference in 1982. In early 1982, Personal Software, Inc. changed the name of the company to VisiCorp. Subsequently Software Arts and VisiCorp had disputes regarding the development and marketing of

VisiCalc. In September 1983 it resulted in litigation, and in early 1984 Software Arts terminated its marketing agreement with VisiCorp. The litigation was settled out-of-court in September 1984 in favor of Software Arts.

Lotus 1-2-3 had significantly impacted the sales of VisiCalc. Also the development costs of VisiOn and the litigation had affected the financial viability of VisiCorp. VisiCorp merged with Palladin Software in November 1984. Then Software Arts was purchased by Lotus Development Corporation for $6.5 million in April 1985. Daniel Bricklin became a consultant for Lotus, then he founded a new software publishing company called Software Garden, Inc.

Other Spreadsheets

See Section 10.4 for the Apple Computer LisaCalc and Section 12.6 for details of Microsoft Electronic Paper, Multiplan and Excel spreadsheets.

SuperCalc was designed by Gary Balleisen and released by a company called Sorcim for the CP/M market. Richard Frank owned the company whose name is micros spelled backwards. Subsequently improvements were incorporated in the release of SuperCalc3 and SuperCalc5 that featured 3-dimensional capabilities. The spreadsheet was acquired by Computer Associates International in 1984.

Randy Wigginton developed Full Impact for the Macintosh in 1989. The program was marketed by Ashton-Tate.

Quattro Pro was a spreadsheet program developed by a company in Hungary and released by Borland International in 1987. Version 2.0 was released in November 1989.

Other spreadsheet programs developed during the 1980's were: Javelin by Javelin Software, pfs:Plan by Software Publishing, SCO Professional by Santa Cruz Operation, T/Maker by Heidi and Peter Roizen, VP-Planner by Paperback Software and WingZ for the Apple Macintosh by Informix. Spreadsheet add-on programs were released during the 1980's to provide additional features. Examples of these are pfs:Graph by Software Publishing and a program called Sideways, which as the name suggests printed a spreadsheet sideways.

13.5 ... Databases

dBASE

The dBASE database program was initially developed by C. Wayne Ratliff in 1979 under the name of Vulcan. However he was not able to market the software successfully.

George Tate and Hal Lashlee founded a company called Software Plus as a discount mail-order software service in August 1980. In late 1980 they signed a marketing agreement with Wayne Ratliff to market his Vulcan database software. The company name was then changed to Ashton-Tate, Inc. The Ashton name does not represent anything and was selected for marketing considerations. Ratliff joined Ashton-Tate later as the chief scientist.

The software was introduced as dBASE II for 8-bit computers with a CP/M operating system in January 1981. There never was a dBASE I, the II implied an improved product. It was one of the earliest full functional relational data base programs for personal computers. The software included capabilities for programming customized requirements. The company offered innovative support services that quickly resulted in its success.

In mid-1983 Ashton-Tate purchased the dBASE II technology and copyright from Wayne Ratliff. The company went public in November 1983. Version 2.4 of dBASE II was released in 1983 with capabilities to run on both an IBM PC computer and a CP/M operating system.

dBASE III for 16-bit computers was released in May 1984. It provided extended functions, pull-down menus and a limited networking capability. dBASE III PLUS was released with built-in multi-user capabilities in November 1985. A dBASE III PLUS LAN PACK was available to share dBASE III files in a network. dBASE Mac was released for the Apple Macintosh computer in September 1987. However, it did not receive good reports.

Oracle

Lawrence J. Ellison, Robert N. Miner and Edward A. Oates founded Software Development Laboratories (SDL) in June 1977. The company had just received a contract to provide software for a mass storage device.

Before the contract was finished the company decided to diversify by developing a packaged software product. This new product would be a relational database system, that was first described by Edgar F. Codd of IBM in the June 1970 issue of the *Communications of the ACM*. It would also incorporate a **S**tructured **Q**uery **L**anguage (SQL), developed by the System R (Relational) group at IBM's Research Laboratory in San Jose, California. SQL was the user interface for the database system. Miner and Bruce Scott were principals in the development of the new program that was named Oracle. The program was developed on a DEC minicomputer and was introduced in 1978. IBM had conceived the relational database system, but SDL had beaten them to the market (IBM did not release a relational system until February 1982). Shortly after the introduction the company name was changed to Relational Software Inc. (RSI) then later to the Oracle Corporation.

RSI's first customer was the Central Intelligence Agency (CIA). However the CIA required the database to run on other operating systems such as IBM's or DEC's VAX. This and other customer requirements resulted in a rewrite of the database program in the C language to make it portable to different computer platforms. In the late 1980's, an easy-to-use version of the Oracle database was developed for the Apple Macintosh computer. This was followed by a version to run on a personal computer using Microsoft Windows. The company also developed network and client-server software for the personal computer market.

The company went public in 1986 and is now the world's largest supplier of database software. Oracle Corporation is the second-largest independent software company after Microsoft.

pfs:File

Software Publishing Corporation (SPC) was founded in 1980 by three Hewlett-Packard associates, Janelle Bedke, Fred M. Gibbons and John D. Page. The initial impetus for establishing the company was by Gibbons. The company president is Gibbons, the vice president of software development is Page and the vice president of marketing is Bedke. Page had created a database for a minicomputer at Hewlett-Packard. He then developed SPC's first program called Personal Filing System, that was abbreviated to pfs:File

The program was developed with the concepts of being simple, easy-to-use and of low cost. The program was written in Pascal with assembler routines for performance critical functions. The program was released in September 1980 for the Apple II computer.

Other programs such as pfs:Graph, pfs:Plan, pfs:Report and pfs:Word have been released using the same design concepts. The programs have also been adapted for various computers such as the IBM PC and Radio Shack models. IBM markets the pfs programs using the IBM Assistant series of labels. See Sections 13.3 and 13.4 for additional details of other Software Publishing products.

Other Databases

Informix Software Inc., is another major database software company founded by Roger Sippl in 1980. The company went public in 1986 and Phillip E. White became the chief executive officer in 1989.

Rupert Lissner developed QuickFile that was an early Apple II database program marketed by Apple Computer in 1980. A version for the Apple IIe was subsequently released named QuickFile IIe.

Jim Button developed a simple inexpensive shareware database program called PC File. He founded a company called ButtonWare around 1983 that was one of the earliest shareware companies.

Mark B. Hoffman and Robert S. Epstein founded Sybase, Inc., in 1984. The company developed relational databases, operating systems and now specializes in client-server software. The company went public in 1991.

Laurent Ribardiére and Marylène Delbourg-Delphis of France developed the 4th Dimension database for the Apple Macintosh computer in 1985. The program was a powerful graphic database and was acquired by Apple Computer. However, due to pressure from Ashton-Tate who were developing dBASE for the Macintosh, Apple Computer decided not to release the program. The developers founded the French company Analyses Conseils Informations (ACI) to market the program. In April 1987, the company formed ACIUS to market the program in the USA and appointed Guy Kawasaki who had been an executive with Apple Computer as president of ACIUS.

In 1987, Borland International bought Ansa Software and its database management system called Paradox. Borland released its version of Paradox in 1989.

13.6 ... Integrated Programs

In the early 1980's, integrated programs that combined features such as a word processor, spreadsheet and database became very popular.

Context MBA

Context MBA was a powerful integrated software package developed by Context Management Systems and released in July 1982. It featured a powerful spreadsheet, communications program, database, graphics and word processor. To improve portability it was developed for the UCSD operating system. However the system was slow, was not user friendly and could not compete with the faster programs such as Lotus 1-2-3. The company subsequently went out of business.

AppleWorks and 3 E-Z Pieces

AppleWorks is an integrated software package developed by Rupert Lissner starting in 1982. It evolved from the QuickFile database program by Lissner and was initially called Apple Pie. The program was written in machine language. It is an integrated word processing, spreadsheet and database software package which was introduced for the Apple IIe in November 1983.

The program was marketed by Apple Computer and sold for $250. AppleWorks became one of the world's best selling programs and enhanced the sale of Apple IIe computers. Its popularity resulted in the formation of the National AppleWorks Users Group (NAUG). An improved Version 2.0 of AppleWorks was released in September 1986. Then Apple Computer's subsidiary Claris Corporation, contracted with Beagle Bros. to develop a major update of the program. This was released as Version 3.0 in March 1989 and a networked version in August.

A similar program called 3 E-Z Pieces was simultaneously developed by Lissner for the Apple III. Lissner sold the marketing rights for this program to Haba Systems.

Subsequently a number of add-on enhancements were made for the AppleWorks program by other companies. Pinpoint Publishing released Pinpoint Desk Accessories in 1985 and Beagle Bros. released MacroWorks in June 1986 and a series of TimeOut modules starting in 1987.

3-Plus-1

3-Plus-1 is an integrated program that was developed by Commodore and included with their PLUS/4 computer which was released in January 1984. The program included a word processor, spreadsheet, business graphics and a file manager.

Symphony and Jazz

Lotus Development Corporation introduced an integrated program called Symphony for the IBM Personal Computer in February 1984. It was a five function integrated package with spreadsheet, business graphics, word processor, database manager and telecommunication capabilities. The project leader on the software development was Raymond Ozzie. However users felt it was too complex and sales were below expectations.

Jazz is a five function integrated program by Lotus similar to Symphony but for use on the Apple Macintosh computer. It was announced in November 1984 for release in March 1985, but the release was delayed until May. The software had a price of $595. However Jazz received mixed reviews. It was reported to be slow and lacked macros as on Lotus 1-2-3. The sales were to some extent affected by the release of the highly successful Microsoft Excel spreadsheet. Lotus reduced the price of Jazz, but it was not a successful product.

Framework

The Framework software was developed by Robert Carr who has both a bachelor's and master's degree in computer science from Stanford University. Carr had previously worked on Context MBA and at the Xerox PARC (Palo Alto Research Center) on software for future products. In 1983, Carr co-founded Forefront Corporation with Marty Mazner. The company was financed by Ashton-Tate in exchange for marketing rights to the software. The software integrated a word processor, data base, spreadsheet, graphics and communication capabilities.

The five function integrated software package named Framework was introduced by Ashton-Tate in July 1984. In 1985 Ashton-Tate acquired Forefront Corporation and Carr became chief scientist. Framework II, a second generation of the software was introduced in September 1985. It included improved functionality and ease of use features.

First Choice

Was released by Software Publishing Corporation (SPC) in 1986. First Choice was an easy to use program that integrated four popular programs into one package.

GS Works and AppleWorks GS

GS Works was initially developed by Styleware for the Apple IIGS computer and was announced in July 1988. The program was subsequently purchased by Claris and released as AppleWorks GS. It is an integrated software package with a Mac-style user interface in six integrated modules. The modules are: a word processor with a spelling checker and thesaurus, database, spreadsheet, graphics with printing and drawing features, page-layout and telecommunications.

AlphaWorks, Enable, Ovation, Q & A and Smart were other integrated programs developed or released in the 1980's.

13.7 ... Miscellaneous

Accounting Programs

Peachtree Software, Inc., started as a retail computer store in 1975 and changed to a software company in 1978. It was one of the first companies to develop accounting software for personal computers. IBM selected Peachtree Software to provide an accounting package for the IBM Personal Computer in 1980. The software was called the Business Accounting Series, that included general ledger, accounts receivable, accounts payable, inventory management and payroll modules. Programs for the IBM PC computer were released in August 1981.

In 1981, Peachtree Software became a wholly-owned subsidiary of Management Science America Inc. Then in May 1985, Peachtree was purchased by Intelligent Systems. Bill Goodhew became the president and chief executive officer. Significant improvements and price reductions were made to the high-end software, that was renamed Peachtree Complete in 1986. Then in 1988, Goodhew, other management and outside investors purchased the company from Intelligent Systems.

BIOS - Basic Input/Output Systems

The Basic Input/Output System (BIOS) for the IBM Personal Computer was developed by David J. Bradley in 1980/81. The BIOS code controls the transfer of information between elements of the hardware system. IBM made the BIOS code proprietary by copyrighting it. This prevented other companies from using the BIOS unless the obtained a license from IBM, or reverse engineered it.

Compaq Computer Corporation was the first company to reverse engineer the functions of the IBM BIOS to obtain compatibility for their portable computer released in November 1982. Then Neil Colvin who founded Phoenix Technologies Ltd., also reverse engineered the

IBM BIOS software. Phoenix released a chip for IBM compatible computers in May 1984. These developments would have a significant impact on the creation of the IBM clone market.

Computer Assisted Drafting (CAD)

John Walker founded Autodesk, Inc., in 1982. The company purchased its main product AutoCAD that was designed by Michael Riddle. Autodesk is a major supplier of computer assisted drafting (CAD) software for use on personal computers by architects and engineers. The software is available in different versions. The company went public in 1985.

Desktop Publishing

Hardware technology that provided the bit-mapped screen and the WYSIWYG (What-You-See-Is-What-You-Get) display of text enabled desktop publishing. The hardware technology was refined at Xerox PARC (Palo Alto Research Center) and implemented in a practical manner on the Macintosh computer. This software made a significant difference in the publishing process by utilizing a relatively inexpensive personal computer. Articles could be readily composed by manipulating text and graphics. The file could then be transferred digitally to a publishing company.

John E. Warnock and Charles M. Geschke founded Adobe Systems Inc., in 1982. Their original product called PostScript was derived from technology that Warnock had developed at the University of Utah and Xerox PARC. Warnock had co-developed at PARC a language called JaM that stands for John and Martin (John Warnock and Martin Newell). This language was the predecessor of PostScript that Adobe introduced in March 1985. The PostScript software contains a page description language that controls the text, graphics, images and color. This facilitates the communication of electronic documents. A special font technology enables the printing of virtually identical characters on various printers with different resolutions. The page-description language also enables a page to be printed with a mix of text and graphics at any resolution. Shortly after the founding of Adobe, Apple Computer made a significant financial investment in the company. Steven Jobs wanted PostScript released for the new Macintosh computer and LaserWriter printer. Adobe Systems became a publicly held company in 1986.

Print Shop is a desktop publishing program developed by Brøderbund Software and announced for Apple computers in May 1984.

PageMaker is a desktop publishing program developed by Aldus initially for the Apple Macintosh computer and LaserWriter printer. Paul Brainerd and colleagues founded Aldus Corporation in 1984 and released PageMaker in July 1985. It became a popular program and contributed significantly to an increase in sales of the Macintosh computer. The Aldus company was acquired by Adobe Systems in 1994.

Ventura Software Inc., was founded in 1985 by three Digital Research employees. The company introduced in 1986 the earliest desktop publishing program for the IBM PC, called Ventura Publisher. Ventura Software was acquired by Xerox Corporation in 1990, who sold it to Corel Corporation in 1993.

Quark, Inc., initially founded as a word processing company, introduced QuarkXPress, a desktop publishing program in 1987. Quark has become a leading supplier of desktop publishing programs.

Timeworks released Publish It!, a desktop publishing program for the Apple II computer in January 1988.

Games

A lot of software has been developed to provide games for personal computers in the 1980's. The following are a few of the more significant companies and their releases.

Zork I was initially developed for mainframe computers by Tim Anderson, Marc Blank, Bruce Daniels and Dave Lebling of MIT in 1977. The game was written using a MIT language called MDL. Albert Vezza who was chief of the programming research group wanted to commercialize some of the capabilities at MIT. Vezza, Joel Berez, the initial developers of Zork and other associates founded

Infocom Inc., in June 1979. The mainframe version of Zork I was then adapted for personal computers and released in December 1980. Joel Berez became president and Personal Software was the initial marketer of Zork. Zork II was released in 1981 and Deadline in 1982.

Douglas G. Carlston and Gary Carlston were principals in the founding of Brøderbund Software, Inc., in February 1980. Brøderbund is a Swedish word for brotherhood. The company released a battle game called Galactic Empire and a game of barter called Galactic Trader for the TRS-80 computer that had been programmed by Douglas Carlston. The games were offered for sale at the West Coast Computer Faire in April 1980. The programs became part of the Galactic Saga series of games. Shortly after the games were converted to run on the Apple II computer. A distribution agreement with a Japanese company called Star Craft allowed Brøderbund to market their games. A number of games such as Alien Rain, Choplifter, Lode Runner, David's Midnight Magic and Space Quarks have been successful. Through the mid 1980's Brøderbund expanded its product line to include items other than games. Brøderbund released an educational program, "Where in the World is Carmen Sandiego?" that teaches geography in 1985. Gary Carlston left the company in 1989.

Ken Williams and his wife Roberta Williams founded On-Line Systems in early 1980. Roberta Williams conceived a game situated in a mysterious house with challenges similar to the Adventure games introduced in the 1970's. Ken Williams developed the software for the game program that included graphics of rooms inside the house. The program was introduced with the name Mystery House for the Apple II computer in May 1980. The game became quite successful and another game called Wizard and the Princess was released shortly after. By 1982, the company name had changed to Sierra On-Line Inc. In 1984, a game called King's Quest was released.

Terry Bradley and Jerry Jewell founded Sirius Software in early 1980. The first entertainment products released by the company were developed by Nasir Gebelli. The first software released was a graphic utilities program called E-Z Draw, then the game programs Both

Barrels, Cyber strike and Star Cruiser followed. Sirius became a major supplier of game software in the early 1980's. However, in the summer of 1984, the company had financial difficulties and became bankrupt.

Bill Budge founded his own company called Budgeco in early 1981, to market a new game he had developed called Raster Blaster. The program was a computer version of a pinball game with innovative graphics. It simulated the bouncing of the steel ball, flippers and the effect of gravity. It became a very successful game program for the Apple II computer. Budge then developed another innovative program called Pinball Construction Set and marketed it through Electronic Arts.

William "Trip" Hawkins who had been a manager at Apple Computer, founded Electronic Arts Inc., in 1982. The company is a creator and distributor of recreational software. Hawkins contracted with independent software developers and created a unique company image by promoting these developers as software artists, similar to musicians, writers and other popular stars. Other company strategies were the direct distribution of its software to retailers, an innovative process for managing creative software development and technology leverage. Two early successful programs were Pinball Construction Set by Bill Budge, and Music Construction Set by Will Harvey. The company started developing video game systems in 1990 and Hawkins left Electronic Arts in 1993 to start another game system company.

Flight Simulator is a popular game designed by Bruce A. Artwick that was initially sold by a company called subLogic. It simulates the flight of a Cessna 182 aircraft. The game is now marketed by Microsoft. A version was released for Windows 95 in November 1996.

Graphics

Mark Pelczarski founded a graphics company called Penguin Software in mid 1981. An early graphics utility called Magic Paintbrush was released for the Apple II computer.

In-A-Vision is a graphics drawing program developed by Micrografx Inc., a company founded by George D. Grayson and J. Paul Grayson in 1982. It was released in July 1985. It was one of the earliest application programs available for Microsoft Windows.

Michael C. J. Cowpland founded the Canadian company called Corel Systems Corporation in April 1985. Corel is an acronym for Cowpland research laboratories. Previously he had co-founded a Canadian telecommunications company called Mitel Corporation and a venture capital company called Bytec Management Corporation. However Mitel had financial difficulties in 1984 and was purchased by British Telecom. Corel Systems originally focused on integrated turnkey systems using a laser printer for word processing and desktop publishing. The company then started assigning resources to the development of graphics software and introduced Version 1.0 of CorelDRAW in January 1989. The product was highly successful and the company went public In November.

Harvard Graphics is a graphics program supplied by Software Publishing, Inc.

Networks

AppleTalk is the network software developed by Apple Computer for Apple computers in January 1985. Related software was LocalTalk and AppleShare.

CP/NET was developed by Digital Research and introduced in late 1980. It connected users of Digital Research's CP/M and MP/M operating systems through the use of an arbitrary network protocol.

GRiD Server is a software package that was released in the early 1980's for communication between different GRiD computers and IBM users.

Novell Data Systems started in 1980 as a manufacturer of personal computer peripherals. Between 1981 and 1982, Safeguard Scientifics, a venture capital firm acquired an 88 percent share of the company. The company name was changed to Novell, Inc., with its incorporation in January 1983 and Raymond Noorda was brought in as chief executive officer. Its early UNIX based software was a network communications program called NetWare, released in 1983. This program was the first to introduce the concept of a file server in a local area network (LAN), that controlled access to shared devices, such as disk drives and printers. Drew Major was the lead architect in developing the program, and became chief scientist at the company. The company became a public corporation in 1985.

IBM released PC Network in the spring of 1985 for use with Version 3.1 of MS DOS from Microsoft.

cc:Mail is a network communications electronic mail program that was created by Lotus Development Corporation.

Other Applications and Companies

Avant-Garde Creations was a unique company founded by Don Fudge and Mary Carol Smith that evolved from a book publishing venture started in 1976. The company specialized initially in psychological self-help programs such as the Creative Life Dynamic series introduced in 1980 that complemented their books. Avant-Garde also developed educational, game and utility programs that enjoyed some success. The company was acquired by David Silver, a venture financier, around 1984 and Tom Measday became president.

Human Engineered Software (HES) is a company that had a rapid rise and fall of fortunes in the early 1980's. The company was founded by Jay Balakrisman in 1980, to market a utility program he had developed. Around 1982, Balakrisman sold his company to USI International, a supplier of microcomputer components who wanted to enter the software market. However, in 1983, USI encountered financial difficulties and sold HES to new investors. HES was a casualty of the software market decline in 1984, and became bankrupt in October.

Spinnaker Software Corporation is another company started to exploit the educational segment of the personal computer software market. It was founded in the early 1980's by William Bowman and C. David Seuss.

In 1978/79 Mitchell Kapor wrote a program called Tiny Troll which did line charts, multiple regressions, statistical analysis and had a text editor. Following this he made an agreement with Personal Software (later VisiCorp) to develop two graphical programs to work with the VisiCalc spreadsheet program. The two programs were VisiPlot for doing charts and VisiTrend a statistics package which were released in April 1981. VisiPlot was initially priced at $199.50. Personal Software purchased the rights to the two programs from Kapor in October 1981 for $1.5 million.

Typing Tutor was designed by Dick Ainsworth and Al Baker. The program was marketed by Microsoft and Kriya Systems in 1982. Sat Tara Singh Khalsa was a principal in the founding of the Kriya Systems.

Gary Hendrix, who was an expert in artificial intelligence, founded Symantec Corporation in 1982. Then in 1983, Gordon E. Eubanks left Digital Research and with Dennis Coleman founded the C & E (Coleman & Eubanks) software company. In 1984, C & E purchased the Symantec Corporation. Gordon Eubanks is the president and chief executive officer of Symantec that went public in 1989.

TK!Solver (TK for Tool Kit) was a program that provided a framework for building and experimenting with expert systems in engineering, scientific and other knowledge disciplines. It was developed by Daniel Bricklin and Bob Frankston and released by Software Arts in February 1983.

Sidekick is a terminate-and-stay-resident (TSR) program introduced by Borland International in mid 1984. Once the program was loaded in memory, it could be recalled at the touch of a key or two while running another program. It was released in June 1984. It had an appointment calendar, calculator, phone dialer, address book and a Word-Star compatible text editor.

Charles B. Wang and Russell M. Artzt started a joint venture with Swiss company Computer Associates (CA) International, Inc., in 1976. Wang purchased control in 1980. CA has become a major worldwide software company by purchasing numerous software companies. Although the company's main focus is on business software for large computers, it has acquired personal computer software products such as EasyWriter word processor in 1983, the SuperCalc spreadsheet in 1984 and the BPI accounting software in 1987.

Quicken is a popular personal finance program that was introduced by a company called Intuit, Inc., in

1984. Intuit was founded by Scott D. Cook and William V. Campbell is the president and chief executive officer.

Roger Wagner Publishing released HyperStudio for the Apple IIGS computer in May 1989. The program used concepts similar to the Macintosh HyperTalk system.

Part IV

1990's -- Current Technology.

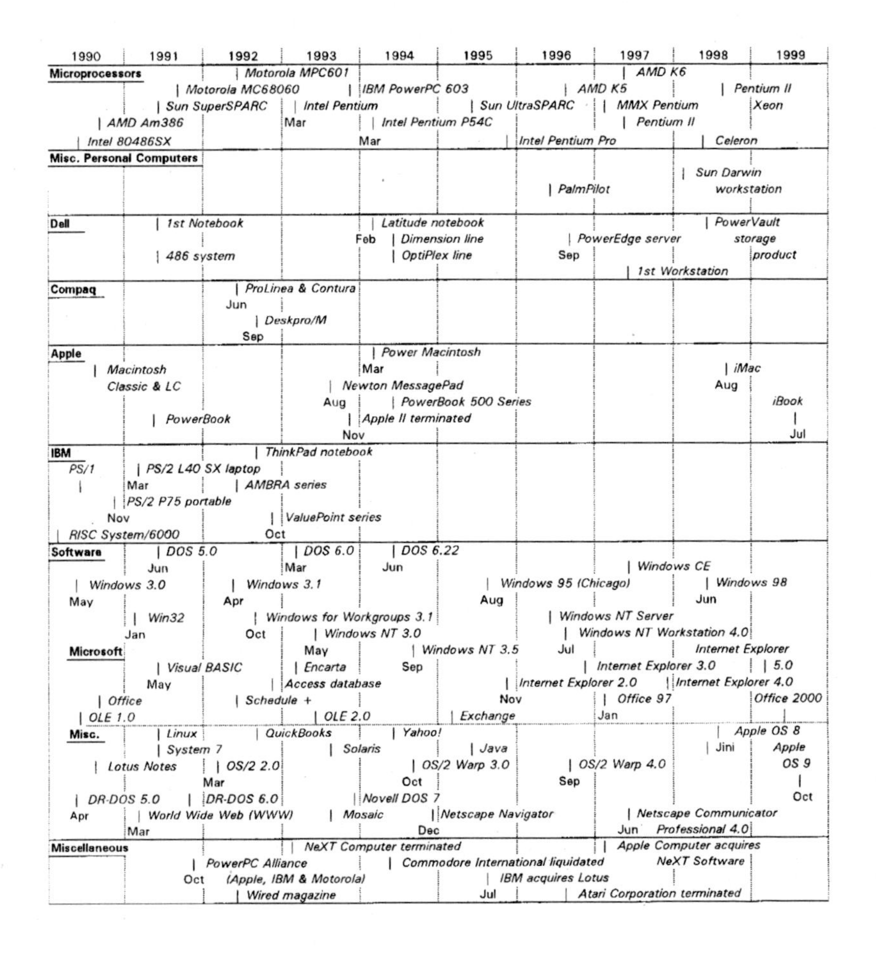

Figure 14.1: A graphical history of personal computers (1990's).

Chapter 14 Hardware in the 1990's

14.1 ... Microprocessors

IBM

IBM introduced a new RISC central processing unit (CPU) for the RISC System/6000 workstation in February 1990. The 32-bit superscalar CPU contained seven to nine VLSI CMOS chips using 1-micron technology with operating speeds of 20, 25 or 30 MHz. The architecture was called a "second generation RISC" by IBM. The CPU contained an instruction/branch unit, fixed point unit, floating point unit, data cache and storage input/output control unit. The instruction set had 184 instructions. Depending on the speed and configuration, the CPU could execute between 28 to over 40 million instructions per second (MIPS).

In October 1991, IBM participated in the formation of the PowerPC Alliance with Apple Computer and Motorola (See Section 19.6). IBM wanted to extend its workstation RISC microprocessor technology to a broader base of personal computers and reduce its dependence on Intel. Production of the PowerPC 601 by IBM began in late 1993. The PowerPC 603 for portable applications was announced in October 1993.

Intel

The 80386SL microprocessor was designed for low power, small size portable PC systems and was introduced in 1990. The chip has a 32-bit internal data path and a memory addressability of 16 megabytes. The microprocessor is available at clock frequencies of 20 and 25 MHz.

The 80486SX microprocessor is similar to the 80486DX except it does not have an integrated floating-point unit. It was introduced in April 1991. The chip has a 32-bit internal data path and a memory addressability of 4 gigabytes. The microprocessor is available at clock frequencies of 16, 20, 25 and 33 MHz.

The 80486DX2 uses a speed doubling technology and was introduced in March 1992. With this technology the microprocessor runs at 66 MHz while interfacing to a low cost 33 MHz system. This boosted computer performance by up to 70 percent without a system redesign. The chip has a 32-bit internal data path and a memory addressability of 4 gigabytes. The microprocessor is available at clock frequencies of 50/66 MHz.

The OverDrive processors were introduced in 1992, as an upgrade strategy for Intel 486 systems. The OverDrive processor is based on the "speed doubling" technology of the 80486DX2. It doubles the internal speed of the CPU while still "talking" to the rest of the system at the same frequency. This boosts overall performance by 70 percent.

The Pentium microprocessor was introduced in March 1993. The name was selected in an employee competition and was registered to prevent similar product designations by competitors. The word Pentium contains the syllable pent, which is the Latin root for five and is also Intel's fifth-generation microprocessor. It has 3.1 million transistors, nearly three times as many as the Intel 486 microprocessor. It uses 0.8 micron BiCMOS technology that combines bipolar (speed) and CMOS (low power consumption) characteristics. It is capable of running many applications five to ten times faster than a 33-MHz 486 unit. It has a 64-bit data bus and at 66-MHz it has a performance of 112 MIPS (Million Instructions Per Second). It utilizes superscalar RISC architecture and has two execution units which can process up to two instructions in a single clock cycle. It also features two Level 1 (L1) 8 KB on-chip caches, one for data and the other for instructions which improves performance. The original Pentium was available at speeds of 60 and 66 MHz. The price at launch was $878. This microprocessor is now available at speeds from 75 to 200 MHz.

Intel introduced the Pentium "P54C" that operated at 3.3 volts in 1994. Then Intel introduced the clock-tripled 80486DX4 with a larger cache in March. A joint venture with Hewlett-Packard to develop a new 64-bit microprocessor was announced in June. This would become the IA-64 microprocessor.

In the fall of 1994, the public became aware of a minor design error in the Pentium microprocessor. The design flaw which was in the floating point unit, caused a mathematical rounding error in a division once every nine billion times. Intel had encountered the problem several months earlier and had established a policy of replacing the chip for those users who were doing a lot of mathematical calculations. Then in December it was reported that IBM was stopping shipment of all computers using the Pentium. The adverse publicity resulting from this and other reports caused Intel to change its replacement policy in late December to include all customers, who wanted the Pentium changed. Intel scrapped all Pentiums that had not been sold. This and the replacement program resulted in a financial loss to Intel of $475 million.

Intel announced the Pentium Pro (initially known as the P6) microprocessor in November 1995. The Pentium Pro contains two chips, a CPU and two sizes of cache in a single package. The CPU has 5.5 million transistors. The chip incorporated a 16 KB Level 1 (L1) cache. The Level 2 (L2) 256K cache has 15.5 million transistors and the 512K version has 31 million transistors. The CPU and cache are in a single package connected by a ultra-high-speed bus. The register size is 32 bits, the data bus 64 bits and the address bus is 32 bits. The microprocessor can process a maximum of three instruction per clock cycle and 300 million instructions per second. Clock speeds were 150, 166, 180 and 200 MHz.

Intel introduced the Pentium processor with MMX technology in January 1997. The MMX processor provided 57 new instructions to improve multimedia program performance. It also included a 32 KB Level 1 (L1) cache. The Pentium II processor was introduced in May 1997. It extended the power of the Pentium Pro by adding MMX technology, dual independent bus architecture and was introduced at processing speeds of 233, 266 and 300 MHz. The chip has 7.5 million transistors. It also featured a new single edge contact cartridge physical configuration.

In October 1997, Intel announced the new IA-64 64-bit microprocessor, code named Merced with introduction planned for 1999 (subsequently changed to 2000). Principals in the joint development with Hewlett-Packard were John Crawford of Intel ad Jerry Huck of HP. It was also announced that the design would use the concept of Explicitly Parallel Instruction Computing (EPIC). It would also be able to run Windows software and HP's version of UNIX.

The Celeron processor with a clock speed of 266 MHz was introduced in April 1998. The Celeron is the same as the Pentium II, but is mounted in a lower-cost module and has no L2 cache. The processor is targeted at the low-cost personal computer market. In August Intel announced two new versions of the Celeron, the 300A and the 333. Both chips had 128 KB of integrated L2 cache and the 333 operated at 333 MHz.

The Pentium II Xeon microprocessor was introduced in August 1998. It was designed for mid- and high-range servers and workstations.

Intel introduced the Pentium III microprocessor, operating at 550 MHz in early 1999. A 600 MHz version was introduced in August. In October, Intel announced it had selected Itanium as the new brand name for the first product in its IA-64 family of processors, formerly code-named Merced.

Motorola

The MC68060 is a 32-bit superscalar microprocessor introduced in 1991. It executes instructions at 100 MIPS, has a 8K byte instruction cache, 8K byte data cache and a floating-point unit. Clock speeds are 50-66 MHz.

Motorola announced the PowerPC 601 microprocessor in 1992. The new microprocessor was developed through the PowerPC Alliance with Apple Computer and IBM (See Section 19.6). This is the first implementation of the PowerPC family of reduced instruction set computing (RISC) microprocessors and is designated MPC601 by Motorola.

The MPC601 is a 32-bit implementation of the 64-bit PowerPC architecture. The microprocessor contains

2.8 million transistors. It is a superscalar processor with the ability to execute three instructions per clock cycle. The MPC601 integrates three instruction units: an integer unit (IU), a branch processing unit (BPU) and a floating point unit (FPU). The microprocessor has a 32K byte cache and is available in 50 and 66 MHz clock speeds. The 50 MHz MPC601 is priced at $380 each and the 66 MHz version lists at $374 for production volumes of 20,000 units.

In 1994, production began of the PowerPC 603 for portable applications, the PowerPC 604 for high performance personal computers and the 64-bit PowerPC 620 for servers and high-end workstations.

Miscellaneous

Advanced Micro Devices (AMD) successfully completed the independent cloned design of the Intel 80386 microprocessor in August 1990. The new processor was named Am386 and was followed by the Am486 clone. AMD then started development of its own microprocessor design using RISC technology that resulted in the release of the K5 microprocessor in 1996, to compete with the Intel Pentium. The K6 microprocessor followed in April 1979 using technology it received after acquiring the NexGen company in 1996. AMD announced the K6-2 with additional features in May 1998.

Sun Microsystems released the SuperSPARC microprocessor that had 3.1 million transistors in 1991. However, the performance was below expectations. It was replaced by the successful UltraSPARC microprocessor in late 1995.

Cyrix is a company that got its start by producing the 80486SLC chip for notebook computers. Digital Equipment Corporation (DEC) developed the 64-bit Alpha 21064 microprocessor in 1992, that had 1.68 million transistors and operated at 200 MHz.

MIPS (purchased by Silicon Graphics) introduced the R8000 microprocessor in June 1994. It was reported to be the world's fastest microprocessor, a supercomputer on a chip. This was followed by the R10000 chip in 1995.

In June 1996, the U.S. Patent and Trademark Office overturned a patent awarded to Gilbert Hyatt for the first microprocessor. The first patent for a microprocessor is now attributed to Gary Boone and Michael Cochran of Texas Instruments.

14.2 ... IBM Computers

Andy Heller managed the RIOS project that began in 1986 to develop a new advanced workstation using RISC (Reduced Instruction Set Computing) microprocessor technology. John Cocke who created the RISC concept at IBM was a principal in the new project. The new computer became the RISC System/6000 family of six advanced workstations that IBM introduced in February 1990.

The entry-level systems were called POWERstation and POWERserver, and used a POWER architecture. POWER is an acronym for Performance Optimization With Enhanced RISC. The 32-bit RISC central processing unit (CPU) was mounted on a card that plugged into the system motherboard. The CPU was available with operating speeds between 20 and 30 MHz that enabled 28 to 40 million instructions per second (MIPS). the six models varied depending on the physical construction, CPU speed, memory and storage capacity. Each model used an enhanced version of the IBM Micro Channel (MCA) bus. IBM also released an enhanced version of the AIX operating system and OSF/Motif software for the workstation. An entry-level system with a 20 MHz CPU, 8 MB of RAM, one 1.4 MB 3.5 inch floppy disk, 120 MB hard disk, a 19 inch 1,280 by 1,024 pixel monochrome display and other accessories had a price of $12,995. The workstations were well received and became effective products in competition with other workstation suppliers.

The PS/1 computer was announced in mid 1990.

IBM introduced the PS/2 Model P75 portable computer in November 1990. The portable computer measured 18 by 12 by 6 inches and weighed 22 pounds. A standard unit utilized an Intel 486 microprocessor operating at 33 MHz, 8 MB of RAM (expandable to 16 MB), 3.5 inch high-density floppy disk drive and a 160 MB hard disk drive. The unit had four MCA expansion slots, a 10 inch diagonal gas-plasma display and a 101 key detachable keyboard. The orange-on-black display supported CGA, EGA, and VGA graphics with up to 16 shades of orange with a resolution of 640 by 480 pixels. The unit also supported XGA graphics with 256 colors with a resolution of 1,024 by 768 pixels on an external monitor. A standard configuration had a base price of $15,990. The portable was not successful due to price and a market change to smaller laptop computers.

Bob Lawten headed a project that started developing a laptop computer in January 1990. The computer design was developed at Boca Raton, Florida and at the IBM Yamato laboratory in Japan. IBM announced the battery operated PS/2 Model L40 SX laptop computer in March 1991. The portable unit measured 12.8 by 2.1 by 10.7 inches and weighed 7.7 pounds. The unit used an Intel 80386SX microprocessor operating at 20 MHz and included a socket for a coprocessor. The base system had 2 MB of RAM, expandable to 6 MB. A 3.5 inch 1.44 MB floppy disk drive and a 60 MB hard disk drive were incorporated into the unit. The display was a 10 inch sidelit supertwist VGA LCD that supported 32 gray scales. The laptop had an 84 key keyboard and a 17 key external numeric keypad. The system used a nickel-cadmium battery and an external power supply was provided. The base system cost $5,995.

Ted Selker, who was director of IBM's ergonomics research at the Almaden Research Laboratory in California, created the TrackPoint pointing device around 1991. It was developed as a means of controlling the cursor on the screen without taking the hands off the keyboard. The TrackPoint is a small pole mounted on the keyboard that converts side pressure to a corresponding movement of the cursor. It was one of the innovative features of IBM's ThinkPad notebook computer.

Between 1990 and 1991, IBM started developing a pen computing type of computer. The project was headed by Kathy Vieth. IBM had done research on handwriting recognition and a pen based operating system. However it chose a pen based operating system from Go Corporation that had been founded by Jerry Kaplan. A pen based computer named ThinkPad was announced in April 1992.

In 1992, IBM introduced two low cost series of computers. The Ambra series was marketed in Britain, Canada and France in June, and the ValuePoint series in the USA in October.

14.3 ... Apple Computers

Disenchanted with the Apple bureaucracy, Steve Sakoman who had headed the Newton project since 1987, resigned from Apple in March 1990. Larry Tesler took over the project in May. In February 1991, Michael Tchao, the product marketing manager, convinced John Sculley to concentrate the project on a less-expensive handheld version of Newton targeted at the consumer market. Sculley envisioned it as a consumer product version of the Knowledge Navigator concept he had described in his 1987 autobiography *Odyssey*. Shortly after, production of the mini Newton with the code name of Junior was approved. Two principals in the product development were Steve Capps and Michael Culbert. It became a new type of consumer oriented handheld computer called a Personal Digital Assistant (PDA). Apple named the new product MessagePad and launched it in August 1993. It had a capability to recognize writing by writing on its 240 by 336 pixel LCD screen with a stylus. It also had an infrared beaming capability for intercommunication between computers. The computer used an ARM 610 microprocessor designed by Advanced RISC Machines (ARM) Ltd. of Cambridge, England. Memory was 4 MB of ROM and 640 K bytes of RAM. The unit measured 7.25 inches long by 4.5 inches wide and 0.75 inches thick, weighed 0.9 pounds and was priced at $699. A number of improved models were released later. However, sales were significantly below expectations. The Newton product line was terminated by Steve Jobs in February 1998.

The Mac LC was released in 1990. Apple discontinued the Apple IIc in November 1990.

In October 1991, Apple participated in the formation of the PowerPC Alliance with IBM and Motorola

(See Section 19.6). Apple wanted a more powerful microprocessor for a new line of Macintosh computers.

Apple discontinued the Apple II product line in November 1993.

In 1994, Apple introduced the Power Macintosh series of computers in March and the PowerBook 500 series of notebook computers in May.

Apple introduced the iMac computer in August 1998. The computer featured a one-piece blue translucent case that incorporated the processor, compact disk drive and 15-inch monitor. The system included a translucent keyboard and a new round translucent mouse. The unit incorporated a 233 MHz PowerPC 750 G3 microprocessor, 32 MB of SDRAM, 4 GB hard disk drive, 24X CD-ROM drive and a 56k modem. A significant omission was that the unit did not include a 3.5 inch floppy disk drive. Apple priced the computer at $1,299. A completely redesigned iMac was introduced in October 1999.

The iBook is a new portable computer introduced in July 1999. It was to be the "iMac to Go' and featured a stylish case, large active-matrix display, long battery life and a PowerPC G3 microprocessor.

14.4 ... Other Computers

Compaq

In response to intense competition from clone manufacturers, Compaq launched a project with the code name of Ruby to develop a low cost personal computer. The project was headed by Richard Swingle. This project resulted in the ProLinea and Contura models being introduced in June 1992 and the ProSignia server computer in October 1992.

The Deskpro/M family of modular computers were introduced in September 1992.

Silicon Graphics (SGI)

SGI introduced the Indigo workstation for the technical market in July 1995. The O2 workstation was introduced to compete with high performance personal computers in October 1996.

U.S. Robotics

U.S. Robotics released a new Personal Digital Assistant (PDA) computer called the PalmPilot in 1996. Principals in the development of the PalmPilot were Jeff Hawkins and Donna Dubinsky. It is a mobile organizer that can interface with a desktop computer. It has become a very successful product.

Chapter 15 Software in the 1990's

15.1 ... Microsoft

Operating Systems and Windows

See Appendix B for a description and release dates of the different versions of DOS.

Microsoft released Windows version 3.0 in May 1990 in New York City with a $10 million promotional campaign. This version provided a new file manager, networking features, more desktop accessories, new screen appearance and new more recognizable icons. It was easier to install and provided an easier user interface for new users. Within four months a million copies were sold. It was a huge success.

Following the release and success of Windows 3.0, Microsoft had additional discussions with IBM in an attempt to improve its relationship with the company and its participation in the continued development of OS/2. In September an agreement was signed that IBM would take over most of the OS/2 development, Microsoft would work on an advanced future version 3.0 of OS/2 and IBM received limited rights to Windows. This resulted in Microsoft shifting programmers to Windows development. However by April 1991, Microsoft had abandoned OS/2 completely, and by 1992 the separation was final.

Object Linking and Embedding (OLE) was a new concept introduced into Microsoft products in 1990. It was also to be incorporated into a new operating system project with the code name of Cairo. Microsoft incorporated OLE Version 1.0 technology into PowerPoint in the summer of 1990 and to Excel in 1991.

In May 1991 Microsoft co-sponsored a new Windows World Exposition Conference.

Version 5.0 of MS-DOS was released in June 1991. This was an upgrade version only available from Microsoft that required a previously installed version of MS-DOS. It was a highly successful release. IBM had released its own version 5.0 but it had a number of problems. The success of the Microsoft products and other problems resulted in a further deterioration in the relationship with IBM during the latter part of 1991.

The highly successful Windows Version 3.1 upgrade with over 1,000 enhancements that included support for Object Linking and Embedding (OLE) and TrueType font technology was released in April 1992. It was in direct competition with the IBM release of OS/2 Version 2 in March. However the new release of Windows was a huge success that resulted in three million copies being shipped in the first six weeks after its introduction.

Microsoft announced Win32 in July 1992. Win32 is an application program interface with a 32-bit flat memory model, multithreading, preemptive multitasking, interprocess communication features and other advanced features.

Windows for Workgroups is a networking program with workgroup capabilities that was released as Version 3.1 in October 1992. It was not successful, and improvements were made that resulted in Version 3.11 being released in November 1993.

Version 6.0 of MS-DOS was released in March 1993 and OLE Version 2.0 in 1993.

Windows NT (New Technology) was announced in May 1993, and the first release was at Version 3.0. Windows NT is an advanced operating system for PC computers. It is a 32-bit system incorporating Win32 concepts with compatibility for applications written for MS DOS, Windows, OS/2 and POSIX. Other features included are: security protection to U.S. Government C-2 level, portability to different microprocessor architectures, symmetric multiprocessing support, built-in networking capabilities and support for international multilingual applications. The system will operate on Intel microprocessors, MIPS workstations and supports the DEC Alpha architecture. The software requires 12 to 16 megabytes of memory and a powerful microprocessor such as an Intel 486 or better. Windows NT is a sophisticated operating system for workstations and file server applications.

In August 1994, the U.S. Patent and Trademark office, approved a Microsoft request to register the label "Windows" as a trademark.

Various improvements were made to the Windows NT software under the code name of Daytona. The hardware requirements were reduced and the system reliability improved. This resulted in an upgrade Version 3.5 being released in September 1994.

The Consumer Products Division released what was called a new "social interface" with the product name of Bob in March 1995. This new easy to use user interface requires Windows, a 486 microprocessor and 8 megabytes of memory. It uses a living room setting metaphor with 12 "intelligent agents" or "friends" and 8 integrated programs. The "room" can be rearranged and customized by the user. An intelligent agent can be selected by the user to act as a guide through different tasks. The agents observe user actions and get to know the user and anticipate the persons needs. The eight integrated programs provide a calendar, checkbook, letter writer, address book, e-mail, financial guide, GeoSafari and household manager. The program has a price of $100 and received mixed reviews.

Chicago was the product code name assigned to a new advanced 32-bit operating system. It evolved from Windows with some MS-DOS code and some features from Windows NT. It incorporated Win32 technology and was targeted at the mass consumer market. Microsoft released an extensive beta test of the new software starting in June 1994. Then after a number of delays the software was released as Windows 95 in August 1995. The program incorporated a new user interface, 255-character file names, preemptive multitasking, multithreading, support for "plug and play" to optimize hardware performance and integrated network connectivity to the new Microsoft Network (MSN). MSN provides on-line communication to commercial services and the Internet. Microsoft estimated that more than 1 million copies of Windows 95 were purchased by customers at retail stores during the first four days after the release.

Windows as a separate graphical user interface for MS-DOS essentially ended with the release of the Windows

95 operating system. The functionality of MS-DOS and the previous Windows graphic user interface had been integrated in the new operating system.

WINPAD is a new operating system being developed for handheld computers.

Windows NT Workstation version 4.0 was released in July 1996. It combined the ease of use of the Windows operating system with the reliability and security of Windows NT. Windows NT Server is a powerful operating system foundation for server applications, such as BackOffice, that was released in 1996.

The Windows CE operating system, is a subset of the Windows family that was released in 1997. It was developed for a broad range of communications, entertainment and mobile devices.

Windows 98 that had a project name of Memphis was released in June 1998. It integrated Internet Explorer version 4 and supported numerous new device types. This was reported to be the last major version of Windows based on the old DOS system. Future versions of Windows will be based on NT technology.

Other Microsoft Product Releases

Microsoft Office was introduced in 1990. Initially it was a discounted suite of applications that consisted of Word, Excel and PowerPoint. The applications contained in the suit were subsequently changed to incorporate a standard user interface and improved integration features using Object Linking and Embedding (OLE) and Dynamic Link Libraries (DLL) technology. Version 4.0 of Office was released in October 1993. It contained Word 6.0, Excel 5.0 and PowerPoint 4.0, plus Mail and Access in the Professional Office edition. Office 95 was released with Windows 95 in August 1995, Office 97 in January 1997 and Office 2000 in June 1999. Microsoft offers various versions of its Office suite and has a dominant position in this market.

Microsoft issues upgrades to its application software to add improvements, new features and correct problems at one to two year intervals. Excel for Windows, Macintosh and OS/2 was improved with the release of Version 3.0 in January 1991, Version 4.0 in

March 1992 and Version 5.0 in October 1993. Word for Windows has also been improved with the release of Version 1.1 that included a grammar checker in November 1991, Version 2.0 in 1992 and Version 6.0 in October 1993. Microsoft moved from Version 2.0 to Version 6.0 to synchronize the Word for Windows version number with the MS-DOS version number. PowerPoint Version 3.0 shipped in May 1993 and Version 4.0 in October 1993.

Microsoft released Visual Basic for Windows in May 1991. Visual Basic is a graphical version of BASIC that simplifies the writing of programs for Windows. It was subsequently released in three editions: Standard, Professional and Enterprise.

Microsoft introduced a personal finance and home banking program called Money in 1991. Schedule + is an appointment, scheduling and list management program that was released in 1992.

Microsoft entered the database segment of the application software market when it acquired Fox Software and its advanced database program called FoxPro for $173 million in March 1992. Then in November, Microsoft announced its own database program called Access at a significant discount to increase market penetration. These actions contributed to the financial difficulties of Borland International, who was a major supplier of database software for the personal computer market.

Microsoft at Work is a software and architecture technology for the connection of office equipment that was announced in June 1993. Equipment such as copiers, fax machines, hand-held devices, printers and telephones would be able to communicate with a personal computer using Windows software. Microsoft released a system for fax machines in January 1995. However, it was not successful.

BackOffice is an integrated series of server applications that enables users to access information from inside and outside an organization. It can be combined with a new system called Microsoft Exchange and the Office suites. Microsoft Exchange was released in beta test in February 1995. It is an extension of

Microsoft Mail and is a groupware type of program that was developed to compete with Lotus Notes.

The company released a final version of Internet Explorer 2.0 for Windows 95 in November 1995. Version 3.0 was released in October 1996, Version 4.0 in September 1997 and Version 5.0 in March 1999.

Multimedia

The CD-ROM division established in 1985, became the multimedia publishing division in March 1992. After many delays and changes in project management, the encyclopedia with code names of Merlin and finally Gandalf shipped with the product name of Encarta in March 1993. Encarta now contains a 29-volume encyclopedia with 26,000 articles, an interactive atlas and an illustrated timeline of world history.

This division has and still is releasing many other multimedia CD-ROM products. Some of these are Ancient Lands, Art Gallery, Atlas, Cinemania, Complete Baseball, Dangerous Creatures, Dinosaurs, Isaac Asimov's The Ultimate Robot, Musical Instruments and disks on a number of musical composers.

New Developments

Microsoft is currently working on a number of systems that suggest a convergence of operating systems, application software, communication technology, personal computers and television. Some of these are the WINPAD operating system, TV set-top device software and Tiger cable-TV project. The Tiger project is a network server system to provide video-on-demand and interactive TV.

Microsoft's vision of "Information At Your Fingertips," "A Computer on Every Desk and In Every Home" and "Windows Everywhere" is becoming a reality.

15.2 ... Apple Computer and IBM

Apple Computer

Apple Computer released HyperCard IIGS in January 1991 and the System 7 operating system for the Macintosh computer in May.

In October 1991, Apple participated in the formation of the PowerPC Alliance with IBM and the creation of joint software companies called Kaleida to develop multimedia applications and Taligent to develop an advance operating system (See Section 19.6). Apple wanted to move in a direction that facilitated the interaction between their systems and IBM.

Apple Computer announced a new strategic plan in September 1994, that would expand the Macintosh technology base. The company had decided to open the Macintosh hardware and software by licensing the operating system to other computer vendors in January 1995. This licensing and an agreement with IBM and Motorola in November 1994, to create a new common hardware reference platform for computers based on the PowerPC microprocessor, was intended to increase market share. Another part of this overall strategy was to offer independent software developers a broader installed base to design applications for the Macintosh platform.

In February 1997, the company acquired NeXT Software, Inc. Apple obtained the NeXT operating system to replace its own future operating system with the code name of Copland that had been having technological problems. The new operating system based on Apple and NeXT software technology would have the code name of Rhapsody. Avidis Tevanian, who had been a principal in the software design at NeXT was placed in charge of the new operating system and became a senior vice president for software engineering.

Apple terminated the program to license the Macintosh operating system to other personal computer vendors and its support of the unified PowerPC platform in September 1997. The company had decided that the benefits of increased market share were more than offset by the costs of the licensing program.

A new operating system called OS 8 was released in July 1998. The new system featured multi-threading, PowerPC processor-native finder, spring-loaded folders, pop-up windows, contextual windows and an Internet Set Up Assistant. This was followed by the release of OS 9 in October 1999 with over fifty new features. Apple indicated that it intended to make a new release of the operating system each year with a major change on OS X (ten) in 2000.

IBM

In early 1990, James Cannavino had discussions with Bill Gates and Steve Ballmer regarding the possibility of Microsoft assuming full responsibility for OS/2 development. However the discussions were not successful. The release of the highly successful Microsoft Windows Version 3.0 in May 1990, had a further detrimental effect on OS/2 sales. Then in September IBM announced that it was taking over most of the responsibility for the development of OS/2. Around this time Cannavino appointed Joseph Guglielmi as a senior marketing executive for OS/2.

Cannavino received approval from the Corporate Management Committee (CMC) to remove Microsoft from any future development of OS/2 in early 1991. This position was announced in April, accompanied by a statement that a new version of OS/2 would run DOS applications better than DOS and Windows applications better than Windows. The announcement also stated that IBM would release the new version of OS/2 by the end of 1991. Coding in the new version was now being changed from assembler to the "C" programming language.

In October 1991, IBM participated in the formation of the PowerPC Alliance with Apple Computer and the formation of joint venture companies called Kaleida and Taligent to develop multimedia applications and an advance operating system (See Section 19.6). IBM wanted to share Apple's expertise in personal computer software development and provide an alternative to OS/2 and Microsoft systems.

Problems accommodating Window applications delayed the release of OS/2 Version 2.0 to March 1992. It included the graphics Presentation Manager user interface. The price had been reduced to a low of $35 for Windows users, $99 for DOS users and $139 for all others. However sales of the new Version 2.0 were well below expectations. Also affecting sales was the lack of application programs from other software companies for the new operating system.

OS/2 Warp Version 3 was released in October 1994.

15.3 ... Other Software

Corel

Corel Corporation introduced new versions of its highly successful graphics program CorelDRAW in the 1990's. Corel initiated a unique marketing strategy for CorelDRAW by marketing several versions of the program simultaneously. Version 2 was introduced in November 1990, Version 3 in May 1992, Versions 4 and 5 in May 1993 and May 1994 respectively, then Version 6 in August 1995. Corel entered the home consumer multimedia CD-ROM market in April 1995. Then after the acquisition of the WordPerfect and related software in February 1996, the company released the Corel WordPerfect suites and applications.

Digital Research

Digital Research released Version 5.0 of DR-DOS in April 1990 followed by Version 6.0 in 1991. Then after Digital Research was purchased by Novell, Inc., in 1992, DR-DOS became Novell DOS.

Linux

Linus Torvalds, a 21-year-old student at Helsinki University in Finland, developed an experimental version of the UNIX operating system in 1991. Torvalds posted the source code on the Internet and named the new operating system Linux. This open-source software was widely disseminated and improved upon by many users. By 1992 it was functioning on Intel processors, had a graphical user interface and had about 1,000 users. By 1995, networking capability had been added, the system had been modified to run on other processors and now had an estimated 500,000 users. In 1998 the number of users was estimated to be 7,500,000.

The dominant commercial supplier of the Linux operating system software, is Red Hat Software Inc., that was founded in early 1995 by Bob Young and a former IBM software engineer named Marc Ewing. The company provides manuals, support and other services for the Linux operating system.

Lotus

Notes is a communications program developed in the mid 1980's by Iris Associates Inc., which was a research group that spun off from the Massachusetts Institute of Technology. Raymond Ozzie was a principal in the development of the software and president of Iris Associates. The software technology was bought by Lotus Development Corporation in 1988 and released with the name of Lotus Notes in 1990. The software enables the connection of multiple personal computers, to share databases, files and provides advanced e-mail capabilities. It is also called groupware. It facilitates collaboration by communication and sharing of information between groups of people. This software was one of the significant reasons for IBM to purchase Lotus Development Corporation in 1995.

Mosaic

In 1993, Marc Andreessen a young undergraduate student and Eric Bina, developed a graphical browser for the World Wide Web (WWW) called Mosaic at the National Center for Supercomputing Applications (NCSA) located at the Urbana-Champaign campus of the University of Illinois. Mosaic provided a more visual form of WWW hypertext presentation, support for images and an intuitive user interface for a non-technical user. The software was developed for use on a UNIX operating system then translated into versions for the Apple Macintosh and IBM PC computer platforms. The program was distributed free over the Internet and received widespread use.

In August 1994, a small company located in Illinois, called Spyglass, Inc., that was founded by Tim Krauskopf in 1990, obtained the exclusive rights to license Mosaic software.

Netscape

When James H. Clark was leaving Silicon Graphics in January 1994, he told a friend that he wanted to start a new high technology company. The friend suggested he contact Marc Andreessen, who had just graduated from the University of Illinois in December 1993, where he codeveloped the Mosaic browser. Clark met Andreessen and after considering various ventures, Clark decided to finance a new company to exploit the commercial possibilities of a Mosaic type browser. Clark founded Mosaic Communications Corporation in April 1994.

Andreessen and Clark had recruited the other key team members from NCSA that developed Mosaic and a few personnel from Silicon Graphics. They then completely recreated the Mosaic browser with additional features, improved performance and stability for the UNIX, Apple Macintosh and PC computer platforms. The company released the beta version of the browser they named Mosaic Navigator in October 1994. This beta release was available free by downloading from the Internet.

In the fall of 1994, the University of Illinois demanded that Clark provide financial compensation for using their technology or intellectual property and to stop using the name Mosaic. This resulted in the company name being changed to Netscape Communications Corporation in November. Then to avoid litigation, an agreement was reached with the university that provided an undisclosed financial settlement in December. The first production version 1.0 of the browser was shipped in December with a new name, Netscape Navigator. "By spring, more than 6 million copies had been downloaded by users all over the world" [134 -page 4].

James L. Barksdale joined the company as president and chief executive officer (CEO) in January 1995 and the company went public in August. Other products that became popular tools for server and authoring Web data were also developed. Navigator 2.0 with integrated e-mail was released in September and version 3.0 with Internet telephone in April 1996. The Communicator Professional with HTML authoring and group calendar was released in June 1997.

Since the founding of the company, Clark had misgivings about future competitive actions by Microsoft. Between September and December of 1994 there was an interchange of communications for Microsoft to license the Netscape browser. However, Microsoft decided to license Mosaic from Spyglass. In the spring of 1995, Clark stated that Microsoft was withholding application programming interface (API) information applicable to Windows 95 that Netscape required for release 2.0 of its browser. In June, Microsoft advised that the API's would be provided if they could obtain an equity position and a seat on the board of Netscape. Clark rejected the offer and obtained legal counsel. This and other uncompetitive activities led to the antitrust action against Microsoft by the Department of Justice in 1977. Clark's concern for the future viability of Netscape led to his consideration of forming an alliance or sale of the company in 1997. America Online completed the acquistion of Netscape Corporation at a cost of $4.3 billion in March 1999.

Novell

After acquiring Digital Research in 1992, a new version of DR-DOS with improved networking capabilities was released as Novell DOS 7 in December 1993. Novell, Inc., sold its WordPerfect word processing program and related suite software to Corel Corporation in January 1996. Corel paid $10.75 million and 9.95 million shares of its common stock for the acquisition.

Sun

In 1991, a small research group was created at Sun Microsystems that conceived the development of a new system for the consumer market. The system would include a portable consumer device and an operating system that could interact with any other system. James Gosling was assigned to develop the operating system that became a project code-named Oak. Project Oak evolved into a platform-independent programming language and operating system for consumer electronic products. In 1994, the language design was repositioned so it could be used to build interactive applications for the Internet. The language was named Java and was introduced in May 1995.

Java is based on the C and C++ languages that has evolved into a general purpose language. It is portable to a variety of hardware platforms and operating systems. It is supposedly a "write once, run anywhere" language. Java is both a programming language and an environment for executing Java programs that has received wide spread use. This use has been increased by Sun posting the language on the Internet for free downloading by programmers. The language has also been licensed by a number of large companies such as Apple Computer, IBM, Microsoft, Netscape and Oracle.

Microsoft licensed the Java programming language technology from Sun in March 1996. Then in October 1997, Sun started litigation against Microsoft regarding its implementation of the language and the compatibility requirements of the license agreement. This action was supported by the court with a preliminary injunction siding largely with Sun in November 1998.

In mid 1998, Sun introduced a sister technology to Java called Jini. Jini was developed by a Sun research group led by William Joy. Jini enables a digital device to be connected into a computer network, identify itself and its parameters. It allows for a group of electronic devices to collaborate and combine to form a complex system.

WordPerfect

WordPerfect Corporation introduced DrawPerfect, a business presentation graphics program in February 1990 and a smaller version of WordPerfect named LetterPerfect in June. In May 1990, the company announced a change in emphasis to release a version of WordPerfect for Microsoft Windows ahead of a version for the IBM OS/2 operating system. The company had misjudged the market acceptance of OS/2 and Windows. WordPerfect for Windows was not released until November 1991 and an OS/2 version was delayed until 1993.

Yahoo!

David Filo and Jerry Yang were graduate students at Stanford University, when they created a web site and a free guide to the World Wide Web (WWW) in early 1994. The guide began as a list of their favorite Web sites and by the summer had tens of thousands visitors daily. As the list grew it was broken into a directory of search categories, then subcategories. Filo and Yang manually designated the categories as compared to computer generated indexes being created elsewhere. This resulted in an intuitive and a more selective means of locating information.

It was in the summer that they selected the name Yahoo! for the search engine. Yahoo! is a whimsical acronym for "Yet Another Hierarchical Officious Oracle." By the fall the number of Yahoo! users had increased dramatically, but they had no revenue.

In the spring of 1995, Filo and Yang approached Don Valentine's venture capital company, Sequoia Capital, and obtained $4 million of finance capital for Yahoo! Inc. Shortly after they started to hire a business team and Timothy Koogle was recruited as president and chief executive officer. Koogle quickly moved to correct the revenue side of the business that resulted in a new look for Yahoo! in August. This included the addition of advertising, a major change in the hierarchy with a reduction in the top level to 14 major search categories and the addition of a Reuters news service. Other services such as weather information and stock quotes were subsequently added.

In the fall of 1995, a second round of investment financing for $40 million was arranged. Two of the strategic investors were Reuters and the Softbank Corporation. In early 1996, Masayoshi Son of Softbank increased his investment in Yahoo! to obtain approximately 30% ownership. The company went public in April 1996 and Filo and Yang each ended up with over 15% of the company. Koogle has built Yahoo! into a powerful portal for e-commerce.

15.4 ... The Road Ahead

Gates book *The Road Ahead*, [89] articulates the future direction of software and new technology as seen by the chairman of the dominant supplier of software for personal computers. Terms such as cable-TV, information highway, information utilities, intelligent agents, interactive TV, multimedia, networks, social interface and video on demand all suggest a future direction for software development. Programming languages dominated the 1970's, application programs and operating systems the 1980's. The 1990's is adding significant capabilities for communication of information. Research to facilitate the use of computer technology will provide an extension of the mass consumer market to include novice home users. Software will be the technology that provides the synthesis to extend personal computing.

Blank page.

Chapter 16 Corporate Activities in the 1990's

Apple Computer

The company entered the 1990's under John Sculley's direction with increasing sales and profits. However, in the following years the company would encounter problems resolving its product strategy. Apple could either limit the use of its proprietary technology and maintain a high profit margin, or license the technology and achieve a greater market share for it. This problem was not helped by an executive group with individual aspirations that lacked corporate cohesiveness. Gilbert Amelio has described the Apple organization as a "dysfunctional culture." This affected the future performance and even the viability of the company.

Litigation by Xerox regarding the similarity of Macintosh graphics systems to the systems developed at PARC (Palo Alto Research Center) was thrown out in early 1990. In March 1991, the Windows/Macintosh graphics litigation with Microsoft was found to be in favor of Apple Computer, only to have most of it thrown out in April 1992.

In early 1991, Apple began negotiations with IBM regarding the possible use of IBM RISC microprocessor technology for a more powerful Macintosh computer. This resulted in an early announcement to form the PowerPC Alliance with IBM and Motorola in July 1991, followed by the final agreement in October (See Section 19.6).

A significant decline of profits in 1993, resulted in John Sculley being replaced by Michael H. Spindler as the chief executive officer in June. Sculley became chairman of the board. Then in October, Sculley resigned from Apple Computer and joined Spectrum Information Technologies as chairman and chief executive officer.

Gilbert F. Amelio, who was the CEO of National Semiconductor Corporation, became a member of the Apple Computer board in November 1994.

Due to a continuation of declining profits and loss of market share, Gilbert Amelio replaced Spindler as the chief executive officer in February 1996. The company lost $816 million in 1996.

Apple acquired NeXT Software Inc., for $427 million in February 1997. The company was purchased to obtain the NeXT operating system. Steven Jobs who had founded NeXT, became a consultant to the chairman of the Apple board.

In the spring of 1997, Apple Computer announced a corporate restructuring and laid off 2,700 employees. Then in May, Apple formed an independent subsidiary to produce and market the money losing Newton hand-held computer.

Although Amelio had implemented many improvements at Apple, the board wanted a new chief executive officer who would increase sales. This resulted in the resignation of Amelio in July. The company lost $1,045 million in 1997. Steven Jobs announced in August, that Microsoft was investing $150 million in the company. Apple agreed to drop its legal dispute about Microsoft Windows versus the Apple Macintosh graphic interfaces. Apple also agreed to promote the use of Microsoft Internet Explorer software on the Macintosh and Microsoft agreed to update and release its popular Office suite of applications for the Macintosh. Significant changes were made to the Apple board of directors with resignations that included vice-chairman Mike Markkula and the appointments of Lawrence Ellison from Oracle Corporation and William Campbell from Intuit.

The company introduced the iMac computer and discontinued its Newton MessagePad and eMate product lines in 1998. Sales of the discontinued products had been significantly below expectations. The company also stated that they wanted to focus their efforts on those products critical to the future success of the company. Apple reported in 1998, that its share of the personal computer market had declined to 4.6% in the USA and to 3.6% worldwide.

The iBook portable computer was introduced in July 1999. At the end of Apple Computer's 1999 fiscal year in September, net sales were $6,134 million and the company had 6,960 regular employees. Steven Jobs was interim chief executive officer and a director of the board.

Apple Computer is a unique company that has created an inspired and devoted following. Its easy-to-use technology, starting with the Apple II and then the Macintosh computer are acclaimed. The company has become an icon, an American success story. conceived by entrepreneurs in a garage who became multimillionaires. Unfortunately it has lost significant market share in a largely IBM compatible market. A lack of consistent leadership has been a handicap. Management have allowed what was the dominant supplier of personal computers, to become a relatively minor participant in the current market. If Apple had made the Macintosh operating system more open, it may have gained a significantly greater share of the market. Its pioneering proprietary technology has lost its initial advantage. The future for the company is somewhat uncertain.

Compaq Computer

In a management reorganization in January 1991, Eckhard Pfeiffer who had previously led Compaq operations in Germany, was appointed chief operating officer. Then in October, the company announced a major restructuring of its operations and a reorganization into distinct product divisions. Pfeiffer was elected president and chief executive officer, replacing co-founder Rod Canion. The company had experienced unsatisfactory financial results that also resulted in a reduction of the number of employees by approximately 14 percent.

Compaq became the world's largest producer of personal computers in 1994 and the world's fifth largest computer company in 1995.

In June 1997, Compaq acquired Tandem Computer Inc. for the equivalent of about $4.1 billion in stock. Tandem is a manufacturer of fail-safe minicomputers and servers for processing online transactions. Also in 1997, Compaq started to make a transition from "build-to-inventory" to a "build-to-order" manufacturing environment.

Compaq announced an agreement to purchase the Digital Equipment Corporation (DEC) for $9.1 billion in January 1998. Compaq Computer stated that the merger would create the "second largest computing company" in the world. The merger was completed in June.

During 1998, Compaq expanded its direct-sell process based on customer build-to-order choice.

In April 1999, Eckhard Pfeiffer left the company. It was reported that Benjamin Rosen and the company's board were not satisfied with the profitability of company. Michael D. Capellas was appointed the new president and chief executive officer in July.

At the end of Compaq's 1999 fiscal year in December, sales were $38.5 billion. Benjamin M. Rosen was chairman of the board and Michael Capellas president and chief executive officer.

Dell Computer

Dell Computer Corporation has become a dominant direct sale provider of build-to-order personal computers. However, Dell had financial difficulties in 1993 due to problems with a new laptop computer and an overaggressive company expansion. Michael Dell reorganized his company and brought in new experienced executives from other companies such as Morton L. Topfer who left Motorola and joined Dell in June 1994.

In 1999, the company stated that it was the "the second-largest manufacturer and marketer of personal computers in the United States and were No. 2 worldwide." At the end of Dell's fiscal year in February 1999, sales were $18.2 billion and the company had more than 24,400 employees. Kevin B. Rollins and Topfer are vice chairmen and Michael Dell is chairman and chief executive officer.

IBM

IBM sold its Lexington, Kentucky keyboard, printer and typewriter division as an independent company that became Lexmark International Group, Inc. in March 1991. Clayton, Dubilier & Rice, an investment firm arranged the financing. Marvin L. Mann, a former IBM manager became the chairman, president and chief executive officer. IBM retained a ten percent share of the new company.

The Lexington division sale helped to minimize a loss of $2.8 billion for IBM in 1991. This was the beginning of profitability problems for the company under the direction of John Akers and Jack Kuehler. In late 1991 IBM announced a major reorganization of the company from a single centralized company into a group of more independent business units.

In October 1991, IBM participated in the formation of the PowerPC Alliance with Apple Computer and Motorola to develop a new RISC microprocessor for personal computers, software for an advance operating system and development of multimedia applications (See Section 19.6).

An intense price war in personal computers, by firms such as Dell Computer, Compaq Computer and the various clone manufacturers, forced IBM to implement an extensive cost reduction program in 1992. This resulted in James Cannavino obtaining special reductions for corporate overhead from the Corporate Management Committee (CMC). Cannavino then appointed Robert J. Corrigan to head the hardware part of the personal computer operations.

In June, IBM formed a separate company known as the Individual Computer Products International (ICPI). This company was formed to market a series of low cost personal computers in Britain, Canada and France using the Ambra brand name. Then in September, James Cannavino announced the creation of the IBM Personal Computer Company (IBM PC Company) with worldwide responsibility for all aspects of the personal computer business. Robert Corrigan was appointed president of the new company. A new series of low cost computers developed in the late summer was introduced in October using the ValuePoint brand name.

In 1992, IBM had a second year of financial losses, $5 billion, a new corporate record. Reorganizations and staff reductions had not returned the company to profitability. IBM also sold Rolm Systems in May, a telecommunications subsidiary it had purchased for $1.3 billion in 1984.

In January 1993, the IBM board announced the resignation of John Akers as chairman and Jack Kuehler as president. Kuehler's resignation was effective in February and Akers in March. The announcement also stated that the company's dividend to shareholders would be reduced for the first time, by 55 percent. In April, Louis V. Gerstner replaced John Akers as chairman of the board and chief executive officer. The position of president was left vacant. Gerstner was from RJR Nabisco and the first CEO who had not progressed to the top through the company. Gerstner started implementing many changes and large write-offs to "right-size" the company that resulted in a net loss of $8 billion in 1993. During 1993, James Cannavino became a senior vice president for strategy and development, and Robert Corrigan became president of the IBM Personal Computer Company.

Personal computer revenue had been growing, but at a slower rate than some of the other competitors. This resulted in G. Richard Thoman, a senior vice-president and group executive being appointed head of the Personal Computer Group in January 1984. Changes implemented by Gerstner, resulted in a financial turnaround for IBM in 1994, a profit of $3 billion. However, the number of employees had gone from a peak of 407,000 in 1986 down to 219,800 in 1994. IBM was a significantly different organization.

IBM acquired Lotus Development Corporation for $3.2 billion ($2.9 billion in cash) in July 1995. Jim Manzi the former chief executive officer of Lotus became a vice-president of IBM. Then in October 1995 Manzi resigned. During 1995, G. R. Thoman became the chief financial officer of IBM, Robert M. Stephenson senior vice president and group executive of the Personal Systems Group and William E. McCracken general manager, sales and service of the IBM PC Company.

Taligent Inc., was dissolved in December 1995 and became an IBM subsidiary.

At the end of IBM's 1999 fiscal year in December, revenue for the year was $87.5 billion and the number of employees was 307,401. Louis V. Gerstner was chairman of the board and chief executive officer.

Intel

Robert Noyce, co-developer of the integrated circuit and co-founder of Fairchild Semiconductor and Intel, died at the relatively young age of 62, in June 1990.

Advanced Micro Devices (AMD) litigation to obtain a license as a second source for the Intel 80386 microprocessor and allegations of antitrust violations continued in the courts during the early 1990's. In February 1992, the court arbitrator awarded license rights for the 80386 to AMD. However, Intel appealed the court ruling to the State Supreme Court, but lost the appeal in 1994. A compromise settlement on all pending litigation between AMD an Intel was reached in January 1995.

Intel began an end-user marketing campaign in May 1991, that used an "*intel inside*" logo. This cooperative advertising campaign with equipment manufacturers, was started to emphasize the Intel brand name to personal computer consumers. It was also targeted at companies like AMD who had developed a clone of the 80386 microprocessor. This also led to the discontinuation of the x86 designation for new microprocessors with the introduction of the Pentium in 1993.

The Peripheral Component Interface (PCI) that had been developed by Intel, was announced in 1993. This technology permitted faster graphics and enhanced computer performance. Intel offered royalty-free licenses on the PCI patents to other companies to promote the new standard.

With the release of the Pentium in 1993, Intel also became a manufacturer of motherboards. This enabled Intel to implement new microprocessor releases and related chips to small companies at a faster rate comparable to the larger personal computer manufacturers.

Intel and Hewlett-Packard announced a joint venture to develop a new microprocessor in June 1994.

The company acquired the outstanding shares of Chips and Technologies, Inc., for approximately $430 million in January 1998. Intel also purchased the semiconductor operations of Digital Equipment Corporation for $585 million in May 1998.

Intel was estimated to have 86.7% share of the microprocessor market in mid 1998. In June the Federal Trade Commission (FTC) file an antitrust complaint against Intel. The FTC alleged that Intel used its dominant position to withhold technical data and threatened to restrict supply of chips to other manufacturers who had intellectual disputes with the company.

At the end of Intel's 1999 fiscal year in December, the company had revenue of $29.4 billion and the number of employees was 70,200. Gordon Moore was chairman emeritus of the board, Andrew Grove chairman of the board and Craig R. Barrett was president and chief executive officer.

Microsoft

In April 1990, the company appointed Michael R. Hallman as president and chief operating officer, to replace Jon Shirley who had decided to retire. Hallman had been at IBM for 20 years and then as president of Boeing Computer Services. Shirley remained on the Board and Hallman assumed operational responsibilities in June. Brad A. Silverberg also joined the company in 1990 to head the Windows and MS-DOS group and became a vice president of Microsoft. Nathan Myhrvold became a vice president responsible for advanced technology and business development in 1990.

In March 1991 arguments on items in the Windows/Macintosh graphics litigation were determined in favor of Apple. Also in March, it became public that the Federal Trade Commission (FTC) was investigating Microsoft for possible antitrust violations. The FTC actually began the inquiry in June 1990. The Commission of the European Communities also launched a similar investigation of Microsoft practices, after receiving complaints from Novell, Inc.

Microsoft had another stock split in 1991, a three for two issue. By the end of 1991 Bill Gates owned about 57 million shares or 33 percent of Microsoft. In October 1992, *Forbes* magazine reported Gates to be the richest person in North America with an estimated worth of about $6.3 billion.

James Allchin joined Microsoft in 1991 and subsequently became a vice president. He is head of the business systems division with responsibilities that include Windows NT and the advanced Cairo project.

After less than two years as the president Michael Hallman left the company. This resulted in a major reorganization of the company into three major groups in March 1992. The three groups reporting through a new Office of the President, were Products, Sales and Support, and Operations. Executive vice president Mike Maples had the Products Group, Steve Ballmer as an executive vice president, the Sales and Support Group and Frank Gaudette was Chief Financial Officer with responsibilities for the Operations Group. Craig Mundie joined the company in 1992 and became head of the advanced consumer division and was appointed a vice president of Microsoft.

In April 1992, the judge favored Microsoft and dismissed a significant part of the Apple Computer charges that Windows software infringed the "look and feel" of the Macintosh graphics system. In August the court made additional judgments in favor of Microsoft. Then in August 1993 the court dismissed all of Apple Computer's remaining infringement claims.

Then in June 1992, Microsoft announced a major agreement with IBM that confirmed their joint development separation. IBM relinquished rights to Windows NT, but was however allowed to use Windows software until September 1993. Microsoft would receive a royalty for each copy of OS/2 sold, but paid IBM a one-time payment for use of certain IBM patents.

In 1993 the Federal Trade Commission litigation started in 1990, was moved to the Department of Justice by the Clinton administration. Richard (Rick) Rashid joined the company in 1993 and became head of research and a vice president of Microsoft. Rashid was an expert in operating systems and chief architect in the development of Mach, a UNIX-based operating system.

Bill Gates married Melinda French, a Microsoft product manager on January 1, 1994 on the Hawaiian island Lanai. In March 1994, Microsoft reached an agreement with Tele-Communications, Inc., to develop an interactive cable-TV system for personal computers. This is a new entry into the information highway market by Microsoft.

Microsoft lost a patent infringement law suit by Stac Electronics in February 1994. The infringement pertained to the DoubleSpace disk compression utility included in MS-DOS Versions 6.0 and 6.2. Microsoft deleted DoubleSpace in Version 6.21 and added DriveSpace in Version 6.22 released in June 1994. Also in June, Microsoft acquired SoftImage, Inc., a leading developer of high performance 2-D and 3-D computer animation and visualization software.

Microsoft signed a consent agreement with the Department of Justice in August 1994, regarding potential antitrust violations. A similar agreement was reached with the European Union. However in February 1995, a Federal judge decided not to ratify the agreement. This was successfully appealed by Microsoft and the Department of Justice later in 1995.

Microsoft signed an agreement in October 1994 to purchase a company called Intuit, Inc. Intuit had a very popular personal finance program called Quicken. However after the intervention of the Department of Justice to block the merger, Microsoft withdrew its offer to purchase the company in May 1995.

In July 1995, Gates made another major reorganization following the retirement of Mike Maples. Five senior executives now directed four operating groups. Steve Ballmer the Sales and Support Group, Robert J. Herbold the Operations Group, and Frank M. (Pete) Higgins and Nathan P. Myhrvold directed the Applications & Content Group and Paul A. Maritz the Platforms Group.

Microsoft released Windows 95 that previously had the code name of Chicago in August 1995. Microsoft also provided capability within Windows 95 to access a new Microsoft Network (MSN). In December, Bill Gates announced the company's commitment to supporting and enhancing the Internet by integrating its software with the public network. Also in December, the company entered into a 50/50 partnership with the NBC television network to create a news/information channel and an interactive online news service for the Microsoft Network (MSN).

In January 1996, Microsoft acquired Vermeer Technologies Inc., and the FrontPage application software. FrontPage is a tool for creating and managing Web documents without programming. In February, the Interactive Media Division was created with responsibilities for applications for children and games, the Microsoft Network on line service and products of the now-dissolved Consumer Products Division. Jeffrey Raikes was promoted to group vice president for sales and marketing in July and in December the Office of the President was replaced by an executive committee.

In June 1997, Microsoft invested $1 billion in Comcast Corporation, the fourth-largest cable television operator in the USA.

WebTV Networks, Inc., was acquired by Microsoft for $425 million in August 1997. WebTV Networks was an online service that enabled consumers to access the Internet through their television via set-top terminals using proprietary technologies.

Microsoft invested $150 million in non voting shares of Apple Computer stock in August 1997. Bill Gates and Steve Jobs described a broad product and technology development agreement between the two companies. Microsoft also agreed to develop future versions of the popular Office suit of programs for the Macintosh computer.

In October 1997, the Department of Justice announced a new investigation of Microsoft. The department stated that the company was violating anti-competitive licensing practices for personal computer manufacturers by tying the use of its Windows 95 operating system to the use of its Internet Explorer software. This investigation, resulted in the launching of an antitrust suit by the Department of Justice and 20 States against Microsoft in May 1998. The antitrust trial opened in October, and the number of States was reduced to 19 when South Carolina withdrew in December.

Microsoft released Windows 98 in June. In July, Gates appointed Steve Ballmer as president of the company.

At the end of Microsoft's 1999 fiscal year in June, net revenue was $19.7 billion. Microsoft is now the world's largest software company.

In September 1999, Microsoft announced it will buy the Visio Corporation for $1.3 billion in stock. Visio's main software is a graphics package for drawing such things as flowcharts, block diagrams and networks. The November 29, 1999 issue of *Forbes* magazine reported Bill Gates net worth to be $93 billion, Paul Allen $26 billion and Steve Ballmer $22 billion. Gates was number 1, Allen number 2 and Ballmer number 3 on the list of the world's top technological billionaires.

Novell

In 1992, Novell acquired Digital Research, Inc. from Gary Kildall who died in July 1994.

Novell acquired the WordPerfect Corporation for $1.4 billion in March 1994 and bought the Quattro Pro spreadsheet from Borland International for $110 million in June. Novell also purchased a license to market the Borland Paradox database for $35 million. Novell made the acquisitions to gain a greater penetration in the business applications market. However, the Novell board became unhappy with the acquisitions and forced Raymond Noorda to leave. He was replaced by Bob Frankenberg who had been at Hewlett-Packard for twenty-five years.

The company released a suit of applications called PerfectOffice in early 1995.

By October 1995 the WordPerfect contribution to Novell sales had declined and was affecting the company's profitability. This resulted in the sale of WordPerfect and related suite software to Canadian Corel Corporation for $10.75 million and shares of Corel stock in February 1996. Corel Corporation was now one of the dominant suppliers of software for personal computers.

The board once again became unhappy with its CEO and in late 1996 Frankenberg resigned. He was replaced by Eric Schmidt who had been the chief technology officer of Sun Microsystems. Significant reorganization and layoffs were made in 1997 to improve profitability.

Sun

In early 1990, Sun decided to consolidate its computer designs on the SPARC microprocessor technology. Its workstations would not use the Motorola microprocessor and the 386i personal computer with the Intel microprocessor was discontinued. The SuperSPARC microprocessor was released in 1991 to mixed reviews and incorporated into Sun's workstations in 1993. The Solaris operating system was also introduced in 1993.

By 1995, Sun was doing very well. The combination of computers with a new UltraSPARC microprocessor released in 1995 with the Solaris operating system made a significant improvement in revenues and more importantly profits. Sun had also become a major source for powerful servers that powered the Internet. Another important event in 1995 with implications for the Internet, was the introduction of the Java programming language.

In 1996, Sun acquired a high-end server product line from Silicon Graphics. The server had been designed by Cray Research using the Sun SPARC microprocessor and the Solaris operating system. Sun also considered acquiring the Apple Computer company in 1996. However, after further consideration it decided not to.

In late 1997, Sun announced that it would develop a version of Solaris operating system for the new Intel Merced microprocessor. This would reduce Intel's dependence on Microsoft and provide a UNIX operating system alternative.

An entry-level line of workstations called Darwin was introduced in January 1988. These low-cost workstations sold for under $3000 without a monitor.

Ed Zander who is the chief operating officer was given the additional title of president in April 1999. Scott McNealy who relinquished the position of president remains the chairman and chief executive officer.

In August 1999, Sun acquired Star Division, a software developer of the StarOffice suit of productivity programs. Sun subsequently announced the development of a version of the suit called StarPortal for the World Wide Web. This is intended to provide a new method of competing with the highly successful Microsoft Office suit of programs.

Sun, and in particular Scott McNealy has contributed to and supported the Department of Justice anti-monopolistic litigation against the Microsoft Corporation. Sun has become not just a major supplier of workstations and computer servers, but also a significant innovator in software and a potential threat to Microsoft in certain segments of the market.

Other Companies

1990

Jean-Louise Gassée with financial backing from AT&T and Seymour Cray founded Be Inc., in 1990. Erich Ringewald who had also been at Apple Computer, was the director of engineering. The company demonstrated a new personal computer called the BeBox in October 1995. The computer system price ranged from $1,600 to $3,000.

In the early 1990's, the Symantec Corporation acquired a number of companies such as the Peter Norton Computing company in 1990 and the rival Central Point software company. Symantec now focuses its product line on communications, networks and is a leading supplier of utilities software.

1991

Borland International acquired Ashton-Tate and the dBASE database application program software for $440 million in 1991.

Starting in 1991, the Digital Equipment Corporation (DEC) encountered financial losses. This resulted in significant staff reductions and company reorganization in 1992. However the losses continued and resulted in the retirement of co-founder Kenneth Olsen in October. He was replaced by Robert B. Palmer who had joined DEC in 1985.

1992

MIPS Computer Systems started having problems around 1990 and was purchased by Silicon Graphics in 1992. Also in 1992, the Corel Systems Corporation, the Canadian developer of CorelDRAW graphics software, changed its name to Corel Corporation.

At WordPerfect Corporation, the two principal owners Alan Ashton and Bruce Bastian decided to implement a management reorganization in March 1992. This resulted in Pete Peterson, the executive vice president who made significant contributions to the development of the corporation, leaving the company shortly after.

Intuit released QuickBooks, an easy-to-use accounting program in 1992. The company went public in 1993.

Carol Bartz who had been a vice president of marketing at Sun Microsystems, was recruited by the board of Autodesk in 1992 to replace founder John Walker. Bartz is now the chief executive officer and chairman of Autodesk. Walker left the company in 1994.

1993

In February 1993, Steven Jobs sold the computer hardware operations of the NeXT Computer company to Canon Inc. Also in 1993, the Tandy/Radio Shack personal computer manufacturing operations were merged with AST Research.

1994

In early 1994, WordStar International (formerly MicroPro International Corporation) had financial difficulties and merged with Softkey Software Products and Spinnaker Software Corporation to become Softkey International, Inc.

Commodore International was liquidated in April 1994. Adobe Systems acquired Aldus Corporation in August 1994 whose main product was the desktop publishing program called PageMaker.

Hayes Microcomputer Products encountered financial problems in November 1994 that resulted in them filing for protection from creditors. The company then reorganized and emerged from Chapter 11 protection in early 1996.

1995

Advanced Micro Devices (AMD) and Intel, reached a compromising settlement in their joint litigation regarding AMD's use of Intel technology in January.

The Zeos personal computer company was purchased by Micron Technology in 1995. Joseph and Ward Parkinson and Doug Pitman had founded Micron Technology, Inc. in 1978 as a semiconductor design consulting company. In 1982, the company started manufacturing semiconductor memory chips. The purchase of Zeos resulted in the formation of Micron Electronics, Inc. that is now a major direct marketer of personal computers.

In December 1995 the joint venture between Apple Computer and IBM to develop an advance operating system by Taligent was dissolved. Taligent became an IBM subsidiary.

1996

Silicon Graphics acquired Cray Research, Inc., a manufacturer of supercomputers, in February 1996 for $767 million. However, it was not a successful acquisition and Silicon Graphics announced it was selling the remains of Cray Research in March 2000.

In July 1996, The Atari Corporation was terminated. AMD acquired the NexGen company in 1996, that had developed advanced microprocessor designs.

In early 1996, Packard Bell Electronics purchased Zenith Data Systems and its personal computer products. Then in July, a merger was formed to integrate Packard Bell Electronics and the NEC corporation's worldwide personal computer operations outside Japan. The merged company became Packard Bell NEC, Inc. Beny Alagem serves as chairman, chief executive officer and president of the new company. The company is reported to be the second largest supplier of personal computers in the USA market.

Since the termination of its computer, the NeXT Computer company has concentrated its efforts on software development. In early 1996 the company changed its name to NeXT Software, Inc. Then in December, Steven Jobs sold the company to Apple Computer.

1997

Acer Inc., acquired the notebook computer operations of Texas Instruments in February 1997. The 3Com Corporation acquired the U.S. Robotics Corporation in June 1997 in an $8.5 billion stock swap.

1998

The Hayes Corporation filed for protection from creditors in October 1998, and subsequently announced it was closing down the company in January 1999.

Borland International changed the company name to Inprise Corporation in April 1998.

1999

Lew Platt, chairman of the board, president and chief executive officer of Hewlett-Packard, announced that the company was being slit into two separate companies in 1999. The medical and instrument business became Agilent Technologies Inc. The computer, printer, software and service business retained the Hewlett-Packard company name. Carleton S. Fiorina became president and chief executive officer of the HP computer company in July.

America Online completed its acquisition of Netscape Corporation for $4.3 billion in March. The acquisition was assisted by a partnership between AOL and Sun Microsystems. Sun agreed to buy Netscape software and AOL agreed to buy Sun products and support the Java programming language.

In November, Packard Bell NEC announced it was closing its USA operations by the end of the year. Cheaper competitive products had resulted in financial losses and quality problems contributed to the closure.

Blank page.

Part V

Bits and Bytes.

Blank page.

Chapter 17 Hardware and Peripherals

17.1 ... Memory

The microprocessor was a significant key to lowering the cost of the personal computer. However the other key and an equally important one was low-cost semiconductor memory. Semiconductor memory started replacing magnetic core memory around 1967.

Their are two types of semiconductor Random Access Memory (RAM). Dynamic RAM (DRAM) requires periodic refresh of the memory contents and Static RAM (SRAM) retains the contents without refresh. Both types of RAM loose their contents when the power is turned off. Read only memory (ROM) retains its contents once it is programmed, even when the power is turned off.

The first commercial 1K metal oxide semiconductor DRAM was the Intel 1103 released in October 1970. This chip had a pivotal role in undercutting the price and replacement of core memory. Intel continued to improve DRAM capacities with the release of the 4K 2107 chip in 1972 and the 16K 2117 chip in 1977. However, competitive challenges from Japanese companies, would have a significant impact on Intel and other North American producers of memory chips.

Japan decided to make a strategic investment in the semiconductor memory industry around in the late 1970's. The effect of this was the first open market release of a 64K DRAM chip by Fujitsu Limited in 1979, and introduction of the first 1-megabit DRAM chip by the Toshiba Corporation in 1985. A number of other factors contributed to the dominance of Japanese manufacturers in the 1980's. Some of these were: a cooperative relationship between various companies in the Japanese industry, illegal use of U.S. technology, superior quality that contributed to lower costs and a significant investment in new facilities to produce memory chips. This resulted in a price war by the Japanese producers to increase their market share through the early 1980's. By 1985 the market situation for North American producers had so deteriorated, that the U.S. Government accused Japan of unfair trading practices and filed an antidumping complaint against the Japanese manufacturers. A semiconductor agreement was signed by the governments of Japan and the United States in 1986. However, by this time it had adversely affected many U.S. companies such as the Intel Corporation that had already decided to withdraw from the DRAM market. The company also withdrew from the EPROM chip market in 1989.

Erasable Programmable Read-Only Memory (EPROM) was invented by Dov Frohman at Intel. The memory contents can be programmed then erased by exposing the chip to ultraviolet light. Intel released the 2K-bit 1702 EPROM chip in September 1971. This alterable storage medium provided a low cost way to store microcomputer programs and became a successful and extremely profitable product for Intel until the mid 1980's.

Flash memory was developed by Toshiba. It provided the non-volatility of EPROM but the memory could be erased electrically. Electrically Erasable Programmable Read-Only Memory (EEPROM) was developed by National Cash Register (NCR) and Westinghouse companies.

17.2 ... Storage Devices

Tape Drives

Paper tape was one of the earliest forms of storage for personal computing. However it normally required a teletype machine for input/output that was too expensive for the average user.

Another early storage medium for personal computers was the magnetic audio cassette tape and the subsequent digital data cassette. Information Terminals Corporation (ITC) was the first producer of high quality data cassettes. The company was founded by J. Reid Anderson in April 1969. Anderson had previously developed acoustic-coupler modems and a prototype for a "smart" computer display terminal. During the development of the computer display terminal, Anderson determined that audio tape cassettes were not a sufficiently reliable storage medium for recording digital data. The audio cassettes did not have a uniform magnetic coating or a precise cassette body that resulted in "dropouts" or lost data. This resulted in the development of a high quality, precision data cassette that ITC started producing in 1970. The company became the dominant supplier of digital data cassettes in the 1970's. ITC introduced a mini cassette for portable data processors and a quarter-inch data cartridge in 1975. A new superior coating media for tapes and disks named Verbatim, was announced in February 1977. The company changed its name to Verbatim Corporation in late 1978, and went public in February 1979.

The 3M company introduced quarter-inch tape drive media in 1971. The capacity of these early drives was only 30 megabytes.

Jerry Ogdin developed the concept of using two tones on magnetic tape to represent digital data. This was implemented in a *Popular Electronics* construction article with the name of HITS (Hobbyists' Interchange Tape System) in September 1975. It was inexpensive and was adapted by many manufacturers. Initially each company had their own formatting standards. However in November 1975 *BYTE* magazine organized a meeting in Kansas City, Missouri of interested companies. The companies agreed to a format that became known as the "Kansas City Standard." This standard facilitated the exchange and use of magnetic tapes on different systems.

Disk Drives

The Beginning at IBM

Hard disk drive technology was developed by IBM in the 1950's as described in Section 1.3. The first Winchester hard disk drive was announced by IBM in March 1973 as the Model 3340 Disk Storage Unit. It was developed as a low-cost drive for small to intermediate computer systems. The term Winchester was used by the engineers due to the storage capacity characteristics and similarities to the name of a popular rifle as described in Section 20.4. A principal in the development was Kenneth E. Haughton who had assumed responsibility for the project in 1969. The drive assembly used a removable sealed cartridge with 14-inch diameter disks and was available in 35 and 70 megabyte storage capacities.

Floppy disk drives were developed at IBM laboratories by David L. Noble during the period of 1967 to 1971. They were initially developed by IBM as a means of storing and shipping microcode for Initial Control Program Load (ICPL) software programs on mainframe computers. The jacket enclosing the diskette was developed to protect the disk during handling and shipping.

The initial eight inch diameter read only units had a product designation of 23FD, a code name of Minnow and shipped in 1971. The diskette on the read only units rotated at 90 revolutions per minute and data was recorded on one side only. The diskette capacity was 81,664 bytes on 32 tracks which were hard sectored with eight holes around the outer edge of the disk.

The eight inch diameter read-write units had a product designation of 33FD, a code name of Igar and shipped in 1973. The diskette on the read-write drive units rotated at 360 revolutions per minute, had a capacity of 242,944 bytes on 77 tracks which was recorded on one side only and used magnetic soft sectoring (no sector holes). The 33FD diskette drive was a success and was used in data entry products which started to replace IBM card systems.

In 1976 the 43FD unit was shipped with data being recorded on both sides and capacity increased to 568,320 bytes on 154 tracks. In 1977 the 53FD double-density unit was shipped with capacity increased to 1,212,416 bytes.

IBM's research and development activities created the Winchester hard disk drives and the floppy disk drives. However other companies entered the market to compete with IBM products and to provide disk drives for other computer systems. Some of these manufacturers were Control Data Corporation (CDC), Conner Peripherals,

Maxtor, Micropolis Corporation, MiniScribe, Quantum Corporation, Seagate Technology, Shugart Associates and Western Digital Corporation. In the mid 1970's, hard disk drives were not suitable for use with microcomputers due to their large size and high cost. However by 1976, inexpensive floppy disk drives became available for personal computers.

Floppy Disk Drives

Alan F. Shugart joined IBM as a customer engineer in 1951. After a number of positions related to memory and storage technology he became manager for direct access storage products. Shugart left IBM in 1969 to become manager of storage products at Memorex. In 1973 Shugart left Memorex and with Finis F. Connor and Donald J. Massaro founded Shugart Associates. The company announced the SA-900 8-inch floppy diskette drive that retailed for $500 in the summer of 1973. After two years Shugart had a dispute regarding capitalization of the company and left. Shugart Associates announced the SA-400 5.25-inch minifloppy disk drive for $390 in December 1976. The drive used a single-sided single-density floppy disk with a capacity of 110 kilobytes. Shugart Associates was acquired by Xerox Corporation in 1977. However it was not profitable and resulted in Xerox terminating Shugart operations in 1985.

The first advertisement for a microcomputer floppy disk drive in the *Byte* magazine appeared in the August 1976 issue. The eight inch drive is described as "iCOM's Frugal Floppy. At $995, your microprocessors best friend." It was produced by iCOM Microperipherals that was a division of the Pertec Computer Corporation. Then in the February 1977 issue of *Byte*, iCOM advertised a 5.25-inch Microfloppy disk drive system for $1,095.

North Star Computers, was another early manufacturer of floppy disk drives for MITS Altair and compatible microcomputers. The company advertised the Micro-Disk System (MDS) in the January 1977 issue of *Byte* magazine. The unit used a Shugart SA-400 mini floppy disk drive and sold for $599 as a kit, or $699 assembled.

Reference Section 5.5 for information on the Apple Disk II floppy disk drive introduced by Apple Computer in 1978. In the late 1970's, other companies such as Alps Electric Company of Japan (who supplied Apple Computer), Sony Corporation and Tandon Corporation entered the floppy disk drive market. During 1982, various Japanese manufacturers offered half-height 5.25 inch floppy disk drives.

Microfloppy disk drives were introduced in the early 1980's for portable computers and to provide a more durable diskette and a less expensive drive assembly. The term floppy was not accurate as the disk was contained in a hard-shell cartridge. It also included an automatic shutter that closed over the recording surface when it was removed from the drive. Initially there were different incompatible disk sizes. Companies such as Hitachi introduced a 3.0 inch drive, Seagate supported a 3.25 inch dive, Sony and Shugart a 3.5 inch drive, Canon a 3.8 inch drive and IBM a 4.0 inch drive. This resulted in the Microfloppy Industry Committee (MIC) being formed in May 1982 to reach a consensus on a common configuration standard. In September a 3.5 inch system was proposed. Sony became a dominant supplier of 3.5 inch disk drives in 1983. The initial Sony disk drive had a storage capacity of 438 kilobytes that was subsequently increased to 1 megabyte (720 kilobytes formatted). Early applications of the 3.5 inch drive were in the Hewlett-Packard HP-150 computer and the Apple Macintosh computer.

Hard Disk Drives

Hard disk drive technology changed significantly in the years following its initial development by IBM. The storage capacity, access time and physical size have been dramatically improved. The original 14-inch diameter disk was reduced to 8-inches in 1978, then to 5.25-inches in 1980, to 3.5-inches in 1984 and to 2.5-inches in 1989.

The first Winchester 8-inch hard disk drive was introduced by Shugart Associates in 1978. Alan Shugart and Finis F. Conner who had been cofounders of Shugart Associates founded Shugart Technology in 1979. A major investment was made in the new company by the Dysan Corporation, a disk manufacturer. Shortly after the company name was changed to Seagate Technology, Inc. The first Winchester 5.25-inch hard disk drive with a storage capacity of 10 megabytes was announced in June 1980. Conner left Seagate and founded his own disk drive company called Conner Peripherals, Inc. in 1985. Seagate Technology became a dominant supplier of disk drives when it acquired the disk operations of Control Data Corporation in 1989. In early 1996, Seagate purchased Conner Peripherals and became the largest U.S. manufacturer of hard disk drives.

David Brown, James Patterson and others founded the Quantum Corporation in 1980. The companies first 8-inch hard disk drive was produced in early 1981 and a 3.5-inch hard disk drive was introduced in 1988. The company is now the second largest manufacturer of hard disk drives.

Western Digital Corporation was founded as a manufacturer of calculators and semiconductors in 1970. In the mid 1980's, the company reorganized and changed product lines to concentrate on storage devices. The company acquired the disk drive operations of Tandon Corporation in 1988. Western Digital is now the third largest U.S. manufacturer of hard disk drives.

Apple Computer's first mass storage system called the ProFile was introduced in September 1981 for the Apple III computer. The unit used Winchester technology, had a 5 megabyte storage capacity and was priced at $3,495. The hard drive within the ProFile unit was the ST-506, a 5.25-inch drive manufactured by Seagate Technology. The drive had a built-in power supply and a Z-8 based controller.

Compact Disk

Compact disk (CD) technology was developed as a joint effort by N.V. Philips of the Netherlands and Sony Corporation of Japan in 1976. This led to a number of specifications to define the disk format standards for the various types of media by the early 1980's (see Section 20.3). Compact Disk - Read Only Memory (CD-ROM) drives with a capacity of 550 megabytes were introduced in the USA in the fall of 1984. However with an initial price of over $2,000 they were expensive. High price, lack of applications and a need for format recording standards inhibited the early proliferation of the device.

Bernoulli Drive

David Bailey and David Norton founded the Iomega Corporation in 1980. Iomega introduced the Bernoulli removable disk drive with a storage capacity of 44 megabytes for personal computers in 1983. It is also known as a Bernoulli Box and features capacity comparable to a hard disk and a removable assembly for portability.

The name Bernoulli is from an eighteenth century mathematician Daniel Bernoulli who described the air dynamics utilized in the drive. The concept enables an extremely close read/write head to disk relationship but also a more tolerant protection from drive-head crashes.

A 100 megabytes Zip drive was introduced in 1995 and became quite popular as a removable high capacity disk storage system. The larger Jaz drive was introduced later, but had a number of problems.

Optical Drives

The first 12 inch diameter optical drives with Write Once, Read Many Times (WORM) recording capabilities were introduced in 1983. This was followed by 5.25 inch drives that were introduced in 1985. The principal feature of the optical drive is its extremely large storage capacity, up to 1 gigabyte. The NeXT computer was one of the earliest applications of the optical disk drive.

17.3 ... Input/Output Devices

Prior to the introduction of low cost monitors, printers and storage devices the teleprinter was a common computer input/output device. The teleprinter had an alphanumeric keyboard for input and a character printer to produce hardcopy output. It could also include a communications interface, magnetic tape unit or a paper tape punch and reader for data storage input/output. The teleprinter was often loosely referred to as a "Teletype" due to the dominant position of the Teletype Corporation in the market.

The other significant supplier of teleprinters was IBM. The market was generally divided between Teletype and IBM compatible teleprinters.

Teletype

The Teletype Corporation produced many different teleprinter models, however popular units included the Models 33 and 35 which were announced in 1963. Three versions of each model were produced with different system designs. The ASR-33 (Automatic Send-Receive) version was priced from $755 to $2,000 depending on the configuration. The printer speed was 10 characters per second.

For the personal computer user a new machine was not only expensive but difficult to obtain. The use of used or rebuilt machines at more affordable prices was more common.

17.4 ... Displays

The September 1972 issue of *Electronic Design* had an article describing how to build a circuit that could display 1,024 ASCII characters on a TV set.

Don Lancaster was an electrical engineer who in the late 1960's started writing articles for *Popular Electronics* and *Radio-Electronics* magazines. One of Lancaster's articles described a project on how to build a decimal counting unit. Then in the September 1973 issue of *Radio-Electronics* Lancaster had an article entitled "TV Typewriter" [433] that described how the computer could be connected to a television set. A TVT-1 prototype was built by Lancaster and sold as a kit for $120 in 1973. The unit could store up to 1,024 characters and display 16 lines of 32 characters. The unit had text editing capabilities and construction details were available for $2.

The VDM-1 Video Display Terminal was a prototype only that was developed by Lee Felsenstein in 1974. It was the first video terminal to be used interactively with a personal computer.

In 1972 the IBM 3270 Information Display system was announced. This provided improved speed and silence of operation. It also facilitated interaction between the user and the computer.

Lear Siegler Inc. (LSI) was an early supplier of "glass teletype" terminals after introducing the LSI ADM-1 terminal at a price of $1,500. Another major supplier of microcomputer monitors was Amdek, founded by Go Sugiura in 1977.

17.5 ... Printers

Wire Matrix

The initial development of wire matrix printing was by Reynold B. Johnson at IBM. The initial concept used a 5 by 7 array of wires to form a character. It was introduced with the Type 26 keypunch in 1949.

In 1954 Burroughs Corporation announced a wire printer producing 100 character lines printing at 1000 lines per minute. In 1955 IBM announced two high speed printers capable of printing 1000 lines per minute. These high speed wire printers experienced numerous problems and were not successful.

In 1969 IBM introduced the Model 2213 seven-wire printer. This printer was unidirectional and printed at a rate of 66 characters per second.

Centronics

Centronics Data Computer Corporation was founded by Robert Howard as a computer systems company. The company designed a dot matrix printer called the Model 101 which was introduced in the spring of 1970. It had a speed of 165 CPS (Characters Per Second) using a 5 by 7 matrix and sold for $2,995. The Micro-1 printer with a print speed of 240 CPS and a price of $595 was released in 1977. Then in 1979, the company introduced the Centronics 700 series that included the Model 779 that was priced at less than $1,000. Centronics was a dominant supplier of dot matrix printers in the 1970's.

Epson

Epson was one of the initial developers of low cost dot matrix printer technology. Epson's technology evolved from a printing device developed to print results from Seiko's quartz watch which was introduced at the 1964 Olympics in Japan. Subsequently a miniature printing device called the EP-101 was marketed by Seiko.

In 1975 Seiko established a subsidiary which they named Epson America, Inc. It was established to market and distribute microcomputer products worldwide. The name Epson was derived from the "son" of the EP-101 printer. Initially the company sold component parts to original equipment manufacturers (OEM's) who manufactured printers under their own brand name. Epson released its own dot-matrix printer, the TX-80 in 1978. This was the first low cost printer for microcomputers and was an immediate success.

The MX series of printers were introduced in 1980. This series was sold for the IBM Personal Computer under an OEM agreement. Subsequently Epson has developed an extensive range of printers using various technologies.

Reference Section 11.7 for Epson computer developments.

IBM

The IBM ProPrinter was introduced in the spring of 1985. It had a speed of 200 CPS, NLQ (Near Letter Quality) and was priced at $549.

Other Printers and Developments

In the late 1970's additional wires were added to the printhead to improve the resolution of dot matrix printers. The early 7-wire heads were changed to include 9, 12, 14, 18 and by the early 1980's the 24-wire head was introduced. These improvements have provided what is called "Near Letter Quality" (NLQ) and "Letter Quality" (LQ) printed output.

Color dot matrix printers became available in the late 1970's. A four color ribbon was used with overprinting to obtain various colors.

C.Itoh Electronics (CIE), Inc. was established in December of 1973. It was an early supplier of low cost printers for personal computers. A low cost, 80-column desktop printer was developed in June 1976. The Apple Computer company marketed the C.Itoh printer under the name of ImageWriter. Apple introduced the ImageWriter in December 1983 at a price of $675.

In the 1980's many other companies started competing in the low cost dot matrix printer market. Some of these were NEC, Okidata and TEC. Currently there is a rapid shift in the market to move from wire matrix printers to ink jet and laser type printers. This is due to the noise, print quality and print speed of the wire matrix printer. Also the decreasing cost of ink jet and laser printers is a significant factor.

Ink Jet

Ink jet printing technology has evolved from a long history of development. However during the 1960/70's research and development accelerated. In 1976 IBM introduced the Model 6640 continuous ink jet printer which set new standards for print quality.

Canon Inc., a Japanese camera company founded in the 1930's, introduced what they called "Bubble Jet" concept of printing in 1978. Then starting in 1978 Hewlett-Packard developed a thermal drop-on-demand concept of printing. Color printing capabilities were introduced in the 1980's.

Hewlett-Packard

The ThinkJet printer was introduced by Hewlett-Packard in 1984. The printer used a disposable printhead with twelve individually controlled chambers that expelled drops of ink from the nozzle. The printer had a speed of 150 CPS with a 11 by 12 dot character and a resolution of 96 dots per inch. The printer ink had some limitations on the type of paper that could be used. The price of the ThinkJet printer was $495.

Laser

Laser printing technology evolved from Chester Charlson's electrophotographic inventions in 1938. It was further developed as a copying technology at Haloid Corporation which became Xerox Corporation in 1961.

Electrophotographic printing is a complex process involving six steps: Charge of a photoconductor (PC) surface, exposure of the PC surface to a light pattern of the print image, movement of the toner to the appropriately charged areas of the PC surface, transfer of the developed image to a sheet of paper, fusing the transferred image to the paper and finally cleaning the PC surface in preparation for the next printing. Most electrophotographic printers use either a gas or diode laser printhead to scan the PC surface. Typical resolutions for laser printers are from 240 to over 800 dots per inch.

The first laser printer was developed by Gary Starkweather at Xerox PARC in 1971. Starkweather modified a Xerox 7000 copier and named the machine "SLOT," an acronym for Scanned Laser Output Terminal. The digital control system and character generator for the printer were developed by Butler Lampson and Ronald Rider in 1972. The combined efforts resulted in a printer named EARS (Ethernet, Alto, Research character generator, Scanned laser output terminal). The EARS printer was used with the Alto computer system network and subsequently became the Xerox 9700 laser printing system.

An inexpensive laser printer was introduced by Canon in 1983. The Canon LPB-CX had a resolution of 300 by 300 dots per inch and a operator changeable disposable cartridge. The Canon engine was sold to Hewlett-Packard and Apple Computer. The engine includes the laser diode, lens and mirror system, photosensitive roller, toner cartridge and paper handler. The Hewlett-Packard printer was named the HP LaserJet and was priced at $2,500. The Apple Computer printer was named the LaserWriter and was priced at $6,000. Hewlett-Packard subsequently became a dominant supplier of laser printers.

Burrell Smith was a principal in the development of the Apple LaserWriter printer. The printer was developed for the Macintosh computer and included a Motorola MC68020 microprocessor. The LaserWriter Plus was introduced in January 1986 and the LaserWriter II family of printers in January 1988.

Thermal Printers

The concept of thermal printing was developed during the 1960/70's using special sensitive paper. Thermal wax transfer printers were introduced by Brother, Toshiba and others in 1982. A resistive ribbon thermal transfer printer was introduced by IBM in 1983.

17.6 ... Peripheral Cards

Manufacturers developed many peripheral or add-on cards for various personal computers. These add-on cards extended and enhanced personal computer capabilities beyond those envisioned by the computer manufacturers. The following is representative of some of the more significant cards.

Creative Technology

Sim Wong Hoo, Chay Kwong Soon and Ng Kai Wa founded the Singapore company Creative Technology Ltd., in 1981. The company started by producing Apple II and IBM PC clones. Subsequently the focus was changed from clones to peripherals with the introduction of the Sound Blaster audio card in 1989. The company is a world leader in the manufacture of sound cards and multimedia accessories. Creative Labs, Inc., is a wholly-owned U.S. subsidiary and Creative Technology became a public company in 1992.

Cromemco Inc.

Cromemco was founded by two Stanford University professors, Harry Garland and Roger Melen in 1975. The first product was an add-on board called "Bytesaver" for the MITS Altair 8800 microcomputer. The board had a 2704 EPROM memory chip that could be programmed to load a monitor program to simplify the startup or "booting" of the computer. A kit cost $195 or $295 assembled. The second board produced by Cromemco was called the "TV Dazzler" and enabled the microcomputer to be connected to a color television set. The board provided a 128-by-128 pixel display. A software program called Kaleidoscope provided an impressive demonstration of the board capabilities. A kit cost $215 or $350 assembled. In October 1976 the company released a Zilog Z-80 board for the MITS Altair 8800 microcomputer. The Z-80 microprocessor was faster and had more extensive instruction set. The Z-80 board cost $195 as a kit or $295 assembled.

Microsoft

Microsoft conceived the concept of an add-on card that would enable their software to run on an Apple II computer. The Apple II computer used a 6502 microprocessor, but most of Microsoft's software had been developed for the Intel series of microprocessors and the CP/M operating system. With the increasing sales of Apple II computers, this segment of the software market was growing.

Microsoft had Tim Paterson of Seattle Computer Products develop what became the Z-80 SoftCard. Microsoft announced the Z-80 SoftCard with a price of $399 in March 1980. Included with the card was the CP/M operating system from Digital Research and two versions of BASIC: MBASIC (which was compatible with Microsoft BASIC-80) and GBASIC with high resolution graphic enhancements. The card was an immediate success.

The RamCard was released around 1981/82 for the Apple II Plus computer to extend the memory by 16K to 64K bytes. This additional memory allowed the computer to run CP/M applications that required 64K bytes.

Processor Technology

Processor Technology Corporation was founded by Robert Marsh and Gary Ingram in April 1975. The first product was a 4K static RAM memory expansion board for the MITS Altair 8800 microcomputer. This computer only had 256 bytes of memory in the standard unit and the 4K memory board produced by MITS was not reliable. The memory board was first advertised at the Homebrew Computer Club in April 1975 and the first order was from Cromemco. It cost $218 as a kit or $280 assembled. A 2K memory board was also available. The company also made other boards for the S-100 bus such as a 2K ROM Board, a 3P+S (parallel/serial) board, VDM-1 (Video Display Module) board designed by Lee Felsenstein and an improved motherboard for the Altair 8800.

Seattle Computer Products

Rod Brock owned Seattle Computer Products, Inc., that supplied memory cards for the S-100 bus computers around 1978. In late 1978 Tim Paterson, an employee of the company started developing a card using the new Intel 8086 microprocessor. The first prototype card was completed in May 1979. It was then demonstrated using the new Microsoft 8086 BASIC interpreter at the June 1979 National Computer Conference in New York City. Production units shipped in November 1979.

Tim Paterson also developed the operating system called QDOS for the CPU card in 1980 (see Section 13.1). This operating system later became MS-DOS.

Other Early Manufacturers

Robert Metcalfe, Greg Shaw and Howard Charney founded 3Com Corporation in 1979. 3Com is an acronym for the three *com*'s in computer, communication and compatibility. The company's main product is communication interface hardware for computer networks. Metcalfe had previously been a principal in the development of the Ethernet communications software at Xerox PARC. 3Com became a public company in 1984, and acquired U.S. Robotics Corporation in 1997.

Applied Engineering released the Transwarp accelerator card which more than tripled the speed of the Apple IIe was released in January 1986. In November 1987 the PC Transporter card was introduced to run MS-DOS programs on an Apple II computer.

The Hercules Card is a display adapter card developed by Hercules Computer Technology to display high resolution text.

North Star Computers advertised in the January 1977 issue of *Byte* magazine, a FPB Model A floating-point board to provide faster mathematical calculations. The board sold for $359 as a kit or $499 assembled. They also developed cassette tape and floppy disk interface boards.

Howard Fulmer founded Parasitic Engineering. The company initially provided add-on boards for the MITS Altair 8800 microcomputer.

SwyftCard is a card developed for the Apple II computer by Jef Raskin at Information Appliance Inc., in the early 1980's. It facilitated a number of convenient operations such as printing, calculations, telecommunications and sold for $89.95.

Vector Graphic is a company operated by Lore Harp and Carol Elly. They manufactured memory and other boards in the late 1970's. The boards were designed by Bob Harp.

IBM PC Cards

Tecmar, Inc. is a company founded by Martin A. Alpert around 1974, that provided add-on cards for the IBM Personal Computer.

In April 1982 the Xedex Corporation announced the Z-80 coprocessor card named "Baby Blue" for the IBM Personal Computer. The card had a Z-80B microprocessor that enabled CP/M programs to be run on an IBM PC.

Some other suppliers were: Quadram's Quadboard, which provided a clock, 64K bytes of additional memory, parallel and serial ports for $595 and the AST Research Combo Card.

17.7 ... Modems

The modem was invented by AT&T in 1960 and one of the earliest hobby modems called the Pennywhistle was described in the March 1976 issue of *Popular Electronics*.

Paul Collard, Casey G. Cowell and Steve Muka founded the U.S. Robotics Corporation in 1975. The first product was an acoustic coupler followed by modems. U.S.

Robotics became a public company in 1991 and was acquired by the 3Com Corporation in 1997.

Dennis C. Hayes and partner Dale Heatherington founded D.C. Hayes Associates Inc., in January of 1978. The company name changed later to the Hayes Microcomputer Products Inc., then to the Hayes Corporation. Although modems were common in the business world, Hayes was an early commercial developer of modems for microcomputers. Hayes introduced the 80-103A Data Communications Adapter modem for professional and hobby communicators in April 1978. The unit was priced at $49.95 for a bare board and $279.95 assembled. The Micromodem 100 was introduced for S-100 bus microcomputers in 1979. It could transmit data at 110 to 300 bbs and had a price of $399. In mid 1981 the Smartmodem 300 was released. Hayes also developed software to facilitate the transfer of information by modem on a phone line.

In late 1979, a company called Novation, Inc., introduced the CAT acoustic modem that was advertised at a price of "less than $199."

The VICMODEM was introduced by Commodore in March 1982 for use with the VIC computer. An interface program named VICTERM was included with the unit. The modem was priced at only $109.95 and included free offers from CompuServe, Dow Jones News and The Source.

17.8 ... Miscellaneous

Bus Systems

A bus system is a set of hardware connections used for power, signal and data transfer between components of a computer system. The bus system is characterized by its size, such as 8-bit or 16-bit and the number of lines or connection points. One of the earliest bus systems for a personal computer was the Altair Bus developed by MITS, Inc., for the Altair 8800 in January 1975. Other manufacturers adapted this bus and it became known as the S-100 Bus.

Subsequently the IEEE established a standard for the S-100 Bus. Then a working group of the IEEE Computer

Society developed the IEEE 696 bus standard. It is an augmentation and extension of the S-100 bus to 16 bits

Southwest Technical Products Corporation (SwTPC) developed the SS-50 bus for the SwTPC 6800 Computer System that they released in November 1975. It was used by a number of other manufacturers in computers using the Motorola 6800 microprocessor, such as the Smoke Signal Broadcasting Chieftain and the Gimix Ghost.

The release of the IBM Personal Computer in August 1981 established another new bus standard, the PC Bus. This was an 8-bit bus with 62 connection lines. IBM then added additional connection lines to the bus with the release of the 16-bit AT computer in August 1984. The AT Bus subsequently became known as the Industry Standard Architecture (ISA) bus.

Micro Channel Architecture (MCA) is a proprietary 32-bit multitasking bus architecture of IBM. It was a design feature of the Personal System/2 (PS/2) family of computers that were released by IBM in April 1987.

The Extended Industry Standard Architecture (EISA) was developed by a consortium of nine companies: AST Research, Compaq, Epson, Hewlett-Packard, NEC, Olivetti, Tandy, Wyse and Zenith and was announced in September 1988. It was developed as an alternative to the IBM MCA (Micro Channel Architecture) bus used on the PS/2 computers and provided some of the MCA features. EISA has a 32-bit data path and maintained compatibility with the earlier ISA architecture.

NuBus is a high-performance expansion bus used in the Apple Macintosh computer that was developed at the Massachusetts Institute of Technology (MIT). SCSI (Small Computer System Interface) that is pronounced "scuzzi," is an input output bus that provides a high-speed interface for connecting personal computers to peripheral devices. The VL Bus is a design established by the Video Electronics Standards Association (VESA) in 1992.

Digitizers

The first digitizer was called the Bit Pad with a 11-inch active area and was advertised by the Summagraphics Corporation in November 1977. The unit had a price of $555.

Floppy Disks

Development of the floppy disk drive by IBM in 1971, created a requirement for floppy disks. Significant suppliers of floppy-disks were 3M, Dysan, Elephant, IBM, ITC (later Verbatim), Maxell, Memorex, Sony and Xidex.

Information Terminals Corporation (ITC), a dominant supplier of digital data cassettes, obtained a license from IBM to manufacture 8-inch floppy disks in June 1973. The company produced its first floppy disks in December and became a dominant supplier of floppy disks. ITC collaborated with Shugart Associates to provide disks for the new 5.25-inch disk drive introduced in December 1976. The single-sided, single-density disks had a storage capacity of 180 kilobytes. Then in July 1978, ITC introduced a 720 kilobyte double-sided, double-density disk. The company name changed to Verbatim Corporation in 1978. Between 1979 and 1980 Verbatim had severe and costly quality problems. A license was obtained from the Sony Corporation to manufacture 3.5-inch diameter hard plastic case microdisks in the spring of 1983. Verbatim also introduced a high-density minidisk with a storage capacity of 1.2 megabytes in 1983 and increased it to 1.44 megabytes in 1986.

Between 1984 and 1985, Verbatim started encountering significant financial difficulties due to increasing competition and falling prices for disks. This resulted in the company being purchased by the Eastman Kodak company in June 1985 for $175 million. But by 1990, Kodak was also having problems and sold Verbatim to Mitsubishi Kasei, a large diversified Japanese company in May. Mitsubishi Kasei changed its name to Mitsubishi Chemical Company in October 1984. Verbatim is still a dominant supplier in the floppy disk market.

Keyboards

An article entitled "A Short History of the Keyboard" in the November 1982 issue of *Byte* magazine describes variations in keyboard layouts.

Another article entitled "Keyboard Karma" in *DIGITAL DELI* [190, pages 267-269] describes the problems the Japanese have with keyboards.

Microsoft introduced a new ergonomic Natural Keyboard in 1994.

Mouse

The mouse concept was invented by Douglas C. Engelbart at the Stanford Research Institute in 1964. Roger Bates and William K. English assisted in the development. The first public demonstration was at the ACM/IEEE Fall Joint Computer Conference in December 1968. Engelbart's mouse was an analog device with a wooden housing that contained a button (subsequently three buttons) and wheels that rotated two potentiometers. The potentiometers converted the movement of the mouse on a surface into electrical signals that controlled the position of the cursor on a terminal screen. The buttons were used for selection and to enter commands.

A digital wheeled type of mouse was developed by Jack Hawley for the Alto research computer at the Xerox PARC (Palo Alto Research Center) in 1972. Also at PARC in 1972, Ronald Rider developed the ball type of mouse that was subsequently changed by Hawley to improve its operation.

The first commercial implementations were on the Xerox Star in 1981, and on the Apple Lisa and Macintosh computers in 1983 and 1984 respectively. The Xerox Star digital mouse used two buttons for control purposes. Apple Computer designed a new digital mouse for the Lisa computer that used a rubber ball with optical scanners to detect motion and one button for control purposes. A degree of controversy exists regarding the number of buttons to include on a mouse. Human factor studies to determine the simplest operation, tend to favor a two button mouse.

A company called Mouse Systems introduced the first commercial mouse for the IBM Personal Computer in 1982. It was a three-button mouse.

Microsoft introduced a mouse with an add-on card for the IBM Personal Computer and a mouse for any MS-DOS computer using the serial port in May 1983 (see Section 12.1). It was priced at $195 with interface software. A new design resembling a bar of soap was released in September 1987. In late 1996, Microsoft announced the Intellimouse priced at $85. The principal new feature of the Intellimouse was an additional miniature wheel located between the left and right buttons that could be used for scrolling in application programs.

Chapter 18 Magazines and Newsletters

18.1 ... The Beginning

Publication of personal computing articles was initially in electronic magazines such as *Popular Electronics*, *QST* and *Radio-Electronics*. Then came the magazines and newsletters devoted to personal computing and microcomputers. Most of these initial publications were not specific to a particular microprocessor or type of microcomputer. The following are some of the more significant publications.

The first publication devoted to personal computing was the Amateur Computer Society *ACS Newsletter*. The editor was Stephen B. Gray who was also the founder of ACS. The first issue was published in August 1966 and the last in December 1976. It was a bi-monthly directed at anyone interested in building and operating a personal computer. The newsletter was a significant source of information on the design and construction of a computer during the time period it was published.

The *PCC Newsletter* was published by Robert L. Albrecht of the People's Computer Company in California. The first issue was published in October 1972. The first issue cover stated it "is a newspaper... about having fun with computers, learning how to use computers, how to buy a minicomputer for yourself your school and books films and tools of the future." The newspaper name changed to the *People's Computers* with a magazine type of format in May-June 1977.

Hal Singer started the *Micro-8 Newsletter* in September 1974. This was a newsletter published by the Micro-8 Computer Users Group, originally the Mark-8 Group for Mark-8 computer users. Another publication started in 1974, was The Computer Hobbyist newsletter. This newsletter had an emphasis on computer circuits and assembly language programs.

The first publication of *Creative Computing* was the November/December 1974 issue. The magazine had an initial emphasis on education and recreational computing. The editor and publisher was David H. Ahl who had previously worked for Digital Equipment Corporation and AT&T in education marketing. Due to financial problems the magazine was subsequently purchased by Ziff-Davis Publishing Company who terminated publishing the magazine in 1985.

The Homebrew Computer Club printed the first issue of its *Newsletter* in March/April 1975. Fred Moore was the first editor.

MITS, Inc., published the first issue of *Computer Notes, A Publication of the Altair Users Group* in June 1975. The editor was David Bunnell who was also the advertising and marketing manager of MITS, Inc.

Wayne Green published the first issue of *BYTE -- the small systems journal* in September 1975. "Computers --the World's Greatest Toy!" headlined the cover of the first issue. The first editor was Carl T. Helmers, who had previously published the *ECS Magazine*. Subsequently Green lost publishing control of *BYTE*, then his former wife sold the magazine to McGraw-Hill, Inc. in 1979. The publication is oriented as a technical magazine for personal computer technology, hardware and software. Byte magazine has made a significant contribution to the personal computer industry. Unfortunately, the editor advised in July 1998 that "This is the last issue of Byte you'll be receiving for a few months." It was also advised that CMP Media Inc. had acquired the magazine and would re-launch it at a later date.

The Southern California Computer Society (SCCS) started a newsletter that became the *SCCS Interface*. The Society published the first issue in December 1975. Subsequently the editor left SCCS and founded the *Interface Age* magazine.

Jim Warren was the first editor of *Dr. Dobb's Journal of Computer Calisthenics and Orthodontia (Running Light Without Overbyte)*. The first issue was January/February 1976. The focus of the magazine was on the dissemination of free or inexpensive software for microcomputers. The first issue had an article describing "Tiny BASIC" for the MITS Altair 8800 microcomputer. The magazine name was subsequently shortened to *Dr. Dobb's Journal* and the content focus on software and programming.

David Bunnell founded the *Personal Computing* magazine and the first issue was January/February 1977. Bunnell had previously worked at MITS, Inc., as editor of *Computer Notes*. Nelson Winkless was the first editor. The magazine is now published by Hayden Publishing. The focus of the magazine is on beginners and intermediate users who want to use the microcomputer as a productivity tool.

Wayne Green published the first issue of *kilobaud --The Computer Hobbyist Magazine* in January 1977. The name of the magazine was changed later to *Microcomputing*. The last issue was published in November 1984.

Other Early Magazines and Newsletters

Hal Chamberlain and some associates started the *Computer Hobbyist* newsletter in November 1974. The initial issue had articles on the Intel 8008 microprocessor.

Carl T. Helmers started the *Experimenters' Computer System (ECS)* magazine and published the first issue in January 1975. Only five issues were printed then Helmers became the editor of *BYTE* magazine.

Erik Sandberg-Diment started *ROM --Computer Applications for Living* magazine in 1977. However only nine issues were printed then it was merged into *Creative Computing*.

Roger Robitaille started a magazine called *SoftSide* about software that began with the October 1978 issue. Mark Pelczarski was an early editor for the magazine. Robitaille published several variations of the magazine. However, he was not successful and the magazines ceased publication in 1984.

Other publications were: *Microprocessors and Microsystems* first published in September 1976, *MicroTeck* --but only two issues were published in 1977 then it ceased publication, *ASCII* is a Japanese magazine started by Akio Gunji and Kazuhiko (Kay) Nishi in 1977. *Popular Computing* is another early magazine published for people who like to compute.

As microprocessors and microcomputers were released during the mid 1970's, magazines and newsletters devoted specifically to those devices were released. The following are some of those early publications.

Dr. Robert Tripp published a magazine called *Micro* that began with the October/November 1977 issue. The articles were usually technical and included many machine language programs. It was devoted to 6502 microprocessor based personal computers. The magazine ceased publication in 1985.

With the introduction of the Tandy Radio Shack TRS-80 in 1977, Wayne Green started a magazine called *80 Microcomputing*.

18.2 ... Apple Publications

Apple II

An Apple II user group in Seattle, Washington called the Apple Pugetsound Program Library Exchange (A.P.P.L.E.) started the newsletter *Call-A.P.P.L.E.* in February 1978. The first editor was Val J. Golding. The publication became a full magazine by 1980. The name of the sponsoring group changed to A.P.P.L.E. Co-op in September 1984 and to TechAlliance in late 1988. With declining sales of the Apple II computer the magazine changed to a quarterly part way through 1989. However with the ninth issue of that year, the magazine ceased publication.

Mike Harvey started the *Nibble* magazine for Apple II computer users in January 1980. The magazine was published by Software Publishing and Research Co. (S.P.A.R.C.). The articles had an orientation to software and programming for beginners and advance readers. Harvey published a *Nibble Mac* edition in 1985. However with declining sales the magazines ceased publication in July 1992.

Softalk is a monthly magazine that was started by Softalk Publishing, Inc., in September 1980. Two principals in the founding of the magazine were Al and

Margot Tommervik. *Softalk* was an informative magazine with a varied content and became a popular publication. It had a unique offer of a six month free subscription to new purchasers of Apple II computers. However, a significant reduction in advertising revenue in 1984, resulted in the termination of the magazine after the August issue.

A+ is a monthly magazine published by Ziff-Davis Publishing Company that started in 1983. It was an Apple II magazine for home and business users. The magazine merged with *In Cider* in 1989.

inCider was a monthly magazine founded by Wayne Green that started in January 1983. In 1989 *A+* magazine merged with *inCider* and became *inCider/A+*. The last issue of the consolidated magazine was July 1993.

Other Apple II Publications

Apple Assembly Line is a newsletter produced by Bob Sander-Cederlof between 1980 and 1988.

Peelings II is a magazine started in 1980, that was devoted to software reviews.

8/16 is a monthly magazine featuring tips and techniques for programmers that was published by Ariel Publishing beginning December 1980. It subsequently became *8/16-Central* in the form of a monthly disk. However, Ariel terminated publishing the magazine in October 1991.

Open-Apple is a newsletter founded by Tom Weishaar in 1985. Weishaar had a column in the *Softalk* magazine and developed software for Beagle Bros. The name of the magazine was changed to *A2-Central* in December 1988.

Softdisk is a disk-based magazine that was founded by Jim Mangham and Al Tommervik in September 1981 as part of the Softalk Publishing company. With the termination of *Softalk* magazine in 1984, Softdisk Inc., evolved. Other magazines created were *Diskworld* for the Macintosh, *Loadstar, On Disk Monthly* and *Softdisk GS*.

AppleWorks Forum is a monthly newsletter published by the National AppleWorks Users Group (NAUG) for AppleWorks users.

Cider Press is a newsletter published by the San Francisco Apple users group.

The Apple II Review magazine began in the fall of 1985, then after five issues the name changed to *Apple IIGS Buyer's Guide*. This magazine ceased publication in the fall of 1990.

II Computing magazine was published from October/November 1985 until February/March 1987.

GS+ is a bimonthly magazine founded by Steven Disbrow and published by EGO Systems for Apple IIGS users in September 1989. It ceased publication in November 1995.

Macintosh

A number of magazines exist for the Macintosh computer. Some of these are *MacWeek*, *Macintosh Today*, *MacBusiness Journal*, and *MACazine* that stopped publication in 1988. However two of the more popular are *MacUser* and *Macworld*.

Macworld is a popular magazine for Macintosh users which was started by David Bunnell. The first issue was coordinated with the introduction of the Macintosh computer in January 1984.

MacUser is a monthly magazine for Macintosh users which is published by Ziff-Davis Publishing Company.

18.3 ... PC Publications

PC Magazine was conceived by David Bunnell and financed by Tony Gold, a founder of Lifeboat Associates. The first editor was David Bunnell and the first issue was published as a bi-monthly publication in March 1982. The magazine was an instant success. However financial considerations resulted in the magazine being sold by Tony Gold to Ziff-Davis Publishing Company in 1982. The magazine became a monthly publication in 1983 and then to twenty-two issues per year in 1984. The sale to Ziff-Davis created differences, that resulted in Bunnell and a group of the staff leaving to start a new magazine called *PC World*. *PC Magazine* now has one of the largest distributions.

PC World is a monthly magazine started by David Bunnell. The first issue was published in January 1983. It is now published by PC World Communications Inc.

PCjr was a magazine published by Ziff-Davis Publishing Company for users of the IBM PCjr computer. The magazine was available for the introduction of the computer in November 1983.

PC Week is a weekly publication started in 1984 by Ziff-Davis Publishing Company for the PC computer industry.

PC Computing is a monthly computer magazine published by Ziff-Davis Publishing Company.

Big Blue Disk is a disk-based magazine introduced by Softdisk Inc., in 1986. The magazine name was subsequently changed to *On Disk Monthly*. The company also published a disk magazine named *Gamer's Edge*.

18.4 ... Other Publications

Datamation started in 1955 and published twice a month by Cahners Publishing. It is a business magazine covering the computer industry.

ComputerWorld started in 1967, it is the oldest computer weekly newspaper with coverage of both mainframes and microcomputers. The publisher is CW Communications.

The *Annals of the History of Computing* quarterly periodical that was initially published by the American Federation of Information Processing Societies (AFIPS) Inc. The publications focus is on the history of computing including the people and companies. The first issue was in July 1979. It was changed to the *IEEE - Annals of the History of Computing* in 1984.

Glenn E. Patch started *Computer Shopper* in November 1979 as a trading paper for used computers and peripherals. It evolved into a tabloid magazine by 1983, devoted to direct sale of computers, related equipment and software. Stanley Veit became editor-in -chief in 1983.

InfoWorld is a weekly publication started in February 1980 and published by CW Communications. It provides news on the microcomputer industry.

Computist is a magazine that began publication in 1981 with the name *Hardcore Computing*. The magazine was initially dedicated to Apple II computer users with some emphasis on breaking copy-protected programs. Then in 1983 the name changed to *Hardcore Computist* and subsequently to just *Computist.* The periodical now covers IBM and Macintosh computers.

Microsoft published the first issue of *the Microsoft Quarterly* around April 1982. The *Microsoft System Journal* is another publication published by Microsoft.

Popular Electronics, an electronic experimenters magazine, pioneered personal computing with the January 1975 issue describing the Altair 8800 microcomputer. The editor was Arthur P. Salsberg and the technical editor was Leslie Solomon. In November 1982, the magazine name was changed to *Computers & Electronics*. However, the transition from an electronics to a computer oriented magazine was not successful. Declining revenues during 1984 resulted in the termination of the magazine, with the last issue in April 1985.

Springer-Verlag published the first issue of *Abacus* in 1983 and the last in 1987. It had numerous articles on the history of computers.

Loadstar is a disk-based magazine published by Softdisk Inc., for Commodore 64 computer users that started in June 1984.

Compute is a monthly magazine published by Compute Publications and targeted at beginning users of personal computers.

Pico is a magazine on portable computing started by Wayne Green.

Scientific American is a magazine that has published a number of articles and special issues on computing. An article by Eric A. Weiss entitled "Scientific American's Snapshot of Software" in *Abacus* Volume 2, Number 2 (Winter , 1985) has a sampling of computer articles from *Scientific American* on page 47. The September 1977 issue on "Microelectronics," the

December 1982 issue on "Personal Computers," the September 1984 issue on "Computer Software" and the January 1998 issue on "The Solid-state Century" are examples of special issues devoted to computer technology.

Time magazine has also had special issues and covers related to the personal computer industry. A special cover and section describing a "miracle chip" and the concerns of a computer society was featured in the February 20, 1978 issue. Since 1927, *Time* magazine featured real individuals for its Man of the Year issue and cover. On the 3rd of January 1983, Time magazine featured the microcomputer on the cover. It was headlined "Machine of the Year: The Computer Moves In." This change by Time magazine symbolized the technological impact that the personal computer had created. Another article featured Andrew Grove in the Dec. 97/Jan. 98 issue. Recent issues in 1997 with special front covers and articles are Bill Gates on Jan. 13th., Steve Jobs on Aug. 18th., and Steve Case on Sep. 22nd.

Wired magazine was co-founded by Nicholas Negroponte in 1992.

18.5 ... Reference

The publication of periodicals and newsletters relating to microcomputers is extensive. A source for additional details is a publication entitled "Microcomputing Periodicals: An Annotated Directory"[126]. The 1985 directory listed over one thousand titles.

For additional information on the *Creative Computing* magazine and publisher David Ahl, reference article by John J. Anderson entitled "Dave Tells Ahl: The History of Creative Computing" [447, pages 66-77].

An overview of some early magazines is contained in "Computer Magazine Madness" by Stanley Veit in *DIGITAL DELI* [190, pages 66-69].

Blank page.

Chapter 19 Other Companies, Organizations and People

19.1 -- Early Organizations

Amateur Computer Society (ACS)

Was formed in May 1966 by Stephen B. Gray. It was created as "a nonprofit group open to anyone interested in building and operating a digital computer that will at least perform automatic multiplication and division, or is of a comparable complexity." A publication entitled *ACS Newsletter* started in August 1966. Grey has stated that "ACS membership never totaled more than a few hundred." The newsletter and the Society terminated in December 1976.

People's Computer Company (PCC)

Robert L. Albrecht founded the People's Computer Company and associated Community Computer Center in the late 1960's. The company published the *PCC Newsletter* starting in October 1972. The cover of the first newsletter stated the "People's Computer Center is a place. ...a place to do things the People's Computer Company talks about. ...a place to play with computers - at modest prices. ... a place to learn how to use computers." It also indicated that "We have a small, friendly computer ...an Edu System 20, a time sharing terminal that connects us to the world and a Tektronix programmable calculator and some simple calculators and books to help you learn and ..."

Community Memory

Lee Felsenstein who belonged to a group called Resource One, helped to organize Community Memory, a public information network in the early 1970's. Felsenstein established the organization and system to humanize the computer interface and bring computing power to the people. It provided free access to a time sharing system. The Community Memory system consisted of remote teletype terminals located in several storefronts located around Berkeley, California. The system operated a bulletin board and enabled people to communicate or leave messages. Felsenstein also wanted to replace the Teletype units with an easy-to-use machine he called "The Tom Swift Terminal."

Homebrew Computer Club

At a number of locations in early 1975 in the Silicon Valley area, a notice was posted reading "Amateur Computer Users Group. Homebrew Computer Club... you name it. Are you building your own computer? Terminal? TV Typewriter? I/O Device? or some other digital black box? Or are you buying time on a time-sharing service? If so you might like to come to a meeting of people with like-minded interests." The notice had been posted by Fred Moore and Gordon French. French was a mechanical engineer and computer hobbyist and both had been associated with Robert Albrecht of the People's Computer Company. The meeting was held at Gordon French's garage in Menlo Park, California on the 5th of March 1975. About thirty people attended the first meeting. Albrecht demonstrated the new Altair computer. Another computer enthusiast Steve Dompier described his visit to MITS Inc., in Albuquerque, New Mexico. Stephen Wozniak also attended this first meeting. From this first meeting the Homebrew Computer Club was formed. The attendance quickly increased and meetings became fortnightly gatherings at the Stanford Linear Accelerator Center (SLAC).

The club was a forum for the interchange of information and became a catalyst for the technological development of the microcomputer. Gordon French was the secretary and librarian. Fred Moore issued the initial newsletters highlighting the meeting activities and other news. A dominant member was Lee Felsenstein who became moderator of the meetings. A number of members of the club became entrepreneurs who established their own companies. Some of these were, Steve Jobs and Stephen Wozniak who founded Apple Computer, Robert Marsh founder of Processor Technology. Adam Osborne sold his books on microprocessors at the club and Felsenstein would design the SOL and Osborne microcomputers.

By 1979/80 the Homebrew Computer Club was past its peak with other user groups being formed with a focus on more specific interests.

Other Early Groups

The Amateur Computer Group of New Jersey was founded in May 1975 with Sol Libes as the first President. Then the Long Island Computer Association of New York formed in 1975. The Southern California Computer Society (SCCS) began a few months after the Homebrew Computer Club in 1975. It was a well organized group which quickly grew and had a membership in the thousands. The editor of their magazine left to publish his own *Interface Age* periodical. In 1977, Jonathan Rotenberg who was 13 years old at the time, founded the Boston Computer Society in 1977. a special interest group called SIGPC on personal computing was formed by the Association for Computing Machinery (ACM). The first chairperson for ACM SIGPC was Portia Isaacson.

Many other computer clubs were formed during the mid 1970's in the USA and other countries. Two *Byte* magazine articles: "A Computer Hobbyist Club Survey" [477] and "Clubs and Newsletters Directory" [478] provide additional details of these early organizations.

Clubs, groups and societies helped to disseminate information on personal computers and computing. However, other early organizations that provided similar input came in various names such as conventions, fairs, festivals and shows.

19.2 -- Conventions, Fairs and Shows

World Altair Computer Convention

The first World Altair Computer Convention (WACC) was organized by David Bunnell of MITS, Inc. It was held in Albuquerque, New Mexico in March 1976. This was the first microcomputer convention and several hundred people attended.

Trenton Computer Festival

The Trenton Computer Festival was the first regional convention. Sol Libes organized it and the Amateur Computer Group of New Jersey sponsored the festival that was held in May 1976.

Personal Computing 76

The first national microcomputer show was organized mainly by John Dilks and co-chaired by James Main. The show was held in Atlantic City, New Jersey in August 1976. The Sol computer and the Apple I computer board were introduced at this show. Attendance was estimated to be about 4,500 people.

Computerfest

Computerfest was a conference for hobbyist computing sponsored by the Midwest Affiliation of Computer Clubs (MASC) that was first held in Cleveland, Ohio in 1976.

Personal Computing Show!

Personal Computing magazine sponsored three shows called the Personal Computer Show! in 1977. The First Western Show was in Los Angeles in March, The First Eastern Show was in Philadelphia in April/May and The Fist New England Show was in Boston in June.

West Coast Computer Faire

Founded by Jim Warren, the first Faire was in April 1977 at the Civic Auditorium in San Francisco, California. Warren was also the first editor of *Dr. Dobb's Journal*. The Faire was oriented to computer hobbyists and personal computer users. Attendance at the first Faire was almost 13,000 with around 180 exhibiters. In 1983 the Faire was sold to Prentice-Hall, Inc.

COMDEX Shows

A show organized by a company called The Interface Group Inc., which was founded by Sheldon Adelson in 1973. COMDEX is an acronym for COMputer Dealers' EXposition. The show is oriented to computer manufacturers , dealers and distributors. The first show was held in December 1979. COMDEX has become one of the largest computer shows in the world. It holds two major shows a year. In winter the show is in Las Vegas and in the spring it is in another major city such as Atlanta, Chicago or Toronto.

COMDEX was purchased in 1995 by the Softbank Corporation, a large Japanese software distribution company. A subsidiary named Softbank COMDEX Inc., now operates the COMDEX shows.

Consumer Electronics Show (CES)

A show for wholesalers and retailers of consumer electronic products. It is held twice a year, Las Vegas in January and Chicago in June.

National Computer Conference (NCC)

It is the largest annual computer show in the data processing industry and is sponsored by the American Federation of Information Processing Societies (AFIPS). The first national show was held in New York in June 1973. Personal computing was recognized as a special theme by having a Personal Computer Fair and Exposition at the June 1977 conference in Dallas, Texas. Following this, microcomputer products became a significant group of the exhibitors.

Other Shows

The first MacWorld Expo was held in Boston in August 1985. CeBIT is one of the largest technology expositions in the world, and is held annually in Hanover, Germany.

19.3 -- Historical Organizations

A significant interest is developing in the history of computing within the history of science and technology. A number of organizations have been organized to support and encourage this specialty.

The American Computer Museum

The American Computer Museum was founded by Barbara and George Keremedjiev in May 1990 and is located in Bozeman, Montana. The museum displays a history of computer technology from ancient Babylonian and Egyptian times to recent personal computers. The museum has items such as: an IBM Tabulating Machine, IBM 1620 transistorized computer, DEC PDP-8 minicomputer, IBM System/360, various personal computers from the 1970's and 1980's and other items of comparative technology. Special displays are introduced periodically such as items from the Smithsonian Institution and a rare mathematical book collection of Erwin Tomash's.

For further information contact Barbara Keremedjiev at The American Computer Museum Ltd., 234 East Babcock street, Bozeman, Montana, USA. MT 59715. The phone number is (406) 587-7545.

Charles Babbage Institute (CBI) -- Center For The History Of Information Processing

The CBI was established in 1977. A principal in the 1977 founding was Erwin Tomash. The first headquarters were established in Palo Alto, California in April 1978. In the fall of 1980 CBI moved to the University of Minnesota.

The Institute has archives of historic materials from pioneers, companies and organizations related to computing. It also has photographic archives, oral histories, reprint series and data bases of computing literature , company developments and information on archival holdings outside the CBI.

A brochure describing the activities and services of the Institute is available. A quarterly *Newsletter* detailing current activities at the CBI and elsewhere relating to the history of computing is also available.

For further information, write to the Charles Babbage Institute, University of Minnesota, 103 Walter Library, 117 Pleasant Street SE, Minneapolis, Minnesota, USA MN 55455. The phone number is (612) 624-5050.

Computer History Association of California

Kip Crosby founded the Computer History Association of California (CHAC) in April 1993. It was an educational organization that studied, preserved and popularized the history of electronic computing in the State of California. The Association published a quarterly newsletter called *The Analytical Engine*. The first issue published was July/September 1993 (Volume 1.1). The association encountered difficulties in 1997 and the last issue of *The Analytical Engine* was Volume 4.1, Winter 1997.

The Computer Museum

The Computer Museum was located in Boston, Massachusetts and opened in late 1984. The museum had a number of early North American and British computers and a collection of manuals, developers notes, technical memoranda, marketing materials etc. A collection of thousands of photographs, hundreds of video film titles and approximately 1,200 books were also at the museum.

In 1999, The Computer Museum closed at its 300 Congress Street, Boston location and joined forces with the museum of Science, Boston. The Computer museum's collection of artifacts resides at The Computer Museum History Center in Moffett Field, California.

The Computer Museum History Center

The museum was established in 1996 and is dedicated to the preservation and celebration of computing history. "It is home to one of the largest collection's of computing artifacts in the world, a collection comprising over 3,000 artifacts, 2,000 films and videotapes, 5,000 photographs, 2,000 linear feet of catalogued documentation and gigabytes of software."

For further information write to The Computer Museum History Center, Building T-12A, Moffett Federal Airfield, Mountain View, California, USA. 94035.

Historical Computer Society

The Historical Computer Society was founded by Tamara Greelish and David A. Greelish who is the director and editor. The society publishes a quarterly magazine called *Historically Brewed*. The first issue was August/September 1993. The magazine headline states "Since 1993 --What's New in What's Old! --The Enthusiast's Magazine of Computer History Nostalgia." The society encountered difficulties in 1996, and the last issue of *Historically Brewed* was Issue #9.

IBM Archives

The International Business Machines (IBM) Corporation established IBM Archives as a separate department in 1974. The Archives primary mission is to preserve materials documenting the history and evolution of IBM and its predecessor companies. The holdings have limited information on items after 1982.

For further information contact IBM Archives, 400 Columbus Avenue, Valhalla, New York, USA. 10595

Intel Museum

Intel Corporation established the Intel Museum in Santa Clara, California in February 1992. The museum naturally concentrates on Intel history and has interactive video and real-time automated displays. Exhibits describe how semiconductor chips are made and used.

For further information contact: Intel Museum, Robert Noyce Building, 2200 Mission College Boulevard, Santa Clara, California, USA. CA 95052-8119. The phone number is (408) 765-0503.

Lawrence Livermore National Laboratory

The Lawrence Livermore National Laboratory in California has a Computer Museum with a collection of early computers. Some of those computers are: Control

Data 660, Cray-1, DEC PDP-8, DEC PDP-10 and Commodore Pet.

Additional information can be obtained from Lawrence Livermore Computer Museum, Pod F North, 1401 Almond Avenue, Livermore, California, USA. CA 94550.

Motorola Museum of Electronics

Motorola founded the museum in September 1991. The museum traces the history of the company and its products. It has historical exhibits, audiovisual displays and interactive computer displays.

For further information contact: Motorola Museum of Electronics, 1297 East Algonquin Road, Schaumburg, Illinois, USA. 60196-1065. The phone number is (847) 576-6559.

National Museum of American History

The Smithsonian Institution's National Museum of American History, has a number of historical computers in its collection. Some of the significant early mainframe computers and minicomputers are: CRAY-1, ENIAC, Harvard Mark 1, UNIVAC and the DEC PDP-8. The personal computer holdings include: Apple I and II computers, Altair 8800 and other S-100 bus microcomputers, IBM PC, Apple Macintosh, Osborne 1, Sun-2 workstation, TRS-80 and Xerox Alto. An Information Age permanent exhibition displays a number of these computers, communications technology and interactive workstations are provided for use by visitors.

For further information contact Jon Eklund, Curator of Computer Technology, American Museum of American History, Smithsonian Institution,, Washington, USA. DC 20560. The phone number is (202) 357-2828.

Stanford University Libraries

The Stanford University Libraries acquired the historical collections of Apple Computer, Inc., in November 1997. The collection includes books, documents, hardware, memorabilia, periodicals, software and videotapes.

Other historical organizations are being started. Some of these are the Computer History Association of Delaware, Computer History Association of Iowa and Cornell University Classic Computer Club.

19.4 -- Retailers and Software Distributors

Arrowhead Computer Co. --The Computer Store

This was the first personal computer store and was founded by Dick Heiser in West Los Angeles, California in July 1975. He was a dealer for Altair microcomputers.

Byte Shop's

Paul Terrell became the MITS Altair 8800 representative for Northern California in 1975 and founded Byte Shop's in December. The Byte Shop store opened in Mountain View, California and by 1976 he had 76 retail stores around the country. Terrell gave Steve Jobs an order for 50 Apple I microcomputer boards in April 1976. This led eventually to the founding of Apple Computer.

CompUSA

Mike Henochowicz and Errol Jacobson founded a software store called Soft Warehouse Inc., in 1984. The company then started opening superstores and selling computer hardware. In 1991, the company went public and changed its name to CompUSA Inc. The two founders left around this time. James F. Halpin became president and chief executive officer of what is now the largest personal computer products retailer in the USA. In 1998, the company started providing custom-built personal computers and purchased the Computer City chain of stores from Tandy.

Computer Mart

Computer Mart was the first computer retail store in New York and was founded by Stanley Veit in February 1976. Computer Mart sold Apple Computer, IMSAI, Processor Technology Sol, SwTPC and other computers. It also had a large selection of magazines and books.

Subsequently a number of other stores opened in other parts of the USA with the same name, but different owners. They had an informal association then a company called XYZ Corporation was formed to enable consolidated purchasing and coordinate assistance. However changing market conditions resulted in the store closing in 1979.

ComputerLand

William H. Millard founded Computer Shack in September 1976 and the first president was Edward Faber. The company franchised personal computer retail stores which at the beginning emphasized the IMSAI 8080 microcomputer. However due to threatened litigation by Tandy Corporation for encroaching on the Radio Shack trade name, the name was changed to ComputerLand in early 1977.

The first franchised store opened in February 1977 in Morristown, New Jersey. By 1984 it was an international chain of approximately 700 franchised computer stores.

However in 1984 litigation over a convertible loan to IMS Associates (a holding company for ComputerLand) resulted in Millard losing 20 percent of the ownership. Millard then relinquished control of ComputerLand and moved to the Pacific island of Saipan in the spring of 1986.

However by 1994 ComputerLand had financial problems. This resulted in the franchise and distribution portions of ComputerLand being sold to the Merisel, Inc., for close to $100 million. Merisel is a dominant distributor of hardware and software.

Computer Store

This was a retail computer store organization founded by Dick Brown and Sid Harrigan in 1975. It had the entire USA East Coast distribution rights for Altair microcomputers

Lifeboat Associates

Lifeboat Associates was founded by Larry Alcoff and Tony Gold. The company became a major distributor of software in the late 1970's. One of the early major products was the CP/M operating system. They also developed modified versions of CP/M for North Star Computers and other floppy disk drive systems. After the release of the IBM PC computer the company also became a distributor for Microsoft MS DOS. Lifeboat renamed the operating system SB-86 (Software Bus 86).

Merisel

Robert Sherwin Leff started a software distribution company called Robwin in April 1980. Robwin is a contraction of Leff's first and middle names. In the summer of that year he formed a partnership with David Wagman and in January 1981 the company name was changed to Softsel Computer Products. Softsel became one of the world's largest distributors of personal computer software during the 1980's. The company acquired Microamerica in 1990, a major hardware distributor, and changed the corporate name to Merisel, Inc.

Personal Software

Dan Fylstra and Peter Jennings founded Personal Software, Inc. in February 1978. The company started initially by marketing a game developed by Jennings called Microchess and other game programs. Personal Software marketed personal computer software in a manner similar to book publishers. He would acquire the rights of software from the developer and add sophisticated marketing and distribution. A significant agreement in 1979, was for the distribution rights for VisiCalc as developed by Daniel Bricklin and Robert Frankston of Software Arts. The Personal Software company name was changed to VisiCorp in early 1982.

Software Plus

Was founded by George Tate and Hal Lashlee in 1980 as a discount mail-order software service.

19.5 -- Networks and Services

Networks

CIE Net

The concept of the CIE (Community Information Exchange) Net was introduced by Mike Wilber at the First West Coast Computer Faire in April 1977. Wilber proposed this early telecommunications network for personal computer users, as a means of exchanging programs and files of data.

Ethernet

A short local area network developed at the Xerox PARC (Palo Alto Research Center) for the Alto research computer in late 1973. Two of the principals in the development were Robert Metcalfe and David Boggs. It is a multi-access broadcast system used to link many computer systems with a single coaxial cable. The control of access is by a system called CSMA/CD (Carrier Sense Multiple Access with Collision Detection). Xerox released Ethernet for use outside the corporation in 1979.

Internet

The Internet evolved from the U.S. Government ARPANET as described in Section 2.6 and a sister network funded by the National Science Foundation (NSF) for academic purposes, called the NSFnet. U.S. Government funding was terminated for the ARPANET in 1989 and for the NSFnet in April 1995. This resulted in the emergence of commercial networks that became loosely known as the Internet.

Internet is a global network of more than 34,000 smaller networks, public and private. The number of users worldwide was estimated to be 304 million by the U.S. Department of Commerce in March 2000. The network uses the TCP/IP (Transport Control Protocol/Interface Program) software specifications. TCP/IP was developed by the U.S. Department of Defense for communications between computers.

ISDN (Integrated Services Digital Network)

ISDN was introduced in the mid 1980's to combine voice and data with about 20 times the throughput.

PCNET

After the West Coast Computer Faire in April 1977, Dave Caulkins organized a group to design a net which became known as PCNET (Personal Computer NETwork).

UseNet

UseNet (for "*Us*ers *net*work") is a collection of UNIX systems that is a public forum for the exchange of ideas and news articles. A group for computer historians worldwide is accessed by: *alt.folklore.computers*.

World Wide Web

Tim Berners-Lee started developing the concepts for the World Wide Web (WWW) for application to the Internet at CERN in Switzerland, in 1989. CERN is an acronym for Conceil Européen pour la Researche Nucléaire, (the international council which started the laboratory). CERN is now known as the European Particle Physics Laboratory. He had previously developed a personal hypertext program called Enquire in 1980. The World Wide Web evolved from a desire to link in a hypertext manner similar to the capabilities of his Enquire program, information resources at CERN and other laboratories around the world. The information resources included graphics, sound, text and video. Berners-Lee designed the hypertext markup language (HTML) for encoding documents, the hypertext transfer protocol (HTTP) and the universal resource locator (URL) for addressing documents (the *WWW.whatever* system). This became the basis for a global hypertext system that Berners-Lee named the World Wide Web (WWW).

The WWW program was released to a limited number of NeXT computer users at CERN in March 1991 and for NeXT users outside CERN in August. The program was then converted for other computer users and demonstrated in San Antonio, Texas in December 1991. This resulted in an increasing number of users during 1992 and the following years.

On-line Services

America Online

Stephen M. Case first got involved with an on-line service for Atari games in 1983, at a company called Control Video Corporation. The company started to encounter financial difficulties in 1984. This resulted in Case and entrepreneur James V. Kimsey obtaining control of the company and changing its name to Quantum Computer Services in 1985. Quantum arranged distribution and marketing agreements with Apple Computer, Commodore, IBM and Tandy to bundle Quantum's on-line service called Q-Link with their computers. In 1989, the company introduced a new service called America Online. Then in October 1991, the company changed its name to America Online (AOL), Inc., and went public in March 1992 with 187,000 subscribers. By 1997, AOL had more than 8 million subscribers and had become a dominant on-line service.

In September 1997, AOL and a telephone company called WorldCom, signed a three-way agreement. This resulted in WorldCom purchasing CompuServe, an exchange whereby AOL obtained the 2 to 3 million CompuServe subscribers and WorldCom obtained AOL's Internet division, and AOL committed to a long-term phone pact with WorldCom. AOL is now the largest on-line service provider with about 14 million subscribers.

AOL acquired Netscape Communications Corporation for $4.2 billion in November 1998. AOL also signed at the same time, a licensing and marketing agreement with Sun Microsystems Inc. The company wanted to dramatically step up its Web presence by using Netscape's Web sites and software.

CompuServe

Jeffrey Wilkins founded CompuServe Corporation as a computer service to an insurance company in 1970. Wilkins expanded the company into a computer time sharing service to provide personal computer users access to large data banks of information in 1978. This new service was named MicroNET. It also became a popular means whereby computer users could exchange information via bulletin boards.

The company was acquired by H & R Block Inc., and the service name changed from MicroNET to CompuServe Information Services in 1980. The company provided a diverse range of services which included: bulletin board, business data, computer technology information, educational reference, electronic mall shopping, entertainment, home and health, money markets, news and weather, sports and travel.

By 1997, CompuServe was encountering difficulties and was losing market share to companies such as America Online and Microsoft. This resulted in the company being sold to WorldCom in September.

Delphi

General Videotex Corporation started an on-line service named Delphi in 1982. To expand the service, General Videotex purchased the *Byte* magazine BIX (Byte Information Exchange) on-line service in January 1992.

Dow Jones News/Retrieval

The Dow Jones News/Retrieval service was provided by Dow Jones and Company Inc.

GEnie

GEnie is an on-line service started by the General Electric Company in 1985. GEnie is an acronym for "General Electric network for information exchange."

Prodigy

The Prodigy Services Company was founded in 1984 by CBS, IBM and Sears, Roebuck and Company. CBS withdrew its investment in Prodigy in 1987.

The Source

The Telecomputing Corporation of America was founded by William von Meister in 1979. The company provided an on-line information service for personal and business users called The Source. The *Reader's Digest* organization purchased the company in 1981. Then The Source was purchased by CompuServe and its subscribers merged that service in 1989.

19.6 -- Associations

ACE (Advance Computing Environment)

ACE was founded in April 1991 by twenty companies led by Compaq, DEC, Microsoft, MIPS Computer Systems and the Santa Cruz Operation. The alliance was formed to develop a new standard for an "advanced computing environment." The consortium's intent was to develop compliant systems that would accommodate Microsoft's object oriented operating system, RISC personal computers and UNIX operating systems. However Compaq left the ACE initiative in April 1992.

APDA (Apple Programmer's and Developer's Association)

Apple Computer and the A.P.P.L.E. (Apple Pugetsound Program Library Exchange) user group sponsored the founding of the Apple Programmer's and Developer's Association (APDA) in August 1985. Apple Computer promoted the founding of APDA initially to disseminate Macintosh computer technical information for outside software development. The organization provides up-to-date technical information and preliminary material from Apple Computer. Apple Computer assumed full control of APDA in December 1988.

DMTF (Desktop Management Task Force)

DMTF was formed in 1992 to develop a common framework for managing PC computer systems. The founders were Digital Equipment Corporation, Hewlett-Packard, IBM, Intel, Microsoft, Novell, SunSoft and SynOptics Communications. The group developed the Desktop Management Interface (DMI) specification for desktop hardware and software in March 1974. It is also developing a Management Information Format (MIF) that will act as a central database for DMI.

EIA (Electronic Industries Association)

The EIA is a North American standards organization for computer equipment. A popular standard is RS-232C for connecting computers to modems and terminals.

ICC (International Color Consortium)

The ICC was formed in March 1994 by Adobe, Agfa, Apple Computer, Kodak, Microsoft, Silicon Graphics, Sun Microsystems and Taligent. The consortium was formed to establish a common device profile format for color.

MIC (Microfloppy Industry Committee)

MIC was an association of over 30 companies that was formed in May 1982 to establish a microfloppy media standard. This committee was responsible for the adoption of the 3.5-inch hard-cartridge disk standard in September 1982.

Microcomputer Industry Trade Association

This association was founded in 1979.

OSF (Open Software Foundation)

The Open Software Foundation (OSF) was an organization of initially seven companies that was formed in May 1988. The foundation which included the Digital Equipment Corporation, Hewlett-Packard and IBM was formed to develop a unified UNIX operating system standard independent of AT&T. It was a reaction to the alliance formed in April 1988 between AT&T and Sun Microsystems to develop a unified UNIX system.

Since the release of UNIX in 1969, many different versions of the operating system had been developed by various organizations. The different versions had unique characteristics that affected the portability of the operating system and inhibited the development of application software.

PowerPC Alliance

Apple Computer, IBM and Motorola made an early announcement of the formation of the PowerPC Alliance in July 1991, followed by a final agreement in October. The companies formed the alliance to jointly develop new emerging technologies. Five of the joint initiatives were microprocessor technology, object-oriented technology, multimedia technology, interconnectivity and networking to provide an open systems environment. IBM and Motorola agreed to jointly develop a broad range of microprocessors based on IBM's POWER architecture.

Apple and IBM agreed to form a joint venture company called Taligent that would develop a new operating system. Joseph M. Guglielmi who had been an IBM executive on the OfficeVision and OS/2 software development, was selected to head the new Taligent company in early 1992. The new operating system would be based on object-oriented design principles incorporated in the Apple Computer Pink project.

Apple and IBM also agreed to form another joint venture company called Kaleida to develop multimedia technologies. Nathaniel Goldhaber who had worked with Apple Computer on new multimedia technology was selected to head the new Kaleida company in mid 1992.

The three companies also agreed to develop a PowerOpen environment project for support of IBM AIX and Macintosh applications. Finally Apple and IBM agreed to develop solutions that would allow their systems to interact more effectively.

Software Publishers Association

The Software Publishers Association (SPA) is an organization representing individual software developers. It was founded to promote and protect the rights of software publishers. It has a special interest in the illicit copying of software products. The SPA merged with another association in January 1999, to form the Software & Information Industry Association.

VESA (Video Electronics Standards Association)

VESA is a group of companies who have organized to establish industry standards for video cards and monitors.

19.7 -- Other Companies and People

The following provides information on companies and people of significance in the personal computer industry, that have not been detailed previously in the book.

Companies

Amazon

Amazon Inc., was founded by Jeffrey Bezos in July 1994 as an online Internet bookstore. The website became operational in July 1995. It has since expanded its products to include items such as: compact disks, computer products and auctions. In 1998 it had sales of $610 million. The November 29, 1999 issue of *Forbes* magazine estimated Bezos's wealth at $7.3 billion. Then in the December 27 issue of *Time* magazine, he was featured on the cover as "Person of the Year."

ATI

Kwok Yuen Ho, Lee Lau and Benny Lau founded ATI Technologies Inc. in Toronto, Canada in 1985. ATI went public in November 1993. Ho is the chief executive officer of the company that is now a leading producer of desktop graphics systems.

Brother

Brother International Corporation is a leading supplier of fax machines, labeling devices, printers and word processors. Hiromi Gunji became the chairman, president and chief executive officer in 1986.

Cisco

Cisco Systems, Inc., was founded in 1984 by Leonard Bosack, Sandra Lerner and three colleagues Kirk Lougheed, Greg Satz and Bill Westfield. The company is a market leader in the networking industry. The products are data routers, network software, servers and switches. The company received financing from venture capitalist Donald Valentine's company Sequoia Capital in 1988. Valentine purchased controlling interest in the company and hired John P. Morgridge as president and chief executive officer. In 1990, the company went public and Bosack and Lerner left the company. Morgridge became chairman in 1995 and John T. Chambers who had joined the company in 1991, succeeded him as president and CEO. Cisco has acquired numerous companies since 1993 and has become a dominant supplier of products for the Internet. This has also resulted in Cisco attaining a market capitalization in 2000, as the most valuable company in the world.

Computer Intelligence

Computer Intelligence Infocorp is a computer market research unit of the Ziff-Davis Publishing Company.

Dataquest

Dataquest is a subsidiary of the Gartner Group Inc. The company is a dominant supplier of information technology research. It provides forecasts, market analysis, statistics and summaries of research to subscribers.

eBay

Pierre Omidyar founded AuctionWeb in the fall of 1995, that shortly after became eBay Inc. In early 1996, Omidyar brought in Jeff Skoll: a friend and Stanford M.B.A. as a partner. Meg Whitman who was a marketing executive, was also recruited as the chief executive officer in early 1998. The company had become very successful and went public in September 1998.

Egghead

Victor Aldhadeff founded Egghead Discount Software in Bellevue, Washington in 1984. Aldhadeff made buying software a more friendly experience. The company subsequently became Egghead, Inc., and went public in 1988. A rapid expansion to 112 stores led to financial losses and Aldhadeff's resignation in 1989. Egghead closed all its physical stores in 1989 and now sells exclusively through the Internet. Terence M. Strom is now the president and chief executive officer.

General Magic

Bill Atkinson, Andy Hertzfeld and Marc Porat founded General Magic in 1990. The company developed an operating system called Magic Cap for personal digital assistants (PDA's).

GT Interactive

Joseph J. Cayre was a principal in the founding of GT Interactive Software Corporation in 1992. The company is a major publisher of game software and became a public company in 1995.

id Software

Adrian Carmack, John Carmack, Tom Hall and John Romeo founded id Software, Inc., in 1990. The company develops game software and uses shareware to release the games.

Kingston Technology

David Sun and John Tu founded Kingston Technology Corporation in 1987. The company is a major supplier of memory enhancement chip clusters and other personal computer peripherals.

Logitech

Daniel Borel and Pierluigi Zappacosta met at Stanford University and founded Logitech International SA in 1981. The company obtained the rights to a Swiss-designed mouse in 1981 and is now a major producer of mice and other computer input devices.

Lycos

Michael Mauldin developed a search engine algorithm for gathering data and information on the World Wide Web. CMG@Ventures bought the rights to Mauldin's technology and founded Lycos Inc. in 1995. The company name is derived from the wolf family of spiders that Mauldin compared his algorithm to. Lycos owns a network of related but separate Web sites and has become a popular portal for Internet users. Robert Davis is the chief executive officer and the company went public in April 1996.

Macromedia

Macromedia, Inc., is a multimedia software company formed in 1992 from the merger of three other companies. John C. Colligan is the chairman, president and chief executive officer.

Micronics

Micronics Computers, Inc., makes motherboards for original equipment manufacturers. Dean Chang, Minsiu Huang and Harvey Wong founded the company in 1986.

Paul Allen Group

Paul Allen, the co-founder of Microsoft, founded the Paul Allen Group in March 1994, as an umbrella organization for his various ventures. Vern Raburn, who had also been at Microsoft, became the president and chief executive officer. One of his ventures is the Asymetrix Corporation that he founded in 1985 to create multimedia development software.

Pixar

Steven P. Jobs founded Pixar Animation Studios after purchasing the computer-graphics division of Lucasfilm for $10 million in 1986. Pixar developed a highly successful computer-animated film called *Toy Story* in 1995. The film was created using software systems also developed by Pixar.

PointCast

PointCast Inc., was founded in 1992 as a company that delivers information and news through the Internet. The company introduced PointCast Network in 1996. The chief executive officer is Christopher R. Hassett.

SAP AG

Systems Applications Products (SAP) is a German software company founded in April 1972. The company was founded by former IBM system engineers Hans-Werner Hector, Dietmar Hopp, Hasso Plattner, Klaus Tschira and Claus Wellenreuther. Its first product was a financial accounting system called R/1. Following this a mainframe version called R/2 was released and subsequently a more extensive version called R/3 was released. SAP went public in 1988 and is now one of the world's largest software companies.

SCI Systems

Olin B. King and two associates founded Space Craft, Inc. in 1961 and subsequently renamed the company SCI Systems, Inc. The company initially specialized in contract engineering and electronic systems for the U.S. space industry. In the 1970's SCI produced sub-assemblies for IBM terminals and in 1981 they began building circuit boards for the IBM Personal Computer. In 1984, the company started producing complete personal computers for other companies that resold them under their own label. SCI Systems is not a well known personal computer company. It is however, a major computer technology company.

Softbank

Softbank Corporation was founded by Masayoshi Son in 1981. The company is the largest distributor of software in Japan. Softbank acquired COMDEX shows in 1995 and Ziff-Davis Publishing Company in early 1996.

Wang Laboratories

An Wang founded Wang Laboratories, Inc., in 1951. The company became one of the largest suppliers of dedicated screen-based word processing systems in the mid 1970's. However, it did not make a successful transition to personal computers, and the appointment of his oldest son as president in 1986 aggravated the problems at the company. The company then began to encounter severe financial difficulties in the late 1980's that resulted in it filing for bankruptcy protection in mid 1992. The company is now focusing on software and systems consulting and management.

People

Alsop

Stewart Alsop is a columnist for the *Fortune* magazine and runs a fall computer industry conference called Agenda. He is also a partner in a venture capital firm.

Brockman

John Brockman and his partner Katinka Matson developed a successful New York literary agency called Brockman, Inc., in the 1970's. In 1983, Brockman announced a transition in his company to also be a software agency. He represented software developers in the marketing of software to major New York publishers. Brockman is also the author of a number of books that includes *Digerati: Encounters with the Cyber Elite* [128].

Dyson

Esther Dyson was a New York security analyst who joined Ben Rosen's investment company and took over the management of his *Rosen Electronics Letter*. In the early 1980's, during the time Ben Rosen became the chairman of Compaq Computer Corporation and a director of Lotus Development Corporation, Dyson purchased the newsletter and renamed it *Release 1.0*. She is editor of the monthly newsletter and runs an annual spring conference called

PC Forum. These and other activities place Dyson in a significant role in the communication and dissemination of information related to the personal computer industry.

Knuth

Donald E. Knuth is a scholar noted for his major contributions to computer technology. His multi-volume text entitled *The Art of Computer Programming* has received wide acclaim.

Negroponte

Nicholas Negroponte is a founder and director of the Massachusetts Institute of Technology Media Laboratory. The laboratory is focusing on the study and experimentation of future forms of human and computer communication. In 1992, Negroponte co-founded the *Wired* magazine of which he is a senior columnist. Negroponte is also the author of *being digital* [203].

Winblad

Ann Winblad is the co-founder of a venture capital company specializing in start-up software firms. She has also been a friend of Bill Gates for a number of years.

Chapter 20 Miscellaneous Items

20.1 ... Bits and Bytes

Early Books

Computer Lib and Dream Machines [204] was a book authored and self-published by Theodor H. Nelson in 1974. It was part of a grass-roots, anti-establishment, computer-power-to-the-people movement in California in the early 1970's. Other organizations related to this movement were the People's Computer Company and Community Memory.

An Introduction to Microcomputers [44] was another early influential book. It was written and self-published by Adam Osborne. It provided a practical introduction to microprocessors.

Knowledge Navigators

Knowledge navigators evolved from the Vannevar Bush future Memex machine, Douglas C. Engelbart's augmentation of man's intellect concepts, Alan Kay's Dynabook concept and Theodor Nelson's hypertext ideas.

Theodor H. Nelson described a "unified system for complex data management and display" called Xanadu in his 1974 book *Computer Lib and Dream Machines*. Nelson also proposed a hypertext text manipulation system.

In 1987, John Sculley proposed "a wonderful fantasy machine called the Knowledge Navigator ..." in his book *Odyssey* [73]. Sculley described it as a tool to "drive through libraries, museums, databases, or institutional archives." This would enable an individual to convert "vast quantities of information into personalized and understandable knowledge."

Another term used by Bill Gates in his book [89] *The Road Ahead* is "spatial navigation." Gates used the term to describe the process of navigating on the information highway.

CD-ROM and Multimedia applications

Multimedia is a technology that can integrate text, graphics and sound in a single document. The use of multimedia in personal computers received significant impetus from the development of CD-ROM technology in the mid 1980's (see Section 20.3).

Two of the earliest companies to develop CD-ROM applications were Cytation and the Activenture Corporation. The early developments were text only. Gary Kildall of Digital Research, also founded Activenture Corporation in 1984 and renamed the company KnowledgeSet Corporation in 1985. Activenture developed a CD-ROM disk based on the *Academic American Encyclopedia* that was offered by Grolier in 1985. Tom Lopez founded the Cytation company in late 1984 and developed a CD-ROM disk called CD-Write that incorporated reference texts. The Cytation company was purchased by Microsoft Corporation in January 1986. These early developments evolved to include multimedia products by companies such as Microsoft (see Sections 12.6 and 15.1).

Other Items

Stephen Wozniak created the mythical Zaltair microcomputer for the First West Coast Computer Faire in April 1977. Wozniak did it as a prank against the established MITS Altair computer. He printed twenty thousand brochures describing an "incredible dream machine" and attributed the computer to Ed Roberts, President of MITS, Inc. It also caused concern for Steve Jobs of Apple Computer who was not aware of the prank.

20.2 ... Reference Sources

Bibliographies

An extensive bibliography of microprocessor literature is contained in the "Architecture of Microprocessors" by Robert C. Stanley, in the *Encyclopedia of Microcomputers* - Volume 1, pages 269 to 281 [236].

A bibliography of various output devices is contained in "Computer Output Devices" by David Bawden, in the *Encyclopedia of Microcomputers* - Volume 3, pages 360 to 362 [236].

History

An *Encyclopedia of Computer History* is a disk stored hyperstack history for use on MS-DOS or Microsoft Windows, that has been developed by Mark Greenia. It is available from Lexikon Services, 3241 Boulder Creek Way, Antelope, California, CA 95842.

20.3 ... Standards and Specifications

Bus Architecture Standards

Reference Section 17.8

Communications

The ITU (International Telecommunications Union) is an international organization that establishes standards for communication devices, such as modems. The V.32 standard is for communication at 9,600 bps. The V.32 bis is an expanded standard for communication at 14.4 Kbps. The V.34 is the latest standard for communication at 28.8 Kbps without compression and up to 115.2 Kbps with compression. K56flex is a 56 Kbps technology developed by Lucent Technologies and Rockwell International. X2 is a technology developed by the U.S. Robotics Corporation and operates at 57 Kbps.

The HTML (HyperText Markup Language) specification was developed by Tim Berners-Lee at CERN (CERN is the European Particle Physics Laboratory) in 1991. It was written as part of the World Wide Web (WWW) to facilitate communication among high-energy physicists. The specification was refined in 1993/94 by Dan Connelly who wrote the Standard Generalized Markup Language (SGML) and Document Type Definition (DTD) specifications for HTML.

Compact Disk

Standard specifications were created by Sony Corporation of Japan and N. V. Philips of the Netherlands following compact disk development in 1976. In 1982 a Red Book specification for CD audio and in 1983 a Yellow Book specification for CD-ROM were released. This was followed by a Green Book for CD-I (Compact Disk - Interactive) specifications detailing requirements for interleaving of audio and video data. An Orange Book has also been issued for specifications on CD-R (Compact Disk - Recordable) drives.

In November 1985 a group of companies met to consider a standard format for organizing data files on CD-ROM. This standard became known as the High Sierra Proposal and was approved as ISO (International Standard Organization) standard 9660.

CD-ROM XA was an audio and graphics standard developed by Sony and Philips in late 1988.

Graphics

IGES (Initial Graphics Exchange Standard) is a standard for defining the format of geometry for CAD (Computer Assisted Drafting) data. It enables the communication of CAD data files between different computer platforms.

Hardware

Apple Computer, IBM and Motorola announced in November 1994 that they would develop a common hardware specification for PowerPC based computers.

Memory

EMS (Expanded Memory Specification) was developed by Intel Corporation. It was first called Above Board then the specification became known as LIM (Lotus, Intel and Microsoft) due to an agreement between those companies in June 1985. This specification was developed to overcome the PC memory limitation of 640K bytes.

XMS (eXtended Memory Specification) was the result of a collaborative effort between AST Research, Intel, Lotus and Microsoft and was introduced in August 1988. XMS is a more sophisticated system than EMS and utilizes a eXtended Memory Manager (XMM) to control the transfer of data in memory above one megabyte.

Video Standards

MDA (Monochrome Display Adapter) was introduced in August 1981 for the IBM PC computer. It had a maximum resolution of 720 by 350 pixels with 1 color.

CGA (Color Graphics Adapter) was introduced by IBM in August 1981 for the IBM PC computer. It had a maximum resolution of 320 by 200 pixels with 4 colors.

MGA (Hercules Monochrome Graphics Adapter) was introduced in 1982 and had a maximum resolution of 720 by 350 pixels with 1 color.

EGA (Enhanced Graphics Adapter) was introduced by IBM in late 1984 for the PC AT computer. It had a maximum resolution of 640 by 350 pixels with 16 colors.

PGA (Professional Graphics Array) was introduced by IBM in 1984 and had a maximum resolution of 640 by 480 pixels with 256 colors.

VGA (Video Graphics Array) was introduced by IBM in April 1987 for the PS/2 series of computers. It has a maximum resolution of 640 by 480 pixels with 16 colors.

MCGA (Multi-Color Graphics Array) was introduced by IBM in 1987 and had a maximum resolution of 640 by 480 pixels with 2 colors.

8514/A was an IBM standard which was introduced in 1987 and had a maximum resolution of 1,024 by 768 pixels with 256 colors.

Super VGA was introduced in 1989 and has a maximum resolution of 800 by 600 pixels with 16 colors.

XGA (Extended Graphics Array) was introduced by IBM in 1990 and has a maximum resolution of 1,024 by 768 pixels with 256 colors.

20.4 ... Terminology: Clarification and Origins

Bit

The first documented use of the term "Bit" was in an internal memo by John W. Tuckey at the AT&T Bell Laboratories in January 1947. It was used in a table defining terms to describe an individual character in the binary system.

The first use of the term bit in a publication was by Claude E. Shannon in the *Bell System Technical Journal*, July 1948 issue. Shannon used the term bit to describe a binary digit in a measuring system related to a mathematical theory of communication. Shannon stated that it was "a word suggested by J. W. Tuckey".

Bug

The first use of the term "Bug" in computer technology is attributed to Grace Murray Hopper during the summer of 1945. During the development of the Harvard Mark II computer an operational failure was caused by a moth getting in one of the computer relays. Subsequently when determining an operational problem it would on occasion be described as "debugging the computer".

There has been some question as to the origin of the problem on the Mark II computer in 1945. Also Fred R. Shapiro has shown in a Commentary of the April 1994 issue of *BYTE* magazine that the terms "bug" and "debugging" had usage prior to 1945.

Byte

The first use of the term "Byte" was on the IBM Stretch computer development in an internal memo written in June of 1956. Initially it referred to any number of parallel bits from one to six. Shortly after August 1956 the Stretch computer design was changed to incorporate 8-bit bytes.

The first published reference using the term byte was in the *IRE Transactions on Electronic Computers*, June 1959 issue. It was stated by W. Buchholz that "The term is coined from bite, but respelled to avoid accidental mutation to bit".

Hypertext

Theodore H. Nelson is attributed as being the originator of the term hypertext. Nelson described it as "non-sequential writing -- text that branches and allows choices to the reader," non-sequential information retrieval and perusal. It is related to his Xanadu text manipulation system.

Microcomputer

Gilbert Hyatt stated in an article "Micro, Micro: Who made the Micro ?" [328] that "I trademarked the name microcomputer" in 1968.

Microprocessor

In an article entitled "A History of Microprocessor Development at Intel" [342], Robert N. Noyce and Marcian E. Hoff stated that "The term "Microprocessor," first came into use at Intel in 1972." Prior to the development of LSI computer chips the term "microprocessor" referred to a processor of a microprogrammed computer.

Minicomputer

The origin of the term minicomputer is attributed to John Leng of Digital Equipment Corporation (DEC). Leng was responsible for establishing a DEC presence in the United Kingdom in the mid 1960's. In reporting sales activity he stated that "Here is the latest minicomputer activity in the land of miniskirts as I drive around in my Mini Minor." The term then became used at DEC and throughout the industry.

Personal Computer

Competing claims have been made in two periodicals. An *IEEE Computer* magazine article states that "Kay and others ... coined the term "personal computer" at Xerox PARC in 1973." *BYTE* magazine claims to have "coined the term, in our May 1976 issue."

Portable Computer

The term portable computer requires clarification because of the way it has evolved into different forms or categories over the years. Generally it is a computer that can be carried by an individual from place to place. Initially it could only be operated from an AC power source, today it is battery operated. The following are the different designations for the various types of portable computers:

The *transportable* computer, also known as a *luggable*, is the earliest form of a portable computer. It weighed fifteen pounds or more and generally ran off an AC power source. The IBM 5100, announced in 1975, is the first commercially produced portable computer (weighed 50 pounds). The Osborne and Compaq portable computers released in 1981 and 1982 are other examples of this type.

The *laptop* computer weighing around seven to fourteen pounds, that could be placed on a persons lap was the next stage in portable computing. It could be operated from a battery or an AC power source. It is also a generic name used by the press for today's lighter notebook computers. The GRiD Compass that was introduced in 1982 is an example of the laptop type of portable

The *notebook* computer is the current popular portable computer and typically weighs between five to eight pounds. The physical size is similar to a paper notebook, with approximate dimensions of nine by twelve by two inches. Of increasing popularity are the "light and thin" versions with a weight of less than five pounds and a thickness of less than 1.5 inches. The NEC UltraLite introduced in 1989 is an early example of this type of portable.

Other variations of even lighter portables are known as the *Ultraportable* with a standard keyboard, and the *Subnotebook* weighing four pounds or less with a smaller keyboard and screen.

The *tablet* is a portable type of computer that does not have a keyboard and uses a pen or stylus for input. The GRiDPad portable announced in 1989 is an early example of this type.

Silicon Valley

The term "Silicon Valley" has been attributed to Don Hoeffler, who was a journalist for *Electronic News*. Hoeffler used the term to refer to the region of Santa Clara Valley, south of Stanford University in a series of articles in January 1971.

Vaporware

The term "Vaporware" came into vogue during 1983/84 to describe software that was announced, highly publicized, long awaited but still not available. Some of this may have been attributed to the Microsoft announcement, prolonged development and delayed release of Windows software. The term was first coined by the *InfoWorld* publication.

Winchester (hard drive)

The term "Winchester" was an early internal code name for a sealed hard disk developed by IBM between 1969 and 1973. It was derived from a disk storage unit development that had two spindles, each with a disk capacity of 30 megabytes. The unit was initially called "30 - 30" and because of the similar designation to a popular rifle from the Winchester Company it became known as the Winchester hard drive.

Wintel

A long-standing relationship between Microsoft and Intel resulted in the term "Wintel"." It refers to Microsoft Widows running on an Intel microprocessor.

Blank page.

Appendix A: Some Technical Details of Various Personal Computers

Apple

Apple II The computer used a MOS 6502 microprocessor operating at 1 MHz with 8K bytes of ROM and 4K bytes of dynamic RAM, expandable to 48K. The integral video system could display 24 rows of 40 characters in upper case only. Each character was a 5 by 7 dot matrix. The display also had a graphics mode with high and low resolutions. In low resolution the 40 horizontal by 48 vertical locations could be in one of 15 colors. In high resolution the screen could display a maximum of 280 horizontal by 192 vertical positions in 4 colors. The video system also included a mixed mode in both low and high resolutions, with graphics and four lines of text at the bottom of the screen.

Apple IIc The computer used a 65C02 microprocessor operating at 1.023 MHz with 16K bytes of ROM and 128K bytes of RAM. The unit had one 5.25-inch 140K byte Alps floppy disk drive. The monitor could display 24 lines of 40 or 80-column text. The video system also had three graphic modes. The low-resolution mode had 40 horizontal by 48 vertical pixels in 16 colors. The high-resolution mode had 280 horizontal by 192 vertical pixels in 6 colors. Then a double-high-resolution mode had 560 horizontal by 192 vertical pixels in 16 colors. The single 5.25-inch Alps half-height floppy disk drive was built-in. The disk controller was the IWM (Integrated Woz Machine) chip as used on the Macintosh computer. The disks were single-sided with 35-tracks and 16-sectors.

Apple IIe The computer used a MOS 6502A microprocessor operating at 1 MHz with 16K bytes of ROM and 64K bytes of RAM, expandable to 128K. The storage system supported six 140K byte 5.25-inch floppy disk drives. The terminal could display 24 lines of 40-column text in both uppercase and lowercase characters. Each character was a 5 by 7 dot matrix. The computer had two standard graphic modes. A low-resolution mode produced 40 horizontal by 48 vertical pixels with 16 colors. The standard high-resolution mode produced 280 horizontal by 192 vertical pixels with 6 colors. The video system retained the Apple II mixed mode in both low and high resolutions, with graphics and four lines of text at the bottom of the screen. Seven slots were available for peripheral boards. The unit included an additional 60-pin auxiliary slot on the motherboard that provided for one of two optional cards.

The two optional cards could either display 80 columns of text or both 80 columns of text display and an extension of memory to 128K bytes. The extended memory 80-column option card and suitable software enabled double-density graphics in both low and high-resolution modes.

Apple IIGS The computer used a Western Design Center W65C816 microprocessor operating at 2.8 MHz with 128K bytes of ROM and 256K bytes of RAM, expandable to 8 megabytes. The microprocessor had a 24-bit address bus and an 8-bit data bus. The unit had two operating modes: a native mode of 2.8 MHz and an Apple IIe emulation mode of 1.02 MHz. The storage system included support for both 3.5-inch 800K byte and 5.25-inch 140K byte floppy disk drives. The terminal could display: 24 lines of 40 or 80-column text. Graphic modes varied from low-resolution with 40 horizontal by 48 vertical pixels in 16 colors, to super-high-resolution with 640 horizontal by 200 vertical pixels in 4 colors. An Ensoniq digital synthesizer chip with 64K bytes of dedicated RAM provided new sound capabilities. The unit had a 44-pin memory expansion slot and seven 50-pin slots for peripherals.

Apple III The computer used a Synertek 6502A microprocessor operating at 2 MHz with 96K bytes of memory, expandable to 128K. The unit included one built-in 5.25-inch, 143K byte floppy-disk drive. The terminal could display 24 lines of 80-column text. Graphic modes were 560 horizontal by 192 vertical pixels in monochrome and 280 horizontal by 192 vertical pixels in 16 colors or 16 shades of gray. The computer had four slots for peripheral cards that had certain compatibility with Apple II cards. An aluminum chassis acted as a radio frequency shield and as a heat sink that eliminated the need for a cooling fan. Three additional disk drives could be daisy-chained from a rear connector. The Apple III floppy-disks had a new 16-sector format (compared with a 13-sector format on the Apple II).

Lisa The computer used a Motorola MC68000 microprocessor operating at 5 MHz with 1 megabyte of RAM. The storage system had two Twiggy 5.25-inch 860K byte floppy disk drives and a separate 5 megabyte Winchester-type hard disk named ProFile. The display was a 12-inch monochrome monitor with a resolution of 720 by 364 pixels. The computer used four additional microprocessors. Two from National Semiconductor controlled the keyboard and mouse, a 6504 controlled the two floppy disk drives and the hard disk controller had a Z-8. The floppy disk drives maintained a constant data density between the outer and inner tracks.

Macintosh The computer used a Motorola MC68000 microprocessor operating at 7.83 MHz with 64K bytes of ROM and 128K bytes of RAM. The storage system had one integral 3.5-inch 400K byte floppy disk drive from Sony. The 3.5-inch disks were single-sided with 80 tracks. The storage system recorded data at a constant rate similar to the Lisa Twiggy drive. A connector provided for an optional second external 3.5-inch floppy disk drive. The display was a 9-inch monochrome monitor with a resolution of 512 horizontal by 342 vertical square pixels. There were no expansion slots, but two high-speed serial ports provided for connection of peripherals. The separate keyboard had 58 keys (59 in the international version), but no function or cursor keys.

Macintosh II The computer used a Motorola MC68020 microprocessor operating at 16 MHz with a floating-point coprocessor and one megabyte of RAM, expandable to 8 MB. The unit had an open NuBus architecture developed at MIT with six slots for plug-in boards. The 13-inch color or 12-inch monochrome monitor could display 640 by 400 pixels as compared to the 512 by 342 pixel display on the previous Macintosh computers. The display system could have either 16 or 256 colors or shades of gray.

Macintosh Portable The computer used a Motorola CMOS 68000 microprocessor operating at 16 MHz with 1 megabyte of RAM expandable to 2 megabytes. The unit included a built-in 3.5-inch 1.4 megabyte floppy-disk drive. The portable had an Active Matrix Liquid Crystal Display with a screen resolution of 640 by 400 pixels.

The computer was 15.25 inches wide by 14.83 inches deep and the height varied from 2 to 4 inches, front to back. The weight without a hard disk drive was 13.75 pounds. The keyboard had 63 keys with a unique arrangement for locating either a trackball pointing device or an 18-key numeric keypad on the left or right hand side of the keyboard. The computer used lead acid batteries with a power management system controlled by a 6502 microprocessor. This provided 8 to 10 hours of operation on a single battery charge.

Atari

The Atari 400 computer used a MOS 6502 microprocessor operating at 1.8 MHz with 8K bytes of RAM, expandable to 16K. The display had 16 lines of 32 characters and a high-resolution mode of 320 by 192 pixels.

Commodore

PET 2001 The PET 2001 computer used a MOS 6502 microprocessor, had 14K bytes of ROM and 4K bytes of RAM, expandable to 8K. The 9-inch black and white CRT could display 25 lines of 40 characters and the 73-key keyboard included a numeric keypad.

PET 4000 The PET 4000 computer used a MOS 6502 microprocessor, had 18K bytes of ROM and 16K bytes of RAM, expandable to 32K. The configuration included a 12-inch green-phosphor display with a capability of showing 25 lines of 40 characters.

Cromemco

Z-1 The Cromemco Z-1 computer used a Zilog Z-80 microprocessor and had 8K bytes of memory.

Z-2 The Cromemco Z-2 computer used the S-100 Bus with 21 slots for plug-in boards.

CTC Datapoint

The CTC Datapoint 2200 computer terminal used a bit-serial processor in TTL logic with shift register memory. The terminal had a keyboard, 12-line screen display and two cassette tape drives.

Digital Group

The Digital Computer System was available with either an AMD 8080A, MOS 6502, Mostek 6800 or Zilog Z-80 microprocessor. The processor board had 2K bytes of RAM. An input/output board and a video interface board that provided a 16 line by 32 character display were part of the system. An optional 8K static RAM board was available to increase the memory capacity.

EPD

The System One computer kit had 82 integrated circuits, 1K bytes of memory, expandable to 8K and used 57 instructions.

HAL

The HAL-4096, was a home-built 16-bit computer with 16 registers and 4K bytes of magnetic core memory from a surplus IBM 1620.

Heath

H8 The H8 computer used an Intel 8080 microprocessor, had 1K bytes of ROM and memory cards were available in 4K-byte increments to 32K. It had a unique 50-pin bus for expansion cards.

H11 The H11 computer used a DEC LSI-11 microcomputer board and had a 4K by 16-bit word memory that could accommodate up to 20K words.

Heath/Zenith-89 The Heath/Zenith-89 computer used a Zilog Z-80 microprocessor operating at 2.048 MHz and had 16K bytes of RAM, expandable to 48K. The 12-inch screen could display 24 lines of 80 characters. The keyboard subsystem and video display used an additional Z-80 microprocessor.

Accessories The H9 video terminal had a 67-key keyboard and the 12-inch CRT could display 12 lines of 80 characters.

Hewlett-Packard

The HP 9831A desktop computer used a HP BPC microprocessor and had 8K bytes of memory, expandable to 32K. The unit had a 32-character LED display, cassette tape drive and a keyboard.

Hyperion

The Hyperion portable computer had an Intel 8088 microprocessor and 256K bytes of RAM. The storage system had two 5.25-inch 320K byte double-density dual-sided floppy-disk drives. The unit had a 7-inch amber-on-gray monitor that could display 25 lines of 80 characters. The 28 pound portable unit included an integrated display, two floppy-disk drives and a detachable keyboard. The operating system was MS-DOS and the unit was priced at $4,950 (Canadian).

IBM

5100 The 5100 portable computer had a built-in keyboard, monitor and magnetic tape cartridge storage system. The unit used an IBM developed microprocessor and memory consisted of 48K bytes of ROM and 16K bytes of RAM, expandable to 64K. The monitor could display 16 lines of 64 characters.

Scamp The Scamp computer used an IBM Palm microcontroller and had 64K bytes of RAM. The unit had an integrated keyboard and a small CRT that could display 16 rows of 64 columns.

System/3 Model 6 This BASIC computer system used monolithic circuit technology and had 8K bytes of silicon chip memory, expandable to 16K. The design included a keyboard for direct data input, a 14-inch diameter disk storage drive with capacities of 2.5 to 9.8 megabytes and a printer. The CRT could display 15 lines of 64 characters.

IMSAI

The IMSAI 8080 computer used the Intel 8080A microprocessor and included 4K bytes of memory. It also had a heavy duty power supply, commercial-grade paddle switches on the front panel and a six-slot S-100 Bus motherboard, expandable to 22 slots.

Kenbak-1

The Kenbak-1 computer incorporated small and medium-scale integrated circuits with a memory capacity of 256 bytes. It had three programming registers, five addressing modes and very limited input/output capabilities.

Mark-8

The Mark-8 computer used an Intel 8008 microprocessor, had 256 bytes of memory, expandable to 16K bytes by adding memory boards. The unit had provision for six circuit board modules.

MITS

The Altair 8800 computer used an Intel 8080 microprocessor and had 256 bytes of memory. The Altair cabinet had space for eighteen cards. The basic unit had two slots, one for the CPU card and one for the 256 byte memory card.

MOS

The KIM-1 computer used a MOS 6502 microprocessor, had 2K bytes of ROM that contained the system executive, and 1K bytes of RAM. The unit included an audio cassette interface, a 23-key keypad and a six-digit LED display.

North Star

The Horizon-1 computer used a Zilog Z-80 microprocessor and had 16K bytes of RAM. It also included one or two 5.25-inch floppy disk drives and a 12-slot S-100 Bus motherboard.

Noval

The Noval 760 computer used an Intel 8080A microprocessor, had 3K bytes of ROM and 16K bytes of RAM.

NRI

The NRI 832 kit had 52 integrated circuits, 32 bytes of memory, used 15 instructions.

Ohio Scientific

The OSI 400 computer used either a Motorola MC6800, MOS 6501 or 6502 microprocessor. The unit had 512 bytes of ROM and up to 1K bytes of programmable memory was available. The board was designed for use with a 48-line expansion bus system.

PolyMorphic

Poly 88 The Poly-88 computer used an Intel 8080 microprocessor, had 512 bytes of RAM and 1K bytes of ROM. It also used the S-100 Bus with four slots, had a cassette interface

System 8813 The System 8813 computer used an Intel 8080 microprocessor and the video could display 16 lines of 64 characters. The main unit could accommodate from one to three floppy-disk drives with 90K capacity each.

Processor Technology

Sol-10 The Sol-10 Terminal Computer used an Intel 8080A microprocessor and had 1K bytes of PROM, 1K bytes of RAM and 1K bytes of video RAM. The computer could generate a 16-line by 64-character video display. The unit had a built-in 85-key keyboard and one slot for a S-100 Bus board.

Sol-20 The Sol-20 computer used an Intel 8080 microprocessor, had 1K bytes of ROM, 8K bytes of RAM and 1K bytes of video RAM. It also had a heavier-duty power supply and five S-100 Bus expansion slots

REE

The Micral computer used an Intel 8008 microprocessor and included 256 bytes of RAM, expandable to one kilobyte. The system used a 60-bit data bus called Pluribus.

Scelbi

The Scelbi-8H computer used an Intel 8008 microprocessor with up to 4K bytes of memory. The Scelbi-8B business computer had up to 16K bytes of memory.

Sphere

Sphere produced three computer models that used the Motorola MC6800 microprocessor; the Hobbyist, Intelligent and BASIC. The Hobbyist had 4K bytes of memory and a keyboard, the Intelligent had more features and the Basic had 20K bytes of memory and full extended BASIC software.

SwTPC

The SwTPC 6800 Computer System used a Motorola MC6800 microprocessor, had 2K bytes of memory, expandable to 16K. A 1K byte ROM chip contained the mini-operating system. The unit also contained an SS-50 bus for eight interface boards.

Tandy/Radio Shack

TRS-80 The TRS-80 computer used a Zilog Z-80 microprocessor, had 4K bytes of ROM and 4K bytes of RAM, expandable to 62K. The system included a separate cassette tape recorder and a 12-inch black and white monitor that could display 16 lines of 64 characters.

A Disk System Expansion Unit had a disk controller which could support up to four drives. The diskette had a capacity of 83K bytes (formatted) with 35 tracks of 2,500 bytes per track divided into ten sectors.

Texas Instruments

The TI-99/4 computer used a TI TMS9900 microprocessor and had 16K bytes of RAM.

Xerox

Alto The Alto computer system had a processor/disk storage cabinet, graphics display unit, keyboard and a mouse. The processor unit had a 16-bit custom-made processor similar to the Data General Nova 1220. The operating speed was 400,000 instructions per second. Memory incorporated an address space of 64K 16-bit words, expandable to 256K. The processor/disk storage cabinet had two 3-megabyte hard-disk drives. The graphics unit featured an 8 inch horizontal by 10 inch vertical black and white display. It used a bit-mapped raster scan with a resolution of 606 pixels horizontally by 808 pixels vertically that could display 60 lines of 90 characters. The computer used a detachable keyboard supplemented by a mouse incorporating three buttons for program control.

Appendix B: Versions of DOS

Version 1.0

This is the version released with the IBM PC computer in August 1981. The program had 4,000 lines of code and required 136K bytes of storage when fully installed.

Versions 1.1 and 1.25

Version 1.1 was released in May 1982 for PC computers with two disk drives and to enable writing on both sides of the diskette. This doubled the diskette capacity from 160K to 320K bytes. Microsoft released MS DOS Version 1.25 for IBM compatible computers.

Version 2.0

Was released in March 1983 for the IBM PC XT computer. Initially IBM only wanted an upgrade to accommodate a hard disk on the new PC XT computer. However Paul Allen decided to enhance the software and add additional features. Some of those new features were hierarchical directories, a limited form of multitasking for printing and certain UNIX features. The floppy disk format was also changed from 8 sectors per track to 9 sectors that increased the storage capacity to 360 K bytes. The program now had 20,000 lines of code.

Version 2.1 (PC-DOS)

Was introduced in November 1983 for the IBM PC Junior.

Version 2.11

Was released in March 1984 for international customers. It supported the variations in date format and comma instead of a period for decimal point notation. The program was translated into many languages and sold worldwide. By June 1984, Microsoft had licensed MS-DOS to 200 manufacturers.

Version 3.0

Was released in August 1984 for the IBM PC AT computer. It accommodated high-density 1.2 megabyte diskettes. It also added features such as: RAM disk, volume names and the ATTRIB command. The program now consisted of 40,000 lines of code.

Version 3.1

Was released in March 1985 to add networking capabilities for IBM PC Network and Microsoft MS-Net.

Version 3.2

Was released in December 1985 to accommodate 3 1/2 inch disk drives.

Version 3.3

Was released in April 1987 to accommodate multiple partitions, improved foreign character support, support for a 32 megabyte hard disk and the IBM PS/2 series of computers.

Version 4.0

Was released in June 1988. It had a graphical user interface or DOS shell for use with a mouse, EMS support and support for hard disk partitions over 32 megabytes. The program required 1.1 MB of storage.

Version 5.0

Was released in June 1991. It featured a MS-DOS kernel which loaded into HMA (high memory), 80386 memory management, a new shell, improved multitasking capabilities, limited switching capabilities, on-line help and a full-screen editor.

Version 6.0

Was released in March 1993. The new features added were: DoubleSpace disk compression, improved memory management, backup and defrag utilities licensed from

Symantec Corporation, anti-virus and undelete utilities licensed from Central Point Software Inc., Multi-configuration support on start-up, an improved SmartDrive, three new commands (Move, Deltree and Choice) and a file transfer utility called Interlink. The program required 8.4 MB of storage.

Version 6.2

Was released in November 1993. DoubleGuard that is a part of DoubleSpace is added to verify data integrity before writing to disk.

Versions 6.21 and 6.22

Version 6.21 was released in February 1994 to delete DoubleSpace after litigation with Stac Electronics. Version 6.22 was released in June 1994 to add a new disk compression program called DriveSpace.

MS-DOS was essentially made obsolete with the introduction of Microsoft Windows 95 in August 1995. Windows 95 integrates the functionality of MS-DOS and the previous Windows graphical interface.

Bibliography

BOOKS

Chapter 1 ... Development of the Computer

1.1 The Original Digital Computers

1. Burkes, Alice R. and Arthur W. Burkes.
   The First Electronic Computer.
   Ann Arbor, Mich.: University of Michigan Press, 1989.
2. Ceruzzi, Paul E.
   Reckoners: The Prehistory of the Digital Computer, from Relays to the Stored Program Concept, 1935-1945.
   Westport, Conn.: Greenwood Press, 1983.
3. Ceruzzi, Paul E.
   A History of Modern Computing.
   Westport, Conn.: Greenwood Press, 1998.
   -- Covers the transition to the microprocessor and personal computer.
4. Goldstine, Herman H.
   The Computer from Pascal to von Neumann.
   Princeton, N.J.: Princeton University Press, 1972.
5. Ifrah, Georges.
   The Universal History of Computing: From the Abacus to the Quantum Computer.
   New York: John Wiley & Sons, 2000.
6. Lavington, Simon S.
   Early British Computers.
   Manchester, England: Manchester University Press, 1980.

7. Lukoff, Herman.
From Dits To Bits: A Personal History of the Electronic Computer.
Portland, Oregon: Robotics Press, 1979.

8. Lundstrom, David E.
A Few Good Men from Univac
Cambridge, Mass.: MIT Press, 1987.

9. Metropolis, N., J. Howlet and Gian-Carlo Rota (Editors).
A History of Computing in the Twentieth Century.
New York: Academic Press, 1980.

10. Mollenhoff, Clark R.
Atanasoff: Forgotten Father of the Computer.
Ames, Iowa: Iowa State University Press, 1988.

11. Moreau, R.
The Computer Comes of Age: The People, the Hardware, and the Software.
Cambridge, Mass.: MIT Press, 1984.

12. Pylyshyn, Zenon W. (Editor).
Perspectives on the Computer Revolution.
Englewood Cliffs, N.J.: Prentice-Hall, 1970.

13. Randell, Brian.
The Origins of Digital Computers: Selected Papers.
New York: Springer-Verlag, 1975.

14. Redmond, Kent C. and Thomas M. Smith
Project Whirlwind: The History of a Pioneer Computer.
Bedford, Mass.: Digital Press, 1981.

15. Ritchie, David.
The Computer Pioneers: The Making of the Modern Computer.
New York: Simon & Schuster, 1986.

16. Shurkin, Joel.
Engines of the Mind: A History of the Computer.
New York: Norton & Company, 1984.

17. Stern, Nancy.
From Eniac to Univac: An Appraisal of the Eckert-Mauchly Computers.
Bedford, Mass.: Digital Press, 1981.

18. Wildes, Karl L. and Nilo A. Lindgren.
A Century of Electrical Engineering and Computer Science at MIT, 1882 - 1982.
Cambridge, Mass.: MIT Press, 1985.
19. Wilkes, Maurice.
Memoirs of a Computer Pioneer.
Cambridge, Mass.: MIT Press, 1985.
20. Williams, Michael R.
A History of Computing Technology.
Englewood Cliffs, N.J.: Prentice-Hall, 1985.

1.2 IBM

21. Bashe, Charles J., Lyle R. Johnson, John H. Palmer and Emerson W. Pugh.
IBM's Early Computers.
Cambridge, Mass.: MIT Press, 1986.
22. Belden, Thomas and Marva Belden.
The Lengthening Shadow: The Life of Thomas J. Watson.
Boston, Mass.: Little, Brown and Company, 1962.
23. Pugh, Emerson W.
Memories That Shaped An Industry: Decisions Leading to IBM System/360.
Cambridge, Mass.: MIT Press, 1984.
24. Pugh, Emerson W., Lyle R. Johnson and John H. Palmer.
IBM's 360 and Early 370 Systems.
Cambridge, Mass.: MIT Press, 1991.
25. Pugh, Emerson W.
Building IBM, Shaping an Industry and Its Technology.
Cambridge, Mass.: MIT Press, 1995.
26. Rodgers, William.
Think: A Biography of the Watsons and IBM.
New York: Stein and Day, 1969.
27. Watson, Thomas J. Jr.
A Business and Its Beliefs: The Ideas That Helped Build IBM.
New York: McGraw-Hill, 1963.
28. Watson, Thomas J. Jr. and Peter Petre.
Father Son & Co.: My Life at IBM and Beyond.
New York: Bantam Books, 1990.

1.3 Technology

29. Braun, Ernest and Stuart Macdonald.
Revolution in Miniature: The History and Impact of Semiconductor Electronics.
Cambridge, Mass.: Cambridge University Press, 1982 (2nd Edition).

30. Gilder, George.
Microcosm: The Quantum Revolution in Economics and Technology.
New York: Simon & Schuster, 1989.
-- "A prescient look inside the expanding universe of economic, social and technological possibilities within the world of the silicon chip."

31. Morris, P.R.
A History of the World Semiconductor Industry.
London, England: Peter Peregrinus, 1990.

32. Reid, T.R.
The Chip: How Two Americans Invented the Microchip and Launched a Revolution.
New York: Simon and Schuster, 1984.

33. Riordan, Michael and Lillian Hoddeson.
Crystal Fire: The Birth of the Information Age.
New York: W.W. Norton & Company, 1997.

1.4 Software

34. Rosen, Saul (Editor).
Programming Systems and Languages.
New York: McGraw-Hill, 1967.

35. Sammet, Jean E.
Programming Languages: History and Fundamentals.
Englewood Cliffs, N.J.: Prentice-Hall, 1969.

36. Wexelblat, Richard L. (Editor).
History of Programming Languages.
New York: Academic Press, 1981.

Chapters 2, 4 & 11 ... Personal Computing, Transition. to Microcomputers & Competitive Computers.

37. Bardini, Thierry.
Bootstrapping: Douglas Engelbart, Coevolution, and the Origins of Personal Computing.
Stanford: Stanford University Press, 2000.

38. Clark, W.A. and C.E. Molnar.
A Description of the LINC.
In "*Computers in Biomedical Research Corporation.*"
Editors: Ralph W. Stacy and Bruce D. Waxman.
New York: Academic Press, 1965; Chap. 2.

39. Evans, Christopher.
The Making of the Micro: A History of the Computer.
London, England: Victor Gollancz, 1981.
-- The "Micro" in the title refers to the microprocessor. The book does not contain any history of the microcomputer. The author died in 1979 prior to completion of the book, which was finished by Tom Stonier.

40. Freiberger, Paul and Michael Swaine.
Fire in the Valley: The Making of the Personal Computer.
Berkeley, Calif.: Osborne/McGraw-Hill, 1984.
-- A second edition was published by McGraw-Hill in 2000 that has been updated, with expanded coverage and numerous photographs.

41. Goldberg, Adele (Editor).
A History of Personal Workstations.
New York: ACM Press, 1988.

42. Kemeny, John G.
Man and the Computer.
New York: Charles Scribner's Sons, 1972.
-- Has a chapter describing Dartmouth Time Sharing System.

43. McCarthy, John.
Time-Sharing Computer Systems.
In: "*Management and the Computer of the Future*".
Greenberger, Martin (Editor).
New York: MIT Press and John Wiley, 1962, pp. 220-248.

44. Osborne, Adam.
An Introduction to Microcomputers. 3 Vols.
Berkeley Calif.: Osborne/McGraw-Hill, 1977.

45. Veit, Stan.
Stan Veit's History of the Personal Computer: From Altair to IBM, A History of the PC Revolution.
Asheville, N.C.: WorldComm, 1993.

Chapters 3 & 8 ... Microprocessors

46. Byman, Jeremy.
Andrew Grove and the Intel Corporation.
Greensboro, NC: Morgan Reynolds, 1999.

47. Jackson, Tim.
Inside Intel: Andy Grove and the Rise of the World's Most Powerful Chip Company.
New York: Dutton, 1997.

48. Kaye, Glynnis Thompson (Editor).
A Revolution in Progress: A History of Intel to Date.
Santa Clara, Calif.: Intel Corporation, 1984.

49. Malone, Michael S.
The Microprocessor: A Biography.
New York: Springer-Verlag, 1995.

50. Noyce, Robert N. and Marcian E. Hoff.
A History of Microprocessor Development at Intel.
Santa Clara, Calif.: Intel Corp., Public AR-173.

51. YU, Albert.
Creating the Digital Future: the Secrets of Consistent Innovation at Intel.
New York: Free Press, 1998.

Chapters 5 & 10 ... Apple Computer

52. Amelio, Gil and William L. Simon.
On the Firing Line: My 500 Days at Apple.
New York: HarperBusiness, 1998.

53. Butcher, Lee.
Accidental Millionaire: The Rise and Fall of Steve Jobs at Apple Computer.
New York: Paragon House, 1988.

54. Carlton, Jim.
Apple: The Inside Story.
New York: Random House, 1997.

55. Deutschman, Alan.
The Second Coming of Steve Jobs.
New York: Broadway Books, 2000.

56. Garr, Doug.
Woz: The Prodigal Son of Silicon Valley.
New York: Avon Books, 1984.

57. Gassée, Jean-Louis.
The Third Apple: Personal Computers and the Cultural Revolution.
Orlando, Florida: Harcourt Brace Jovanovich, 1985.

58. Greenberg, Keith Elliot.
Steven Jobs & Stephen Wozniak: Creating the Apple Computer.
Woodbridge, Conn.: Blackbirch Press, 1994.
-- A juvenile book for young readers.

59. Kawasaki, Guy.
The Macintosh Way.
New York: HarperPerennial, 1990.

60. Kendall, Martha E.
Steve Wozniak: Inventor of the Apple Computer.
New York: Walker & Co., 1994.

61. Kunkel, Paul.
AppleDesign: The Work of the Apple Industrial Design Group.
New York: Graphics Inc., 1997.

62. LeVitus, Bob and Michael Fraase.
Guide to the Macintosh Underground: Mac Culture From the Inside.
Indianapolis, In.: Hayden Books, 1993.

63. Levy, Steven.
Insanely Great: The Life and Times of Macintosh, the Computer That Changed Everything.
New York: Viking, 1994.

64. Linzmayer, Owen W.
The Mac Bathroom Reader.
Alameda, Calif.: Sybex, 1994.

65. Linzmayer, Owen W.
Apple Confidential: The Real Story of Apple Computer, Inc.
San Francisco, Calif.: No Starch Press, 1999.

66. Malone, Michael S.
Infinite Loop: How Apple, the World's Most Insanely Great Computer Company went Insane.
New York: Doubleday, 1999.

67. Menuez, Doug, Marcos Kounalakis and Paul Saffo.
Defying Gravity: The Making of Newton.
Hillsboro, Oregon: Beyond Words Publishing, 1993.
-- A photo-journalistic depiction of the Newton development.

68. Moritz, Michael.
The Little Kingdom: The Private Story of Apple Computer.
New York: William Morrow, 1984.

69. Price, Rob.
So Far: The First Ten Years of a Vision.
Cupertino, Calif.: Apple Computer, Inc., 1987.

70. Rose, Frank.
West of Eden: The End of Innocence at Apple Computer.
New York: Viking, 1989.

71. Rozakis, Laurie.
Steven Jobs: Computer Genius.
Vero Beach, Florida: Rourke Enterprises, 1993.
-- A juvenile book for young readers.

72. Schmucker, Kurt J.
The Complete Book of LISA.
New York: Harper & Row, 1984.

73. Sculley, John with John A. Byrne.
Odyssey: Pepsi to Apple ... A Journey of Adventure, Ideas and the Future.
New York: Harper & Row, 1987

74. Thygeson, Gordon.
Apple T-Shirts: A Yearbook of History at Apple Computer.
Scotts Valley, Calif.: Pomo Publishing, 1998.

75. Various.
Maclopedia: The ultimate reference on everything Macintosh!
Indianapolis, In.: Hayden Books, 1996.

76. Weyhrich, Steven.
Apple II History.
Self Published, Zonker Software, 1991.

77. Young, Jeffrey S.
Steve Jobs: The Journey is the Reward.
Glenview, Illinois: Scott, Foresman, 1988.

Chapters 6 & 12 ... Microsoft

78. Andrews, Paul.
How the Web Was Won: Microsoft from Windows to the Web: The Inside Story of How Bill Gates and His Band of Internet Idealists Transformed a Software Empire.
New York: Broadway Books, 1999.

79. Auletta, Ken.
World War 3.0: Microsoft and its Enemies.
New York: Random House, 2001.

80. Bick, Julie.
All I really Need to Know in Business I Learned at Microsoft.
New York: Pocket Books, 1997.
-- A personal guide to business management.

81. Boyd, Aaron.
Smart Money: The Story of Bill Gates.
Greensboro, N.C.: Morgan Reynolds, 1995.
-- A juvenile book for young readers.

82. Brinkley, Joel and Steve Lohr.
U.S. V. Microsoft: The Inside Story of the Landmark Case.
New York: McGraw-Hill, 2001.

83. Cusumano, Michael A. and Richard W. Selby.
Microsoft Secrets: How the World's Most Powerful Software Company Creates Technology, Shapes Markets, and Manages People.
New York: Free Press, 1995.

84. Dearlove, Des.
Business the Bill Gates Way: 10 Secrets of the World's Richest Business Leader.
New York: Amacon, American Management Association, 1999.

85. Dickinson, Joan D.
Bill Gates: Billionaire Computer Genius.
Springfield, N.J.: Enslow Publishers, 1997.
-- A juvenile book for young readers.

86. Edstrom, Jennifer and Marlin Eller.
Barbarians Led by Bill Gates -- Microsoft from the Inside: How the World's Richest Corporation Wields its Power.
New York: Henry Holt and Company, 1998.

87. Ferry, Steven.
The Building of Microsoft.
Mankato, MN.: Smart Apple Media, 1999.

88. Foreman, Michael.
Bill Gates, Software Billionaire.
Parsippany, NJ.: Crestwood House, 1999.

89. Gates, Bill with Nathan Myhrvold and Peter Rinearson.
The Road Ahead.
New York: Viking, 1995.

90. Gates, Bill with Collins Hemingway.
Business @ The Speed of Thought: Using a Digital Nervous System.
New York: Warner books, 1999.

91. Gatlin, Jonathan,
Bill Gates: The Path to the Future.
New York: Avon Books, 1999.

92. Heilemann, John.
Pride Before the Fall: The Trials of Bill Gates and the End of the Microsoft Era.
New York: HarperCollins, 2001.

93. Ichbiah, Daniel and Susan L. Knepper.
The Making of Microsoft.
Rocklin, Calif.: Prima Publishing, 1991.

94. Lewis, Ted G.
Microsoft Rising ... and other tales of Silicon Valley.
Los Alamitos, Calif.: IEEE Computer Society, 1999.
-- Mainly a reprint of *IEEE Computer* periodical articles, with minimal Microsoft content.

95. Liebowitz, Stan J. and Stephen E. Margolis.
Winners, Losers and Microsoft: Competition and Antitrust in High Technology.
Oakland, Calif.: The Independent Institute, 1999.

96. Lowe, Janet.
Bill Gates Speaks: Insight from the World's Greatest Entrepreneur.
New York: John Wiley & Sons, 1998.

97. Manes, Stephen and Paul Andrews.
Gates: How Microsoft's Mogul Reinvented an Industry and Made Himself The Richest Man in America.
New York: Doubleday, 1993.

98. Marshall, David.
Bill Gates and Microsoft.
Watford, England: Exley, 1994.

99. Microsoft Staff.
Inside Out! Microsoft - In our own words.
New York: Warner Books, 2000.
-- 25th Microsoft Anniversary book, 1975 - 2000.

100. Moody, Fred.
I Sing the Body Electronic: A Year with Microsoft on the Multimedia Frontier.
New York: Viking, 1995.

101. Rivlin, Gary.
The Plot to Get Bill Gates: An Irreverent Investigation of the World's Richest Man...and the People Who Hate Him.
New York: Times Business/Random House, 1999.

102. Rohm, Wendy Goldman.
The Microsoft File: The Secret Case Against Bill Gates.
New York: Times Business/Random House, 1998.

103. Stross, Randall E.
The Microsoft Way: The Real Story of How the Company Outsmarts Its Competition.
Reading, Mass.: Addison-Wesley, 1996.

104. Tsang, Cheryl D.
Microsoft First Generation: The Success Secrets of the Visionaries Who Launched a Technological Empire.
New York: John Wiley & Sons, 1999.

105. Wallace, James and Jim Erickson.
Hard Drive: Bill Gates and the Making of the Microsoft Empire.
New York: John Wiley & Sons, 1992.

106. Wallace, James.
Overdrive: Bill Gates and the Race to Control Cyberspace.
New York: John Wiley & Sons, 1997.

107. Woog, Adam.
Bill Gates.
San Diego: Lucent Books, 1999.

108. Zachary, G. Pascal.
Show-Stopper! The Breakneck Race to Create Windows NT and the Next Generation at Microsoft.
New York: Free Press, 1994.

Chapters 7, 13 & 15 ... Other Software

109. Birnes, William J. (Editor).
Personal Computer Programming Encyclopedia.
New York: McGraw-Hill, 1989.

110. Brooks, Frederick P.
The Mythical Man-Month: Essays on Software Engineering.
Reading, Mass.: Addison-Wesley, 1975.

111. Gosling, James., David S.H. Rosenthal, and Michele J. Arden.
The NeWS Book: An Introduction to the Network/extensible Window System.
New York: Springer-Verlag, 1989.
-- Chapter 3 has a history of Windows development with an emphasis on the relationship to Sun workstations.

112. Grauer, Robert T. and Paul K. Sugrue.
Microcomputer Applications.
New York: McGraw-Hill, 1989.

113. Hoch, Detlev J., Cyriac R. Roeding, Gert Purkert, Sandro K. Linder with Ralph Müller.
Secrets of Software Success: Management Insights from 100 Software Firms Around the World.
Boston, Mass.: Harvard Business School Press, 1999.

114. Hsu, Jeffrey.
Microcomputer Programming Languages.
Hasbrouck Heights, N.J.: Hayden Book Co, 1986.

115. Kemeny, John G. and Thomas E. Kurtz.
Back to BASIC: The History, Competition, and Future of the Language.
Reading, Mass.: Addison-Wesley, 1985.

116. Raymond, Eric S.
The Cathedral & the Bazaar: Musings on Linux and Open Source by an Accidental Revolutionary.
Sebastopol, Calif.: O'Reilly & Associates, 1999.
--A history of hackers and the development of open-source software.

117. Wayner, Peter.
Free For All: How Linux and the Free Software Movement Undercut the High-Tech Titans.
New York: HarperBusiness, 2000.

Chapter 9 ... The IBM Corporation

118. Carroll, Paul.
Big Blues: The Unmaking of IBM.
New York: Crown Publishers, 1993.

119. Chposky, James and Ted Leonis.
Blue Magic: The People, Power and Politics Behind The IBM Personal Computer.
New York: Facts on File Publications, 1988.

120. Dell, Deborah A. and F. Gerry Purdy.
ThinkPad: A Different Shade of Blue: Building A Successful IBM Brand.
Indianapolis: Sams (A Division of Macmillan Computer Publishing), 1999.

121. Ferguson, Charles H. and Charles R. Morris.
Computer Wars: The Fall of IBM and the Future of Global Technology.
New York: Random House, 1993.

122. Garr, Doug.
IBM Redux: Lou Gerstner & the Business Turnaround of the Decade.
New York: HarperBusiness, 1999.

123. Heller, Robert.
The Fate of IBM.
London, England: Little, Brown and Company (UK), 1994.

124. Slater, Robert.
Saving Big Blue: Leadership Lessons & Turnaround Tactics of IBM's Lou Gerstner.
New York: McGraw-Hill, 1999.

Chapter 17 ... Peripherals

125. Durbeck, Robert C. and Sol Sherr.
Output Hardcopy Devices.
San Diego, Calif.: Academic Press, 1988.

Chapter 18 ... Magazines and Newsletters

126. Shirinian, George.
Microcomputing Periodicals: An Annotated Directory.
Toronto, Canada: George Shirinian, 1985.

Chapter 19 ... Other Companies, Organizations & People

127. Adamson, Ian and Richard Kennedy.
Sinclair and the 'Sunrise' Technology.
Harmondsworth, England: Penguin Books, 1986.

128. Brockman, John.
DIGERATI: Encounters with the Cyber Elite.
San Francisco, Calif.: HardWired, 1996.

129. Bronson, Po.
The Nudist on the Late Shift: and Other True Tales of Silicon Valley.
New York: Random House, 1999.

130. Brown, Kenneth A.
Inventors at Work: Interviews with 16 Notable American Inventors.
Redmond, Washington: Tempus Books, 1988.
-- Includes interviews with Marcian E. "Ted" Hoff and Steve Wozniak

131. Bunnell, David with Adam Brate.
Making the CISCO Connection: The Story Behind the Real Internet Superpower.
New York: John Wiley & Sons, 2000.

132. Caddes, Carolyn.
Portraits of Success: Impressions of Silicon Valley Pioneers.
Palo Alto, Calif.: Tioga Publishing, 1986.

133. Carlston, Douglas G.
Software People: An Insider's Look at the Personal Computer Software Industry.
New York: Simon & Schuster, 1985.

134. Clark, Jim with Owen Edwards.
Netscape Time: The Making of the Billion-Dollar Start-Up That Took On Microsoft
New York: St. Martin's Press, 1999.

135. Cohen, Scott.
Zap! The Rise and Fall of Atari.
New York: McGraw-Hill, 1984.

136. Cringely, Robert X.
Accidental Empires: How the Boys of Silicon Valley Make Their Millions, Battle Foreign Competition, and Still Can't Get a Date.
Reading, Mass.: Addison-Wesley, 1992.

137. Cusumano, Michael A. and David B. Yoffie.
Competing On Internet Time: Lessons from Netscape and Its Battle with Microsoft.
New York: The Free Press/Simon & Schuster, 1998.

138. Dale, Rodney.
The Sinclair Story.
London, England: Duckworth & Co, 1985.

139. Datapro Research.
Who's Who in Microcomputing 1984-85.
Delran, N.J.: Datapro Research Corporation, 1984.

140. Dell, Michael with Catherine Fredman.
Direct from Dell: Strategies That Revolutionized an Industry.
New York: HarperBusiness, 1999.

141. Editors (Senior): George Sutton, James R. Talbot and Alan Chai.
Hoover's Guide to Computer Companies: 2nd Edition.
Austin, Texas: Hoover's Business Press, 1996.

142. Ehrbar, Al.
The Verbatim Story: The First Twenty-Five Years.
Lyme, Conn.: Greenwich Publishing Group, 1995.

143. Hall, Mark and John Barry.
Sunburst: The Ascent of Sun Microsystems.
Chicago, Illinois: Contemporary Books, 1990.

144. Hiltzik, Michael.
Dealers of Lightning: XEROX PARC and the Dawn of the Computer.
New York: HarperBusiness, 1999.

145. Jager, Rama D. and Rafael Ortiz.
In the Company of Giants: Candid Conversations with the Visionaries of the Digital World.
New York: McGraw-Hill, 1997.

146. James, Geoffrey.
Success Secrets from Silicon Valley: How to make your teams more effective.
Formerly Titled:
Business Wisdom of the Electronic Elite
New York: Time Business/Random House, 1998.
-- A business management guide.

147. Jessen, Kenneth Christian.
How it All Began: Hewlett-Packard's Loveland Facility.
Loveland, CO.: J.V. Publications, 1999.

148. Kaplan, David A.
The Silicon Boys: and Their Valley of Dreams.
New York: William Morrow & Co, 1999.

149. Kaplan, Jerry.
Startup: A Silicon Valley Adventure.
New York: Houghton Mifflin, 1995.

150. Kenney, Charles C.
Riding The Runaway Horse: The Rise and Decline of Wang.
Boston: Little, Brown and Company, 1992.
-- Has a chapter on "The Personal Computer Revolution."

151. Kenney, Martin and John Seely Brown (Editors).
Understanding Silicon Valley: The Anatomy of an Entrepreneurial Region.
Stanford: Stanford University Press, 2000.

152. Lammers, Susan.
Programmers at Work: Interviews with 19 Programmers Who Shaped the Computer Industry.
Redmond, Washington: Tempus Books, 1986.

153. Laver, Ross.
Random Excess: The Wild Ride of Michael Cowpland and Corel.
New York: Viking, 1998.

154. Lee, J.A.N.
Computer Pioneers.
IEEE Computer Society Press, 1995.

155. Levering, Robert., Michael Katz and Milton Moskowitz.
The Computer Entrepreneurs: Who's Making it Big and How in America's Upstart Industry.
New York: Nal Books, 1984.

156. Lewis, Michael.
The New New Thing: A Silicon Valley Story.
New York: Norton & Company, 1999.
-- A biographical story of Jim Clark (Silicon Graphics and Netscape)

157. Levy, Steven.
Hackers: Heroes of the Computer Revolution.
New York: Doubleday/Anchor Press, 1984.

158. Littman, Jonathan.
Once Upon a Time in ComputerLand: The Amazing, Billion Dollar Tale of Bill Millard's ComputerLand Empire.
Los Angeles, Calif.: Price Stern Sloan, 1987.

159. Mahon, Thomas.
Charged Bodies: People Power and Paradox in Silicon Valley.
New York: Nal Books, 1985.

160. Malone, Michael S.
The Big Score: The Billion-Dollar Story of Silicon Valley.
New York: Doubleday, 1985.

161. Osborne, Adam and John Dvorak.
Hypergrowth: The Rise and Fall of Osborne Computer Corporation.
Berkeley, Calif.: Idthekkethan Publishing, 1984.

162. Packard, David.
The HP Way: How Bill Hewlett and I Built Our Company.
New York: HarperBusiness, 1995.

163. Pearson, Jamie Parker (Editor).
Digital at Work: Snapshots from the first thirty-five years.
Burlington, Mass.: Digital Press, 1992.

164. Peterson, W.E. Pete.
AlmostPerfect: How a Bunch of Regular Guys Built WordPerfect Corporation.
Rocklin, Calif.: Prima Publishing, 1994.

165. Quittner, Joshua and Michelle Slatalla.
Speeding the Net: The Inside Story of Netscape and How it Challenged Microsoft.
New York: Atlantic Press, 1998.

166. Read, Stuart.
The Oracle Edge: How Oracle Corporation's Take-No-Prisoners Strategy Has Created an $8 Billion Software Powerhouse.
Holbrook, Mass.: Adams Media, 1999.
-- Mainly a business management text.

167. Rifkin, Glenn and George Harrar.
The Ultimate Entrepreneur: The Story of Ken Olsen and Digital Equipment Corporation.
Chicago, Illinois: Contemporary Books, 1988.

168. Sigismund, Charles G.
Champions of Silicon Valley: Visionary Thinking from Today's Technology Pioneers.
New York: John Wiley & Sons, 2000.

169. Slater, Robert.
Portraits in Silicon.
Cambridge, Mass.: MIT Press, 1989.

170. Smith, Douglas K. and Robert C. Alexander.
Fumbling The Future: How Xerox Invented, Then Ignored, The First Personal Computer.
New York: William Morrow, 1988.

171. Southwick, Karen.
Silicon Gold Rush: The Next Generation of High-Tech Stars Rewrites the Rules of Business.
New York: John Wiley & Sons, 1999.
-- A business management book about the high-tech industry.

172. Southwick, Karen.
High Noon: The Inside Story of Scott McNealy and the Rise of Sun Microsystems.
New York: John Wiley & Sons, 1999.

173. Spector, Robert.
amazon.com: Get Big Fast.
New York: HarperBusiness, 2000.

174. Spencer, Donald D.
Great Men and Women of Computing, 2nd Edition.
Ormond beach, Florida: Camelot, 1999.

175. Stross, Randall E.
Steve Jobs & The NeXT Big Thing.
New York: Atheneum, 1993.

176. Stross, Randall E.
eBoys: The First Inside Account of Venture Capitalists at Work.
New York: Crown Business, 2000.

177. Swisher, Kara.
aol.com: How Steve Case Beat Bill Gates, Nailed the Netheads, and Made Millions in the War for the Web.
New York: Random House/Time Books, 1998.

178. Texas Instruments.
50 Years of Innovation: The History of Texas Instruments -- A Story of people and their ideas.
Dallas, Texas: Texas Instruments, 1980.
-- A company publication.

179. Thomas, David.
Knights of the New Technology: The Inside Story of Canada's Computer Elite.
Toronto, Canada: Key Porter Books, 1983.

180. Tomczyk, Michael S.
The Home Computer Wars: An Insider's Account of Commodore and Jack Tramiel.
Greensboro, North Carolina: Compute! Publications, 1984.

181. Voyer, Roger and Patti Ryan.
The New Innovators: How Canadians Are Shaping the Knowledge Based Economy.
Toronto, Canada: James Lorimer, 1994.

182. Wilson, Mike.
The Difference Between God and Larry Ellison: Inside Oracle Corporation.
New York: William Morrow, 1997.

183. Young, Jeffrey.
Forbes Greatest Technology Stories: Inspiring Tales of the Entrepreneurs and Inventors Who Revolutionized Modern Business.
New York: John Wiley & Sons, 1998.

184. Young, Robert and Wendy Rohm.
Under the Radar: How Red Hat Changed the Software Business -- and Took Microsoft by Surprise.
Scottsdale, Arizona: Coriolis Group, 1999.

Other ... Bits and Bytes

General

185. Ahl, David H. (Editor).
The Best of Creative Computing: Volumes 1 and 2.
Morristown, New Jersey: Creative Computing Press,6 1976.
-- Vol. 1 consists of material from the first six issues.

186. Ahl, David H. and Carl T. Helmers, Jr. (Editors).
The Best of Byte: Volume 1.
Morristown, New Jersey: Creative Computing Press, 1977.
-- Includes material from the first twelve issues.

187. Augarten, Stan.
BIT by BIT: An Illustrated History of Computers.
New York: Ticknor & Fields, 1984.

188. Bowker, R.R.
Microcomputer Market Place 1985: A Comprehensive Directory of the Microcomputer Industry.
New York: R.R. Bowker, 1985.

189. Campbell-Kelly, Martin and William Aspray.
Computer: A History of the Information Machine.
New York: BasicBooks, 1996.

190. Ditlea, Steve (Editor).
DIGITAL DELI: The comprehensive, user-lovable menu of computer lore, culture, lifestyles and fancy.
New York: Workman Publishing, 1984.

191. Dvorak, John C.
Dvorak Predicts: An Insider's Look at the Computer Industry.
Berkeley, California: Osborne/McGraw-Hill, 1994.
-- The book is now somewhat dated and parts of the book appeared in various periodicals.

192. Dyson, Esther
Release 2.0: A design for living in the digital age.
New York: Broadway Books, 1997.

193. Glass, Robert L.
Computing Calamities: Lessons Learned from Products, Projects, and Companies that Failed.
Upper Saddle River, N.J.: Prentice Hall PTR, 1999.
-- Contains information on problems at companies such as Atari, Commodore, Novell and Wang.

194. Goody, Roy W.
The Intelligent Microcomputer.
Chicago, Illinois: Science Research Associates, 1986.

195. Greelish, David.
Historically Brewed: Our First Year.
Jacksonville, Florida: HCS Press, 1994.

196. Gupta, Amar. and Hoo-min D. Toong (Editors).
Insights into Personal Computers.
New York: IEEE Press, 1985.

197. Haddock, Dr. Thomas F.
A Collector's Guide To Personal Computers and Pocket Calculators.
Florence, Alabama: Books Americana, 1993.

198. Hanson, Dirk.
The New Alchemists: Silicon Valley and the Microelectronics Revolution.
Boston, Mass.: Little, Brown and Company, 1982.

199. Hyman, Michael.
PC Roadkill.
Foster City, Calif.: IDG Books, 1995.

200. Kidder, Tracy.
The Soul of a New Machine.
New York: Avon Books, 1981.

201. Kidwell, Peggy A. and Paul E. Ceruzzi.
Landmarks in Digital Computing: A Smithsonian Pictorial History.
Washington, D.C.: Smithsonian Institution Press, 1994.

202. Mims, Forrest M. III.
Siliconnections: Coming of Age in the Electronic Era.
New York: McGraw-Hill, 1986.

203. Negroponte, Nicholas.
being digital.
New York: Knopf, Inc., 1995.

204. Nelson, Theodor H.
Computer Lib and Dream Machines.
South Bend, In.: Self Published in 1974.

205. Nyce, James M. and Paul Kahn (Editors).
From Memex to Hypertext: Vannevar Bush and the Mind's Machine.
San Diego, Calif.: Academic Press, 1991.

206. Osborne, Adam.
Running Wild: The Next Industrial Revolution.
Berkeley, Calif.: Osborne/McGraw-Hill, 1979

207. Palfreman, Jon and Doron Swade.
The Dream Machine: Exploring the Computer Age.
London, England: BBC Books, 1991.
-- The book is derived from a television mini-series.

208. Ranade, Jay and Alan Nash (Editors).
The Best of BYTE: Two decades on the Leading Edge.
New York: McGraw-Hill, 1994.

209. Rogers, Everett M. and Judith K. Larsen.
Silicon Valley Fever: Growth of High Technology Culture.
New York: Basic Books, 1984.

210. Smolan, Rick and Jennifer Erwitt.
One Digital Day: How the Microchip is Changing Our World.
New York: Times Books/Random House, 1998.
-- The book was sponsored by the Intel Corporation. It is a photographic documentation of the microprocessor's vast influence world-wide.

211. Stork, David G. (Editor).
HAL's Legacy: 2001's Computer as Dream and Reality.
Cambridge, Mass.: MIT Press, 1997.

212. Stumpf, Kevin.
A Guide to Collecting Computers and Computer Collectibles: History Practice and Techniques.
Kitchener, Ontario, Canada: Self published, 1998.

Internet and the World Wide Web

213. Abbate, Janet.
Inventing the Internet.
Cambridge: MIT Press, 1999.

214. Berners-Lee, Tim with Mark Fischetti
Weaving the Web: The Original Design and Ultimate Destiny of the WORLD WIDE WEB by its Inventor.
New York: Harper SanFrancisco, 1999.

215. Ferguson, Charles H.
High Stakes, No Prisoners: A Winner's Tale of Greed and Glory in the Internet Wars.
New York: Time Books, 1999.

216. Hafner, Katie and John Markoff.
Cyberpunk: Outlaws and Hackers on the Computer Frontier.
New York: Simon & Schuster, 1991.
-- Stories of three hackers who create havoc on computer networks.

217. Hafner, Katie and Lyon, Mathew.
Where Wizards Stay Up Late: The Origins of the Internet.
New York: Simon & Schuster, 1996.

218. Jefferis, David.
Cyber Space: Virtual Reality and the World Wide Web.
New York: Crabtree Pub., 1999.

219. Naughton, John.
A Brief History of the Future: The Origins and Destiny of the Internet.
London: Weidenfeld & Nicolson, 1999.

220. Randall, Neil.
the soul of the Internet: Net Gods, Netizens and The Wiring of the World.
London, United Kingdom: International Thomson Computer Press, 1997.

221. Reid, Robert H.
Architects of the Web: 1000 Days that Built the Future of Business.
New York: John Wiley & Sons, 1997.

222. Salus, Peter H.
Casting the Net: From ARPANET to Internet and Beyond.
Reading, Mass.: Addison-Wesley, 1995.

223. Segaller, Stephen.
NERDS 2.0.1: A Brief History of the Internet
New York: TV Books, 1998.
-- A companion book to the PBS television series of the same title.

224. Stauffer, David.
Business the AOL Way: Secrets of the World's Most Successful Web Company.
Oxford: Capstone, 2000.

225. Stoll, Clifford.
Silicon Snake Oil: Second Thoughts on the Information Highway.
New York: Doubleday, 1995.

226. Wolinsky, Art.
The History of the Internet and the World Wide Web.
Springfield, New Jersey: Enslow Publishers, 1999.

Reference

227. Bowker, R.R.
Bowker's Complete Sourcebook of Personal Computing 1985.
New York: R.R. Bowker, 1984.

228. Cortada, James W.
Historical Dictionary of Data Processing: Biographies.
Westport, Conn.: Greenwood Press, 1987.

229. Cortada, James W.
Historical Dictionary of Data Processing: Organizations.
Westport, Conn.: Greenwood Press, 1987.

230. Cortada, James W.
Historical Dictionary of Data Processing: Technology.
Westport, Conn.: Greenwood Press, 1987.

231. Cortada, James W.
A Bibliographic Guide to the History of Computing, Computers, and the Information Processing Industry.
Westport, Conn.: Greenwood Press, 1990.

232. Cortada, James W.
A Bibliographic Guide to the History of Computer Applications, 1950-1990.
Westport, Conn.: Greenwood Press, 1996.

233. Editors of *The Red Herring* periodical.
The Red Herring Guide to the Digital Universe: The inside look at the technology business -- from Silicon Valley to Hollywood.
New York: Warner Books, 1996.

234. Glossbrenner, Alfred and Emily.
Computer Sourcebook.
New York: Random House, 1977.

235. Juliussen, Egil, Portia Isaacson and Luanne Kruse (Editors).
Computer Industry Almanac.
Dallas, Texas: Computer Industry Almanac, 1987.

236. Kent, Allen and James G. Williams (Executive Editors).
Encyclopedia of Microcomputers (Volumes 1 to 15).
New York: Marcel Dekker, 1988/95.

237. Petska-Juliussen, Karen and Dr. Egil Juliussen.
The 8th Annual Computer Industry Almanac.
Austin, Texas: The Reference Press, 1996.

238. Sayre Van.
MicroSource: Where to Find Answers to Questions About Microcomputers.
Littleton, Colorado: Libraries Unlimited, 1986.

Periodicals

This section is a historical reference source of periodical articles. The articles provide additional detailed information on personal computer developments. The product articles are generally an initial review of the product shortly after its introduction.

Chapter 1 ... Development of the Computer.

239. Randell, Brian.
"The Origins of Computer Programming."
IEEE Annals of the History of Computing
(Vol. 16, No.4, 1994), pp. 6-14.

240. Reid-Green, Keith S.
"A Short History of Computing."
Byte (July 1978), pp. 84-94.

241. Ridenour, Louis N.
"Computer Memories."
Scientific American (Vol. 192, No.6, June 1955)
-- An overview of computer memory technology in the mid 1950's.

242. Stern, Nancy.
"Who Invented the First Electronic Digital Computer?"
Abacus (Vol. 1, No. 1, 1983), pp. 7-15.

Chapters 2, 4 & 11 ... Personal Computing, Transition to Microcomputers & Competitive Computers.

Commodore

243. Dickerman, Harold.
"The Commodore 8032 Business System."
BYTE (August 1982), pp. 366-376.

244. Fylstra, Dan.
"User's Report: The PET 2001."
BYTE (March 1978), pp. 114-127.

245. Perry, Tekla S. and Paul Wallich.
"Design case history: the Commodore 64."
IEEE Spectrum (March 1985), pp. 48-58.

246. Williams, Gregg.
"The Commodore VIC 20 Microcomputer: A Low-Cost, High-Performance Consumer Computer."
BYTE (January 1981), pp. 94-102.

Compaq

247. Dahmke, Mark.
"The Compaq Computer: A portable and affordable alternative to the IBM Personal Computer."
BYTE (January 1983), pp. 30-36.

Epson

248. Ramsey, David.
"Epson's HX-20 and Texas Instruments' CC-40."
BYTE (September 1983), pp. 193-206.

249. Williams, Gregg.
"The Epson HX-20: The First Byte-sized Computer."
BYTE (April 1982), pp. 104-106.

250. Williams, Greg.
"The Epson QX-10/Valdocs System."
BYTE (September 1982), pp. 54-57.

Graphics Software

251. Bissell, Don.
"The Father of Computer Graphics: Today's graphics systems owe their existence to an innovative graduate school project called Sketchpad."
Byte (June 1990), pp. 380-381.

252. Editor.
"The CAD Revolution: A 20-Year Saga."
Compressed Air Magazine (October/November 1993), pp. 40-44.

253. Sutherland, Ivan E.
Sketchpad: A Man-Machine Graphical Communication System."
AFIPS Conference Proceedings (Vol. 23), 1963 Spring Joint Computer Conference, pp. 329-346.

254. Sutherland, Ivan E.
"Computer Graphics."
Datamation (May 1966), pp. 22-27.

Heathkit

255. BYTE Staff.
"The New Heathkit Computer Line."
BYTE (August 1977), pp. 86-88.
256. Dahmke, Mark.
"The Heath H-89 Computer."
BYTE (August 1980), pp. 46-56.
257. Poduska, Paul R.
"Building the Heath H8 Computer."
BYTE (March 1979), pp. 12-13 & 124-140.

Hewlett-Packard

258. Archer, Rowland.
"The HP-75 Portable Computer."
BYTE (September 1983), pp. 178-186.
259. Morgan, Christopher P. (Editor-in-Chief).
"Hewlett-Packard's New Personal Computer: The HP-85."
BYTE (March 1980), pp. 60-66.

History

260. Ahl, David H.
"The First Decade of Personal Computing."
Creative Computing (November 1984), pp. 30-45.
261. Allan, Roy A.
"What Was The First Personal Computer?"
The Analytical Engine (Volume 3.3, May 1996), pp. 42-46.
262. Buchholz, Werner (Editor).
"Was the First Microcomputer Built in France?"
Annals of the History of Computing (Vol.10, No.2, 1988), page 142.
-- Editor discusses Micral R2E microcomputer.
263. Ceruzzi, Paul.
"From Scientific Instrument to Everyday Appliance: The Emergence of Personal Computers, 1970 -1977."
History and Technology (Vol. 13, No. 1, 1996), pp. 1-31.

264. Editor.
"Early Small Computers."
Annals of the History of Computing (11, No. 1, 1989), pp. 53-54.
-- The editor discusses Kenbak computer.

265. Garland, Harry.
"design innovations in personal computers."
Computer (March 1977), pp. 24-27.

266. Gray, Stephen B.
"The Early Days of Personal Computers."
Creative Computing (November 1984), pp. 6-14.

267. Isaacson, Portia.
"Personal Computing: An Idea Whose Time Has Come."
Byte (February 1977), pp. 4 & 140-143.
-- Describes evolution of new trend in computing.

268. Layer, Harold A.
"Microcomputer History and Prehistory: An Archaeological Beginning."
Annals of the History of Computing, (Vol. 11, No 2, 1989), pp. 127-130.

269. Libes, Sol.
"The First Ten Years of Amateur Computing."
BYTE (July 1978), pp. 64-71.

270. Miller, Michael J.
"Looking Back: A history of the Technology that changed our world."
PC Magazine (March 25, 1997), pp. 108-136.
-- Part of *PC Magazine* Fifteen Years anniversary issue.

271. Pfaffenberger, Bryan.
"The Social Meaning of the Personal Computer: Or, Why the Personal Computer Revolution was no Revolution."
Anthropological Quarterly (January 1988), pp. 39-47.

272. Press, Larry.
"Personal Computing: Where Did it Come From ?"
Abacus (Vol.1, No. 1, 1983), pp. 56-60.

273. Press, Larry.
"Before the Altair: The History of Personal Computing."
Communications of the ACM (Vol. 36, No. 9, September 1993), pp. 27-33.

274. Sheldon, Kenneth M.
"Micro Edsels: A look back at 15 years of the good, the bad, and marketing bombs of the microcomputer revolution."
Byte (February 1990), pp. 245-248.

275. Warren, Jim.
"personal and hobby computing: an overview."
IEEE Computer (March 1977), pp. 10-22.

276. Wood, Lamont.
"The Man Who Invented the PC."
Invention & Technology (Vol. 10, No. 2, Fall 1994), page 64.

IBM

277. Editors Bytes/Bits column (Byte, December 1975).
"Welcome IBM, to personal computing."
Reprint in *BYTE* (November 1983), page 137.
-- Reviews IBM 5100 portable computer.

278. Littman, Jonathan.
"The First Portable Computer."
PC World (October 1983), pp. 294-300.
-- Describes IBM SCAMP computer.

Miscellaneous

279. Baker, Robert.
"Product Description: OSI."
Byte (January 1977), pp. 94-95.
-- Describes Ohio Scientific(OSI) models 300 & 400 microcomputer boards

280. Ciarcia, Steve.
"Try This Computer on for Size."
BYTE (March 1977), pp. 114-129
-- Describes Digital Group microcomputer.

281. Crosby, Kip.
"Dawn of the Micro: Intel's Intellecs."
The Analytical Engine (Jan. - Mar. 1994), pp. 11-14.

282. Editors.
"Computer! Build this microcomputer yourself. Add it to the TV Typewriter for a complete computer system of your own."
Radio-Electronics (July 1974), pp. 29-33.
-- Describes construction details of Mark-8 computer.

283. Fager, Roger and John Bohr.
"The Kaypro II."
BYTE (September 1983), pp. 212-224.

284. Fiegel, Curtis.
"What a Concept: A View of the Corvus Computer."
BYTE (May 1983), pp. 134-150.

285. Gray, Stephen B.
"Building Your Own Computer."
Computers and Automation (December 1971), pp. 25-31.

286. Harmon, Tom.
"The SwTPC 6809 Microcomputer System."
BYTE (January 1981), pp. 216-222.

287. Hudson, Richard L.
"French Entrepreneur Labors in Obscurity Despite His Big Feat: Truong's Company Invented The First Microcomputer ..."
Wall Street Journal (September 18, 1985), pp. 1 & 27.

288. Hauck, Lane T.
"System Description: The Noval 760."
Byte (September 1977), pp. 102-108.

289. Infield, Glenn.
"A Computer in the Basement?"
Popular Mechanics (April 1968), pp. 77-229.
-- Author describes a home-built ECHO-IV computer.

290. Jones, Douglas W.
"The DEC PDP-8 Story: The First Line of Truly Small Computers."
Historically Brewed (Issues #7 to #9, 1994-1996).
Part I: The Beginning, (Issue #7, 1994), pp. 7-9.
Part II: The Minicomputer Revolution (Issue #8, 1995), pp. 7-10.
Part III: The Concluding Years, (Issue #9, 1996), pp. 11-14.

291. Krause, Llaus.
"Exidy Sorcerer."
The Analytical Engine (April - June 1994), pp. 22-23.

292. Lemmons, Phil.
"Victor Victorious: The Victor 9000 Computer."
BYTE (November 1982), pp. 216-254.

293. Nadeau, Michael E.
"The Littlest Zenith."
Byte (August 1989), pp. 94-96.

294. Scharf, Steve.
"LOBO MAX."
The Analytical Engine (April - June 1994), page 22.

295. Tomayko, James E. (Editor).
"Electronic Computer for Home Operation (ECHO): The First Home Computer."
IEEE Annals of the History of Computing (Vol. 16, No. 3, 1994), pp. 59-61.

296. Toong, Hoo-min D. and Amar Gupta.
"Personal Computers."
Scientific American (December 1982), pp. 87-107.

297. Uttal, Bro.
"Sudden Shake-up in Home Computers."
Fortune (July 11, 1983), pp. 105-106.

298. Van Name, Mark L. and Bill Catchings.
"The Painlessly Portable PC."
Byte (August 1989), pp. 161-164.
-- Describes NEC UltraLite portable computer.

MITS

299. Greelish, David A.
"A Talk With the Creator?!
-- An Interview with Ed Roberts."
Historically Brewed (Issue #9, 1996), pp. 5-10.

300. Mims, Forest M.
"The Altair Story."
Creative Computing (November 1984), pp. 17-27.

301. Roberts, H. Edward and William Yates.
"Altair 8800: The Most Powerful Minicomputer."
Popular Electronics (January 1975), pp. 33-38.

NeXT

302. Thompson, Tom and Nick Baran.
"The NeXT Computer."
BYTE (November 1988), pp. 158-175.

303. Thompson, Tom and Ben Smith.
"Sizing Up the Cube: The long-awaited NeXT cube offers advanced features but only fair performance."
Byte (January 1990), pp. 169-176.

Osborne

304. Dahmke, Mark.
"The Osborne 1."
BYTE (June 1982), pp. 348-362.

305. Pournelle, Jerry.
"The Osborne Executive and Executive II: Adam Osborne's Improved Portable Computers."
BYTE (May 1983), pp. 38-44.

306. Uttal, Bro.
"A Computer Gadfly's Triumph,"
Fortune (March 8, 1982), pp. 74-76.
-- Describes Adam Osborne's introduction of the Osborne 1 portable computer.

Processor Technology

307. Barbour, Dennis E.
"Users Report: The SOL-20."
BYTE (April 1978), pp. 126-130.

308. Bumpous, Robert.
"A User's Reaction to the SOL-10 Computer."
BYTE (January 1978), pp. 86-93.

Radio Shack

309. Fylstra, Dan.
"The Radio Shack TRS-80: An Owner's Report."
BYTE (April 1978), pp. 49-60.
310. Kelly, Mahlon G.
"The Radio Shack TRS-80 Model 100."
BYTE (September 1983), pp. 139-162.
311. Malloy, Rich.
"Little Big Computer: The TRS-80 Model 100 Portable Computer."
BYTE (May 1983), pp. 14-34.
312. Miastkowski, Stan (Editor).
"Three New Computers from Radio Shack."
BYTE (October 1980), pp. 172-180.
313. Worthy, Ford S.
"Here Come the Go Anywhere Computers."
Fortune (October 17, 1983), pp. 97-98.
-- Describes TRS-80 Model 100 and other portable computers.

Sinclair

314. Garrett, Billy.
"The Timex/Sinclair 1000."
BYTE (January 1983), pp. 364-370
315. McCallum, John C.
"The Sinclair Research ZX80."
BYTE (January 1981), pp. 94-102.

Texas Instruments

316. Haas Mark.
"The Texas Instruments Professional Computer."
Byte (December 1983), pp. 286-324.
317. Uttal, Bro.
"TI's Home Computer Can't Get in the Door."
Fortune (June 16, 1980), pp. 139-140.
-- Describes marketing problems of the TI-99/4 computer.

Time Sharing

318. Fano, R.M. and F.J. Corbato.
"TIME-SHARING ON COMPUTERS."
Scientific American (September 1966), pp. 128-312.

Xerox

319. Johnson, Jeff., Teresa L. Roberts, William Verplank, David C. Smith, Charles H. Irby, Marian Beard and Kevin Mackey.
"The Xerox Star: A Retrospective."
Computer (September 1989), pp. 11-29.

320. Kay, Alan and Adele Goldberg.
"Personal Dynamic Media."
Computer (March 1977), pp. 31-41.
-- Authors describe Dynabook, Alto and Smalltalk.

321. Ryan, Bob.
"Dynabook Revisited with Alan Kay."
BYTE (February 1991), pp. 203-208.

322. Smith, David Canfield, Charles Irby, Ralph Kimball, Bill Verplank and Eric Harslem.
"Designing the Star User Interface."
BYTE (April 1982), pp. 242-282.

323. Wadlow, Thomas A.
"The Xerox Alto Computer."
BYTE (September 1981), pp. 58-68.

Chapters 3, 8 & 14.1 ... Microprocessors

324. Antonoff, Michael.
"Gilbert Who?: An obscure inventor's patent may rewrite microprocessor history."
Popular Science (February 1991), pp. 70-73.

325. Baskett, Forest and John L. Hennessy.
"Microprocessors: From Desktops to Supercomputers."
Science (Vol. 261: 13 August 1993), pp. 864-871.

326. Bylinsky, Gene.
"The Second Computer Revolution."
Fortune (February 11, 1980), pp. 230-236.
-- Details impact of microprocessor.

327. Bylinsky, Gene.
"Intel's Biggest Shrinking Job Yet."
Fortune (May 3, 1982), pp. 250-256.
-- Describes development of 432 microprocessor.

328. *BYTE* Staff.
"Micro, Micro: Who Made The Micro ?"
BYTE (January 1991), pp. 305-312.

329. Diefendorff, Keith.
"History of the *PowerPC Architecture*."
Communications of the ACM (June 1994), pp. 28-33.

330. Editors.
"The Microprocessor at 25: Milestones of a Quarter Century."
PC Magazine (December 17, 1996), pp. 147-149.

331. Faggin, Federico.
"The Birth of the Microprocessor."
BYTE (March 1992), pp. 145-150.

332. Frenzel, Lou
"How to Choose a Microprocessor."
Byte (July 1978), pp. 124-139.
-- Reviews microprocessors and applicable microcomputers.

333. Garetz, Mark.
"Evolution of the Microprocessor."
BYTE (September 1985), pp. 209-215.

334. Gwennap, Linley.
"Birth of a Chip: In only 25 Years, the microprocessor has become the life-support system of the modern world."
Byte (December 1996), pp. 77-82.

335. Halsema, A.I.
"A Preview of the Motorola 6800."
BYTE (April 1979), pp. 170-174.

336. Mazor, Stanley.
"Microprocessor and Microcomputer: Invention and Evolution."
The Analytical Engine (Vol. 3.4, Fall 1996), pp. 6-13.

337. Moore, Charles R. and Russell C. Stanphill.
"The PowerPC Alliance."
Communications of the ACM (June 1994), pp. 25-27.

338. Moore, Gordon E.
"Cramming More Components Onto Integrated Circuits."
Electronics (Vol. 38, no.8), April 19, 1965, pp. 114-117.

339. Moore, Gordon E.
"Intel: Memories and the Microprocessor."
Daedalus (Vol. 125: No 2, 1996), pp. 55-80.

340. Morse, Stephen P., William B. Pohlman, & Bruce W. Ravenel.
"The Intel 8086 Microprocessor: A 16-bit Evolution of the 8080."
IEEE Computer (June 1978), pp. 18-27.

341. Morse, Stephen P., Bruce W. Ravenel, Stanley Mazor and William B. Pohlman.
"Intel Microprocessors: 8008 to 8086."
IEEE Computer (October 1980), pp. 42-60.

342. Noyce, Robert N. and Marcian E. Hoff.
"A History of Microprocessor Development at Intel."
IEEE Micro (February 1981), pp. 8-21.

343. Rampil, Ira.
"Preview of the Z-8000."
BYTE (March 1979), pp. 80-91.

344. Ritter, Terry and Joel Boney.
"A Microprocessor for the Revolution: The 6809."
BYTE (January 1979), pp. 14-42.

345. Schlender, Brent.
"Killer chip."
Fortune (November 10, 1997), pp. 70-80.
-- Previews the Intel Merced microprocessor.

Chapters 5 & 10 ... Apple Corporation in the 1970's & 1980's

Apple II and III:

346. Duprau, Jeanne and Molly Tyson.
"The Making of the Apple IIGS."
A+ Magazine (November 1986), pp. 57-74.

347. Edwards, John.
"Apple IIe."
Popular Computing (March 1983), pp. 108-190.

348. Helmers, Carl.
"A Nybble on the Apple."
Byte (April 1977), page 10.
-- Editor of *Byte* previews Apple II computer.

349. Helmers, Carl.
"An Apple to BYTE."
BYTE (March 1978), pp. 18-46.
-- An early review of the Apple II computer.

350. Little, Gary B.
"A Close Look At Recent IIe Enhancements."
A+ Magazine (August 1985), pp. 45-54.

351. Little, Gary, B.
"A Technical Overview of the Apple IIGS."
A+ Magazine (November 1986), pp. 45-52.

352. Markoff, John.
"The Apple IIc Personal Computer."
BYTE (May 1984) pp. 276-284.

353. Moore, Robin.
"The Apple III and Its New Profile."
BYTE (September 1982), pp. 92-132.

354. Moore, Robin.
"Apple's Enhanced Computer, the Apple IIe."
BYTE (February 1983), pp. 68-86.

355. Morgan, Christopher.
"The Apple III."
BYTE (July 1980), pp. 50-54.

356. Williams, Greg.
"C is for Crunch,"
BYTE (December 1984), pp. A75-A121.
-- Author interviews Peter Quinn, who was design manager for the Apple IIe and IIc computers.

357. Williams, Greg and Richard Grehan.
"The Apple IIGS."
BYTE,(October 1986), pp. 84-98.

358. Wozniak, Stephen.
"System Description: The Apple II."
BYTE (May 1977), pp. 34-43.
-- A technical description of the initial Apple II computer by the designer.

Lisa and Macintosh:

359. Craig, David T.
"The Apple Lisa Computer: A Retrospective."
The Analytical Engine (Vol. 2.1, July-September 1994), pp. 18-31.
-- Contains an extensive list of references.

360. Lemmons, Phil.
"An Interview: The Macintosh Design Team."
BYTE (February 1984), pp. 58-80.

361. Markoff, John and Ezra Shapiro.
"Macintosh's Other Designers."
BYTE (August 1984), pp. 347-356.
-- Includes interview with Jef Raskin.

362. Morgan, Chris, Greg Williams and Phil Lemmons.
"An Interview with Wayne Rosing, Bruce Daniels and Larry Tesler: A behind-the-scenes look at the development of Apple's Lisa."
BYTE (February 1983), pp. 90-114.

363. Nace, Ted.
"The Macintosh Family Tree."
Macworld (November 1984), pp. 134-141.

364. Nulty, Peter.
"Apple's Bid to Stay in the Big Time,"
Fortune (February 7, 1983), pp. 36-41.
-- Description of the technical and marketing strategies for Lisa computer.

365. Press, Larry.
"Apple Announces the Mac II and the Mac I 1/6."
Abacus (Vol. 4, No. 4, Summer 1987), pp. 58-62.

366. Raskin, Jef.
"The MAC and Me: 15 Years of Life with the Macintosh."
The Analytical Engine
Part 1: (Vol. 2.4, August 1995), pp. 9-22.
Part 2: (Vol. 3.3, May 1996), pp. 21-33.

367. Schlender, Brent.
"Steve Jobs Apple Gets Way Cooler."
Fortune (January 24, 2000), pp. 66-76.

368. Tesler, Larry.
"The Legacy of the Lisa."
Macworld (September 1985), pp. 17-22.

369. Van Nouhuys, Dirk.
"Apple 32 Past, Present, and Future."
A+ (July 1984), pp. 76-84.

370. Webster, Bruce F.
"The Macintosh."
BYTE (August 1984), pp. 238-251.
-- A system review of the Apple Macintosh computer.

371. Williams, Greg.
"The Lisa Computer System."
BYTE (February 1983), pp. 33-50.

372. Williams, Greg.
"The Apple Macintosh Computer."
BYTE (February 1984), pp. 30-54.

373. Williams, Greg.
"Apple Announces the Lisa 2."
BYTE (February 1984), pp. 84-85.

374. Webster, Bruce F.
"The Macintosh."
BYTE (August 1984), pp. 238-251.

Miscellaneous

375. A+ Magazine.
"Back In Time: Apple's ten-year history has been an eventful one. here are some highlights."
A+ Magazine (February 1987), pp. 48-49.

376. Hogan, Thom.
"APPLE: The First Ten Years: History, Part I."
A+ Magazine (January 1987), pp. 43-46.
"APPLE: The First Ten Years: History, Part II."
A+ Magazine (February 1987), pp. 45-46.

377. Raleigh, Lisa.
"Woz on the Last 10 Years."
A+ Magazine (January 1987), pp. 39-41.

378. Schnatmeier, Vanessa.
"In Search of Early Apples."
A+ Magazine (August 1986), pp. A67-A70.

379. Williams, Greg. and Rob Moore.
"The Apple Story Part I: Early History."
BYTE (December 1984), pp. 67-71.
"The Apple Story Part II: More History and the Apple III." *BYTE* (January 1985), pp. 167-174
-- Informative interviews with Stephen Wozniak.

Chapters 6 & 12 ... Microsoft in the 1970's & 1980's

380. Isaacson, Walter.
"In Search of the Real Bill Gates."
Time (13 January, 1997), pp. 30-42.

381. Paterson, Tim.
"An Inside Look at MS-DOS: The design decisions behind the popular operating system."
BYTE (June 1983), pp. 230-252.

382. Schlender, Brent.
"Bill Gates & Paul Allen Talk."
Fortune (October 2, 1995), pp. 68-86.

383. Simonyi, Charles and Martin Heller.
"The Hungarian Revolution."
BYTE (August 1991), pp. 131-138.

384. Stein, Joel.
"Image is Everything: Bill Gates bid for a digital empire may pay off someday, but for now the King of Content can only scheme."
Time (11 November 1996), pp. TD32-TD38.
-- Describes Corbis company founded by Bill Gates.

385. Udell, Jon.
"Three's the One: Windows 3.0 carries DOS into the 1990's."
Byte (June 1990), pp. 122-128.

386. Udell, Jon.
"Windows 3.1 Is Ready to Roll."
Byte (April 1992), pp. 34-36.

387. Uttal, Bro.
"Inside the Deal that made Bill Gates $350,000,000,"
Fortune (July 21, 1986), pp. 23-33.

Chapters 7 & 13 ... Software in the 1970's & 1980's

Application programs

388. Cmar, Karen A.
"AppleWorks: An Integrated Office Product."
BYTE (December 1984), pp. A18-A22.

389. Lemmons, Phil.
"A Guided Tour of Visi On."
BYTE (June 1983), pp. 256-278.

390. Licklider, Tracy Robnett.
"Ten Years of Rows and Columns: From a 16K-byte VisiCalc to multimegabyte packages, spreadsheets have come a long way in a decade."
Byte (December 1989), pp. 324-331.

391. Ramsdell, Robert E.
"The Flexibility of VisiPlot."
BYTE (February 1982), pp. 32-36.

Languages

392. Bowles, Kenneth.
"UCSD PASCAL: A Nearly Machine Independent Software System."
BYTE (May 1978), pp. 46 & 170-173.

393. Editor
"Design of Tiny BASIC."
Dr. Dobb's Journal (Vol. 1, No. 1, January 1976).

394. Gates, Bill.
"The 25th Birthday of BASIC."
Byte (October 1989), pp. 268-276.

395. Kay, Alan C.
"The Early History of Smalltalk."
ACM SIGPLAN Notices (March 1993), page 87.

396. Lehman, John A.
"PL/I for Microcomputers."
BYTE (May 1982), pp. 246-250.

397. Lockwood, Russ.
"The Genealogy of BASIC."
Creative Computing (November 1984), pp. 86-87.

398. Moore, Charles H.
"The Evolution of FORTH, an Unusual Language."
Byte (August 1980), pp. 76-90

399. Rosner, Richard.
"A Review of Tom Pitman's Tiny BASIC."
Byte (April 1977), pp. 34-38.

400. Tesler, Larry.
"The Smalltalk Environment."
BYTE (August 1981), pp. 90-147.

401. Woteki, Thomas H. and Paul A. Sand.
"Four Implementations of Pascal."
BYTE (March 1982), pp. 316-353.

402. Wozniak, Stephen.
"SWEET16: The 6502 Dream Machine."
BYTE (November 1977), pp. 150-159.

Operating Systems and User Interfaces

403. Hayes, Frank and Nick Baran.
"A Guide to GUI's: Graphical user interfaces make computers easy to use; keeping them all straight is the hard part."
Byte (July 1989), pp. 250-257.

404. Hayes, Frank.
"From TTY to VUI: As computers become more complex, using them becomes easier and easier."
Byte (April 1990), pp. 205-211.
-- Describes development of graphic user interfaces.

405. Kildall, Gary.
"CP/M: A Family of 8 and 16-Bit Operating Systems."
BYTE (June 1981), pp. 216-232.

406. Perry, Tekla S. and John Voelcker.
"Of mice and menus: designing the user-friendly interface."
IEEE Spectrum (September 1989), pp. 46-51.

Chapter 9 ... The IBM Corporation

407. Archer, Rowland.
"The IBM PC XT and DOS 2.00."
BYTE (November 1983), pp. 294-304.

408. Bond, George.
"The IBM PC Network."
BYTE (October 1984), page 111.

409. Bradley, David J.
"The Creation of the IBM PC."
BYTE (September 1990), pp. 414-420.

410. Curran, Lawrence J. and Richard S. Shuford.
"IBM's Estridge: The President of IBM's Entry Systems Division talks about Standards, the PC's simplicity, and a desire not to be different."
BYTE (November 1983), pp. 88-97.

411. Editors.
"Introduction: The Array of IBM Personal Computers."
BYTE (Fall 1984), pp. 9-26.

412. Fisher, Anne B.
"Winners (and Losers) from IBM's PC JR."
Fortune (November 28, 1983), pp. 44-48.

413. Gens, Frank and Chris Christiansen.
"Could 1,000,000 IBM Users Be Wrong?"
BYTE (November 1983), pp. 135-141.

414. Henry, G.G.
"IBM small-system architecture and design: Past, present and future."
IBM Systems Journal (Vol. 25, Nos. 3/4, 1986), pp. 321-333.
-- Provides details of the PC RT computer system.

415. Killen, Michael.
"IBM Forecast: Market Dominance."
BYTE (Fall 1984), pp. 31-38.
-- Reviews mainly the IBM 3270 PC and PC/XT Model 3270.

416. Lemmons, Phil.
"The IBM Personal Computer: First Impressions."
BYTE (October 1981), pp. 27-34.

417. Malloy, Rich.
"Two New Office Products from IBM."
BYTE (December 1983), page 594.
-- Reviews IBM 3270 PC and PC/XT 370.

418. Malloy, Rich.
"IBM Announces the PCjr,"
BYTE (December 1983), page 358.

419. Malloy, Rich., G. Michael Vose and Tom Cluwe.
"The IBM PC AT."
BYTE (October 1984), pp. 108-111.

420. Mitchell, Robert.
"IBM and AT&T Enter the Fray of 386SX Notebook Computers."
BYTE (August 1991), pp. 252-254.
-- Describes IBM's first battery operated laptop computer and competing product from AT&T.

421. Morgan, Chris.
"IBM's "Secret" Computer: The 9000."
BYTE (January 1983), pp. 100-106.

422. Sandler, Corey.
" IBM: Colossus of Armonk."
Creative Computing (November 1984), pp. 298-302.

423. Williams, Gregg.
"A Closer Look at the IBM Personal Computer."
BYTE (January 1982), pp. 36-68.

424. Vose, Michael G. and Richard S. Shuford.
"A Closer Look at the IBM PCjr."
BYTE (March 1984), pp. 320-332.

Chapters 14 ... Hardware in the 1990's

425. Alford, Roger C.
"The Fastest Portable: IBM's P75 Road Warrier."
BYTE (April 1991), pp. 265-268.
-- Reviews IBM PS/2 Model P75 portable computer.

426. Reinhardt, Andy and Ben Smith.
"Sizzling RISC Systems from IBM."
BYTE (April 1990), pp. 124-128.
-- Describes RISC System/6000 workstations.

Chapters 15, Software in the 1990's

427. McHugh, Josh.
"For the love of hacking."
Forbes (August 10, 1998), pp. 94-100.
-- Describes free-software and Linux operating system.

Chapters 16 ... Corporate Activities in the 1990's

428. Brant, Richard with Julia Flynn and Amy Cortese.
"Microsoft Hits the Gas:
Its Bidding to Lead the Info Highway Pack."
Business Week (March 21, 1994), pp. 34-35.

429. Hof, Robert D.
"The Sad Saga of Silicon Graphics: What went wrong at the company that once made everybody say: "Gee Whiz."
Business Week (August 4, 1997), pp. 66-72.

Chapter 17 ... Hardware and Peripherals

430. Christensen, Clayton M.
"The Rigid Disk Drive Industry: A History of Commercial and Technological Turbulence."
Business History Review (Winter 1993), pp. 531-588.

431. Gaskin, Robert R.
"Paper, Magnets and Light: The long history of data storage devices is intertwined with the more recent, meteoric rise of personal computers."
Byte (November 1989), pp. 391-399.

432. Jarrett, Thomas.
"The New Microfloppy Standards."
BYTE (September 1983), pp. 166-176.

433. Lancaster, Don.
"TV Typewriter."
Radio-Electronics (September 1973), pp. 43-50.

434. Lebow, Max.
"Tele-Vic: Commodore Breaks the $100 Price Barrier for Modems."
BYTE (March 1982), pp. 240-246.

435. Lemmons, Phil.
"A Short History of the Keyboard."
BYTE (November 1982), pp. 386-387.

436. Levy, Steven.
"Of Mice and Men: The Mouse is but a small part of Doug Engelbart's larger quest."
Popular Computing (May 1984), pp. 70, 75-78.

437. Mayadas, A.F., R.C. Durbeck, W.D. Hinsberg and J.M. McCrossin.
"The evolution of printers and displays."
IBM Systems Journal (Vol. 25, Nos. 3/4,1986), pp. 399-416.

438. Mendelson, Edward.
"Microsoft Does a Wheelie: With a wheel between its buttons, the IntelliMouse goes where no other mouse has gone before."
PC Magazine (17 December 1996), page 65.

439. Morgan, Chris.
"A Look at Shugart's New Fixed Disk Drive."
BYTE (June 1978), pp. 174-176.

440. Nulty, Peter.
"Big Memories for Little Computers."
Fortune (February 8, 1982), pp. 50-56.
-- Describes development of hard disk drives by Seagate and Shugart.

441. Pelczarski, Mark.
"Microsoft SoftCard."
BYTE (November 1981), pp. 152-162.

442. Peters, Chris.
"The Microsoft Mouse."
BYTE (July 1983), pp. 130-138.

443. Schnatmeier, Vanessa.
"A Modern Mouse Story."
A+ (July 1984), pp. 32-35.

444. *Scientific American.*
"The Solid-state Century: the past present and future of the transistor."
Scientific American (Special issue, January 1998).
-- Includes articles on integrated circuit and microprocessor technology.

445. Whang, Min-Hur and Joe Kua.
"Join the EISA Evolution: The EISA bus is breaking up that old "Gang of Nine"."
Byte (May 1990), pp. 241-247.

446. Wieselman, Irving L. and Erwin Tomash.
"Marks on Paper: Part 1 and Part 2. A Historical Survey of Computer Output Printing."
Annals of the History of Computing,
-- Part 1: Volume 13, Number 1, 1991.
-- Part 2: Volume 13, Number 2, 1991, pp. 203-222.

Chapter 18 ... Magazines and Newsletters

447. Anderson, John L.
"Dave Tells Ahl: The History of Creative Computing."
Creative Computing (November 1984), pp. 66-77.

448. The Staff of inCider/A+.
"100 issues of inCider: A Look Back."
inCider/A+ (April 1991), pp. 36-39.

Chapter 19 ... Other Organizations, Companies & People

449. Davidson, Clive.
"The Man who made Computers Personal."
New Scientist (June 19, 1993), pp. 32-35.
-- A biographical article on Alan Kay.

450. Fraker, Susan.
"How DEC Got Decked."
Fortune (December 12, 1983), pp. 83-92.
-- Describes DEC bureaucratic and marketing problems with its early personal computers.

451. Grover, Mary Beth.
"The Seagate Saga."
Forbes (May 4, 1998), pp. 158-159.

452. Halfhill, Tom R.
"R.I.P. Commodore 1954 - 1994."
BYTE (August 1994), page 252.

453. Jacob, Rahul.
"The Resurrection of Michael Dell."
Fortune (September 18, 1995), pp. 117-128.

454. Kirkpatrick, David.
"Over the Horizon with Paul Allen."
Fortune (July 11, 1994), pp. 68-75.

455. Kraar, Louis.
"Acer's Edge: PCs To Go."
Fortune (October 30, 1995), pp. 186-204.

456. Lesser, Hartley.
"Exec Avant-Garde: The Dynamic Zone."
Softalk (November 1983), pp. 66-70.

457. Petre, Peter D.
"Mass-Marketing the Computer."
Fortune (October 31, 1983), pp. 60-67.

458. Petre, Peter.
"The Man Who Keeps the Bloom on Lotus: Mitch Kapor, a child of the Sixties, has nurtured Lotus Development Corp into the world's largest independent software company."
Fortune (June 10, 1985), pp. 136-146.

459. Ramo, Joshua Cooper.
"How AOL Lost the Battles but Won the War."
Time (September 22, 1997), pp. 42-48.

460. Ramo, Joshua Cooper.
"Andrew S. Grove: A Survivor's Tale."
Time (December 29, 1997 - January 5, 1998), pp. 30-46.
-- This was a special "Man of the Year" issue featuring Andrew Grove.

461. Schlender, Brent.
"The Adventures of Scott McNealy: Javaman Pow!"
Fortune (October 13, 1997), pp. 70-78.

462. Serwer, Andy.
"Michael Dell Rocks."
Fortune (May 11, 1998), pp. 58-70.

463. Tazelaar, Jane Morrill (Editor).
"BYTE, 15th. anniversary SUMMIT: 63 of the world's Most Influential People in Personal Computing Predict the Future, Analyze the Present."
BYTE (September 1990), pp. 218-366.
-- Provides a short biographical background of the participants.

464. Uston, Ken.
"Behind The Scenes At Brøderbund: A Family Affair."
Creative Computing (September 1984), pp. 157-162.

465. Uttal, Bro.
"Xerox Xooms Toward the Office of the Future."
Fortune (May 18, 1981), pp. 44-52.

466. Uttal, Bro.
"The Man Who Markets Silicon Valley."
Fortune (December 13, 1982), pp. 133-144.
-- Describes success of Regis McKenna and his company.

467. Uttal, Bro.
"Sudden Shake-up in Home Computers."
Fortune (July 11, 1983), pp. 105-106.

468. Uttal, Bro.
"The Lab That Ran away From Xerox."
Fortune (September 5, 1983), pp. 97-102.
-- Describes Xerox's failure to market the results of research at its Palo Alto Research Center (PARC).

469. Weiss, Eric A.
"The Computer Museum."
Abacus (Vol. 2, No. 4, Summer 1985), pp. 60-65.

470. Whitmore, Sam.
"Electronic Arts."
Cider (May 1984), pp. 35-37.

471. Wolfe, Tom.
"The Tinkerings of Robert Noyce: How the sun rose on the Silicon Valley."
Esquire (December 1983), pp. 346-374.
-- An informative biographical article on Robert Noyce

472. Wright, Robert.
"The Man Who Invented the Web: Tim Berners-Lee started a revolution, but it didn't go exactly as planned."
Time (May 19, 1997), pp. 44-48.
-- describes development of the World Wide Web

473. Yuln, Matt.
"Exec Electronic Arts: Software Construction Company."
Softalk (August 1984), pp. 36-40.

Chapter 20 ... Miscellaneous Items

474. Ahl, David H. (Editor, In-Chief)
"Tenth Anniversary Issue of Creative Computing."
Creative Computing (November 1984), Vol. 10, No 11.

475. Buchholz, W.
"The Word "Byte" comes of Age..."
BYTE, (February 1977), page 144.

476. Bouchard, Judith W. (Editor-in-Chief).
Micro Computer Abstracts (Formerly *Micro Computer Index*).
-- A quarterly journal of microcomputing abstracts.

477. Caulkins, David.
"A Computer Hobbyist Club Survey."
Byte (January 1977), pp. 116-118.

478. Editors.
"Clubs and Newsletters Directory."
Byte (October 1979), page 210.

479. Friedrich, Otto.
"Machine of the Year."
Time (January 3, 1983), pp. 12-24.

480. Helmers, Carl.
"The Era of Off-the-Shelve Personal Computers Has Arrived."
BYTE (January 1980), pp. 6-98.

481. Nelson, Ted.
"On the Xanadu Project."
BYTE (September 1990), pp. 298-299.

482. Press, Larry.
"Is There Such a Thing as a Personal Computer."
Abacus (Vol. 1, No. 2, Winter 1984), pp. 69-71.

483. Press, Larry.
"The ACM Conference on the History of Personal Workstations."
Abacus (Vol. 4, No. 1, Fall 1986), pp. 65-70.

484. Shapiro, Fred R.
"The First Bug."
BYTE (April 1994), page 308.

485. Tomayko, James E. (Editor).
"Origin of the Term Bit."
"Origin of Word Byte and The First Bug."
Annals of the History of Computing,
(Volume 10, Number 4, 1989), pp. 336-340.

Historical Timelines

486. Ahl, David H.
"Ascent of the Personal Computer."
Creative Computing (November 1984), pp. 80-82.

487. *BYTE* Staff.
"10 Years of BYTE: Special anniversary Supplement."
BYTE (September 1985), pp. 198-208.

488. Halfhill, Tom R.
"Apple's Technology Milestones."
BYTE (December 1994), pp. 52-60.

489. Smarte, Gene and Andrew Reinhardt.
"1975-1990: 15 Years of Bits, Bytes and Other Great Moments."
BYTE (September 1990), pp. 369-400.

490. Williams, Gregg and Mark Welch.
"A Microcomputer Timeline."
BYTE (September 1985), pp. 197-208.

Blank page

Index

Legend: Chap.#/Page# of Chap.

-- Numerals --

3 E-Z Pieces software, 13/20
3-Plus-1 software. *See* Commodore
3Com Corporation, 12/15, 12/27, 16/17, 17/18, 17/20
3M company, 17/5, 17/22
3P+S board. *See* Processor Technology
4K BASIC. *See* Microsoft/Prog. Languages
4th Dimension. *See* ACI
8/16 magazine, 18/5
8/16-Central, 18/5
8K BASIC. *See* Microsoft/Prog. Languages
20SC hard drive. *See* Apple Computer/Accessories
64 computer. *See* Commodore
80 Microcomputing magazine, 18/4
80-103A modem. *See* Hayes
86-DOS. *See* Seattle Computer
128EX/2 computer. *See* Video Technology
386i personal computer. *See* Sun Microsystems
432 microprocessor. *See* Intel/Microprocessors
603/4 Electronic Multiplier. *See* IBM/Computer (mainframe)
660 computer. *See* Control Data
700 series printers. *See* Centronics
820 computer. *See* Xerox/Compu
1101, 1102 & 1103 memory chips. *See* Intel/Misc.
1201 microprocessor. *See* Intel/Microprocessors
1702 memory chip. *See* Intel/Misc.
1802 microprocessor. *See* RCA
2020 computer. *See* Dec/Comp.
2107 and 2117 memory chips. *See* Intel/Misc.
2650 microprocessor. *See* Signetics
4004 microprocessor. *See* Intel/Microprocessors
5100 series of computers. *See* IBM/Computers (Personal)
6085 workstation. *See* Xerox/Computers
6100 CPU. *See* Intersil
6501 and 6502 microprocessor. *See* MOS
6502 BASIC. *See* Microsoft/Prog. Languages
7000 copier. *See* Xerox/Misc.
8000 microprocessors. *See* Intel/Microprocessors
8010 "Star" Information System. *See* Xerox/Comp.
8080 and 8086 BASIC. *See* Microsoft/Prog. Languages
8514/A standard, 20/6
9700 laser printing system. *See* Xerox/Misc.
16032 and 32032 micro/p. *See* National Semiconductor
65802 and 65816 micro/p. *See* Western Design Center
68000 series of micro/p. *See* Motorola
80000 series of micro/p. *See* Intel/Microprocessors
88000 micro/p. *See* Motorola

--A--

A Programming lang. *See* APL
A+ magazine, 18/5
A.P.P.L.E. (Apple Pugetsound Program Library Exchange)
  user group, 18/4, 19/17
  Call-A.P.P.L.E. magazine, 18/4
A2-Central newsletter, 18/5
Abacus magazine, 18/8
ABC (Atanasoff-Berry Computer), 1/5
ABIOS (Advance Basic Input/Output System) chip. *See* IBM/Software
Above Board specification. *See* Intel/Misc.
Academic American Encyclopedia, 20/2
Access database. *See* Microsoft/Applic. Programs
Access Portable Computer, 11/12
Accounting programs, 9/8, 13/22
ACE (Advance Computing Environment), 19/17
Ace 100 computer. *See* Franklin Computer
Acer Inc., 11/13, 16/17
  Acer America Corp., 11/13
ACI (Analyses Conseils Informations) company, 13/19

ACIUS company, 13/19
4th Dimension database, 13/19
ACM (Association for Computing Machinery), 1/14, 2/10, 17/23, 19/3
Communications of the ACM, 1/14, 13/17
Journal, 1/14
SIGPC, 19/3
Acorn prototype computer. *See* IBM/Computers (Personal)
Acorn Computers Ltd., 11/13
Acorn computer, 11/13
BBC Models A and B, 11/13
BBC BASIC, 11/13
Acorn, Al, 5/8
Acoustic coupler, 17/19
Acoustic delay line memory, 1/8
ACS and *ACS Newsletter*. *See* Amateur Computer Society
Activenture Corporation, 20/2
Adam computer. *See* Coleco Industries
Adams, Scott, 7/10
Adelson, Sheldon, 19/5
Adobe Systems Inc., 10/6, 13/23-24, 16/15, 19/18
PostScript, 10/6, 13/23-24
Advance Computing Environment. *See* ACE
Advanced BASIC (BASICA). *See* Microsoft/Programming Lang.
Advanced DOS. *See* Microsoft/Operating Systems
Advanced Micro Devices (AMD), Inc., 3/15, 8/5, 8/9, 14/7, 16/7, 16/16
Am386 microprocessor, 14/7
Am486 microprocessor, 14/7
AMD 8080A microprocessor, 4/20
K5 microprocessor, 14/7
K6 microprocessor, 14/7
K6-2 microprocessor, 14/7
Advanced Research Projects Agency. *See* ARPA
Advanced RISC Machines (ARM) Ltd., 14/10
ARM 610 microprocessor, 14/10
Advanced Systems Division *See* Xerox/Miscellaneous
Advantage Professional Word Processor. *See* MultiMate International
Advantage series computers. *See* North Star Computers
Adventure game. *See* Games and Microsoft/Applications
Adventure International, 7/10
Adventure Land game, 7/10
Laser Ball game, 7/11
Fire Copter game, 7/11
Pirate Adventure game, 7/11
AFIPS. *See* American Federation of Information Processing Societies
Agenda conference, 19/25
Agents, 15/15
Agfa company, 19/18
Agilent Technologies Inc., 16/17
Agrawal, Anant, 8/8, 11/26
Ahl, David H., 4/1, 18/2, 18/9
Aiken, Howard H., 1/3-4 1/7
Ainsworth, Dick, 13/29
AIX (Advanced Interactive Executive). *See* IBM/Software
Akers, John F., 9/23-24, 16/5-6
Alagem, Beny, 11/22-23, 16/16
Albrecht, Robert L., 18/1, 19/1-2
Alcoff, Larry, 19/11
Aldhadeff, Victor, 19/22
Aldus Corporation, 13/24, 16/15
PageMaker, 10/7, 13/24, 16/15
Alex. Brown & Sons, 12/8
ALGOL (Algorithmic Language), 2/5-6, 5/2, 7/5
ALGOL 30. *See* Dartmouth College
ALGOL 60, 7/5
Burroughs ALGOL. *See* Burroughs
Alien Rain game. *See* Brøderbund Software.
Allchin, James, 16/9
Allen, Paul G.
Microsoft, 4/11, 6/1-14, 12/1, 12/4, 12/8, 12/12, 12/14, 16/12
Other companies, 4/10, 10/19
Paul Allen Group, 19/23
Allison, Dennis, 7/4
Alpert, Martin A., 17/19
Alpha computer project (HP 3000). *See* Hewlett-Packard
Alpha microprocessor. *See* DEC/Miscellaneous

Alpha project (HP3000). *See* Hewlett-Packard
AlphaWorks integrated program, 13/22
Alps Electric Company, 5/14, 10/11, 10/15, 10/21, 12/5, 17/8
Alsop, Stewart, 19/25
Altair 680b and 8800 computers. *See* MITS
Altair BASIC and Bus. *See* MITS
Alto computers. *See* Xerox/Computers
Am386 and Am486 microprocessors. *See* Advanced Micro Devices
Amateur Computer Group of New Jersey, 19/4
Amateur Computer Society (ACS), 2/14, 4/3, 18/1, 19/1
  ACS Newsletter, 2/14, 4/3, 18/1, 19/1
Amateur computing, 2/14-16
Amazon Inc., 19/20
Ambra computers. *See* IBM/Computers (Personal)
AMD. *See* Advanced Micro Devices
AMD 8080A microprocessor. *See* Advanced Micro Devices
Amdek company, 17/12
Amelio, Gilbert F., 16/1-2
America Online (AOL), Inc., 15/12, 16/17, 19/15
  America Online service, 19/15
American Computer Museum. *See* The American Computer Museum
American Federation of Information Processing Societies Inc., (AFIPS), 18/7, 19/5
American National Standards Institute. *See* ANSI
American Telephone & Telegraph Corporation. *See* AT&T
Ami Pro word processor. *See* Lotus Development
Amiga Computer Corporation, 11/5
Amiga computer. *See* Commodore International
An Introduction To *Microcomputers*, 11/6, 20/1
Analyses Conseils Informations. *See* ACI
Analytical Engine. *See* Computer History Association of California
Anderson, Harlan E., 1/15-16
Anderson, John J. 18/9
Anderson, J. Reid, 17/4, 17/16
Anderson, Tim, 13/24
Andreessen, Marc, 15/10-11
Annals of the History of Computing. *See* IEEE
Ansa Software company, 13/18
ANSI (American National Standards Institute), 2/12, 4/19, 6/11, 7/3, 13/7
  ANSI BASIC, 6/10
  G subset of PL/I, 2/12, 13/9
  "Full" BASIC, 7/4, 13/7
  Minimal BASIC, 7/4
A-0 compiler, 1/12
AOL. *See* America Online
APDA (Apple Programmer's and Developer's Association). *See* Apple Computer /Miscellaneous
APL (A Programming Language), 4/4, 4/12, 4/19
  Also see Microsoft /Prog. Lang.
Apollo Computer company, 9/17, 11/21, 11/26
Apollo II computer, 11/14
Apple Assembly Line newsletter, 18/5
Apple Computer Company, 5/6, 5/9
Apple Computer, Inc.
  1970's, 5/1-17;
  1980's, 10/1-26;
  1990's, 14/10-11, 16/1-3
  IBM, 9/2, 9/23, 16/15
  Markets, 4/21, 11/1, 19/15
  Microsoft, 6/10-11, 12/2-3, 12/6, 12/10, 12/20-21, 12/27, 16/8-9, 16/11
  Miscellaneous, 2/11, 13/5-6, 17/8-9, 17/14, 17/15
  NeXT, 11/11, 16/16
  Other companies and organ's, 3/11, 4/6, 4/17, 16/13-14, 16/17, 19/2, 19/9-10, 19/17, 20/2
  PowerPC Alliance, 14/3, 14/6, 16/5, 16/16, 19/19, 20/5
  Software, 7/3-5, 7/7-8, 7/10, 11/13, 13/3-4, 13/12-13, 13/18-20, 13/24, 13/27, 15/7-8

Accessories:
- 20SC hard drive, 10/12
- Disk II drive, 5/12, 5/14, 5/17, 7/3, 10/1, 10/5, 10/12, 10/21, 13/3, 17/8
- DuoDisk, 10/5
- ImageWriter printer, 10/5, 10/21, 13/20, 17/14
- LaserWriter printer, 10/6, 13/24, 17/16
- LaserWriter II printer, 17/16
- LaserWriter Plus printer, 10/7, 10/24, 17/16
- ProFile hard disk, 10/11, 10/15, 10/17, 17/9
- Scribe printer, 10/13
- Silentype printer, 5/14
- Twiggy floppy disk drive, 5/14 10/15, 10/17, 10/21
- UniDisk, 10/6

Computers:
- Apple I Board, 4/10, 5/4-7, 5/11, 19/4, 19/9-10
- Apple II,
  - Apple corporate, 5/1, 10/1-2, 10/5, 10/11, 10/18, 14/11, 16/3
  - Clones, 11/13-14, 17/16
  - Development of, 5/10-15
  - Magazines, 18/5, 18/8
  - Markets, 4/1, 4/15, 5/14, 9/1
  - Microsoft, 6/9, 6/14, 12/1, 12/21, 12/23
  - Miscellaneous, 3/14, 9/13, 10/6, 11/16, 17/17-19, 19/9
  - Software, 7/5, 7/8-9, 7/10-11, 10/5, 10/9, 10/11, 10/14, 10/23, 13/3, 13/7-9, 13/12, 13/18, 13/24-27
- Apple II clones, 11/13-14, 17/16
- Apple II Plus, 4/1, 4/16, 5/14, 10/12, 17/17
- Apple IIc, 10/5, 10/7, 10/12-13, 14/9
- Apple IIc Plus, 10/8, 10/13
- Apple IIe, 10/5, 10/6, 10/11-14, 13/18, 13/20, 17/18
- Apple IIGS, 8/6, 10/7-8, 10/12-14, 13/4, 13/21, 13/30, 18/6
- Apple IIx project, 10/3, 10/13
- Apple III,
  - Development of, 5/14-15, 10/2, 10/9-11
  - Miscellaneous, 10/5, 10/15, 10/18, 17/9
  - Software, 13/3, 13/8, 13/12, 13/20
- Apple III Plus, 10/5, 10/11, 17/12
- Apple 32 SuperMicro product line, 10/16
- Cortland project (Apple IIGS), 10/13
- eMate, 16/2
- Junior Newton project, 14/10
- Lisa,
  - Apple corporate, 10/1-2, 10/5-6, 10/9-11, 10/17
  - Development of, 5/15-16, 10/14-17
  - Miscellaneous, 3/11, 10/19-21, 10/23, 12/18, 17/23
  - Software, 10/19, 13/5
- Lisa 2, 10/2, 10/5, 10/11, 10/17
  - Lisa 2/5, 10/17
  - Lisa 2/10, 10/17-18
- Macintosh,
  - Apple corporate, 10/1, 10/5-6, 16/1-3
  - Development of, 5/16, 10/18-23
  - Magazines, 18/6, 18/8
  - Microsoft, 12/2-3, 12/5-7, 12/10, 12/17-19, 12/23-24, 12/26-28, 15/4, 16/8-9, 16/11
  - Miscellaneous, 3/11, 10/8-11, 10/17-18, 11/11-12, 14/11, 16/3, 17/8, 17/16, 17/21, 17/23, 19/9, 19/17, 19/19
  - Software, 10/7, 10/14, 13/5-6, 13/11, 13/15-17, 13/19-20, 13/23-24, 13/30, 15/7, 15/10-11
  - Fat Mac, 10/23
  - iBook portable, 14/11, 16/2
  - iMac, 14/11, 16/2
  - Mac LC, 14/9

Macintosh II, 10/8, 10/24-25, 13/7
Macintosh IIci, 10/8, 10/25
Macintosh IIcx, 10/8, 10/25
Macintosh IIx, 8/6, 10/8, 10/25
Macintosh Plus, 10/7, 10/24
Macintosh Portable, 10/8, 10/26
Macintosh SE, 10/8, 10/24
Macintosh SE/30, 10/25
Macintosh XL, 10/18
Power Macintosh, 14/11
PowerBook 500 Series portable, 14/11
TurboMac, 10/23
Newton MessagePad (PDA), 10/8, 14/10, 16/2
Phoenix project (Apple IIGS), 10/13
Rambo project (Apple IIGS), 10/13
Sara project (Apple III), 5/14, 10/9
Miscellaneous:
Apple II Forever conference, 10/12
Apple Fellow, 10/5, 11/11
Apple Programmer's and Developer's Association (APDA), 19/16-17
Apple University Consortium (AUC), 10/23
AppleBus, 10/18
AppleFest, 10/3
AppleWorld, 10/24
IWM (Integrated Woz Machine), 10/21
MacWorld Expo, 10/8, 19/5
Mega II chip, 10/13-14
NuBus architecture, 10/25, 17/21
Software:
Application Programs:
Apple Pie (AppleWorks), 13/19
Apple Writer, 5/14, 7/7, 10/5, 13/12
Apple Writer II, 13/12
Apple Writer IIe, 10/12
AppleLink, 10/7
AppleNet, 10/17
AppleShare, 13/27
AppleTalk Personal Network, 10/6, 13/27
AppleWorks, 10/5, 13/20, 18/6
AppleWorks GS, 10/8, 13/21
Boot 13 utility, 7/3
Breakout game, 5/7, 7/10
FileServer, 10/6
FST's (File System Transactors), 13/4
GS Works, 13/21
HyperCard, 10/8
HyperCard IIGS, 15/7
HyperTalk, 10/8, 13/30
Lisa Office System, 10/16, 10/18
LisaCalc, 10/16, 13/15
LisaDraw, 10/16
LisaGraph, 10/16
LisaGuide, 10/16
LisaList, 10/17
LisaProject, 10/17
LisaTerminal, 10/17
LisaWrite, 10/16
LocalTalk, 13/27
Lunar Lander game, 7/10
MacDraw, 10/23
Macintosh Office, 10/6
MacPaint, 10/20. 10/22
MacProject, 10/23
MacSketch, 10/20
MacTerminal, 10/23
MacWrite, 10/22, 13/13
Penny Arcade game, 7/10
QuickFile, 13/18-19
QuickFile IIe, 10/12, 13/18
QuickDraw, 10/12, 10/16, 10/22
QuickDraw II, 10/14
Switcher, 10/6
Toolbox, 10/20, 10/22
Operating Systems:
A/UX operating system, 10/25
Blue project (System 7), 10/25
Copland project, 15/7
Desktop Manager, 10/16
DOS (Disk Operating System), 5/12, 7/3
Finder program, 10/20, 10/22
GS/OS operating system, 10/14, 13/4
MacWorks operating system, 10/18
MultiFinder operating

system, 10/8
OS 8, 9 and X, 15/8
Pink project, 10/25, 19/19
ProDOS (Professional Disk Operating System), 10/5, 10/13, 13/3-4, 13/6
ProDOS 8, 10/14, 13/4
ProDOS 16, 10/14, 13/4
Rhapsody project, 15/7
Sophisticated Operating System (SOS), 10/10, 13/3
System 7, 10/25, 15/7
Window manager, 10/16
Programming Languages:
Apple BASIC, 5/5, 5/7, 5/10
Apple Business BASIC, 10/10
Apple FORTRAN, 10/2
Apple II Pascal, 5/14, 7/3
Applesoft BASIC, 5/12, 6/10, 10/12, 10/13-14, 10/23
Applesoft Extended BASIC, 5/14
Assembler/Debugger, 10/23
Integer BASIC, 5/10, 6/10
Logo, 10/23, 13/5
MacBASIC, 10/20, 10/23, 12/7, 12/21
Pascal, 10/10, 10/23

Apple II Review. *See The Apple II Review*
Apple IIGS Buyer's Guide magazine, 18/6
Apple Programmer's and Developer's Association. *See* Apple/Miscellaneous
Apple Pugetsound Program Library Exchange user group. *See* A.P.P.L.E.
Apple Writer. *See* Apple Computer/Software
AppleBus, AppleFest and AppleWorld. *See* Apple Computer/Miscellaneous
AppleWorks Forum newsletter. *See* National AppleWorks Users Group
Application programs, 12/22
Applicon company, 2/10
Applied Computer Technology, 4/20
Applied Engineering, 17/18
PC Transporter card, 17/18
Transwarp card, 17/18
APX (Advance Processor Architecture). *See* Intel
Aquarius computer. *See* Mattel
ARC (Augmented Research Center). *See* Stanford Research Institute
ARC (Automatic Relay Computer), 1/9
ARC (Average Response Computer). *See* MIT
Ariel Publishing, 18/5
ARITH-MATIC software, 1/12
ARM. *See* Advanced RISC Machines
ARPA (Advanced Research Projects Agency), 2/3-4, 2/12-13
IPTO (Information Processing Techniques Office), 2/3-4, 2/9, 2/13
ARPANET, 2/4, 2/12-13, 4/6, 7/10, 19/13
Arrowhead Computer Co., 19/10
Artificial intelligence, 13/29
Artwick, Bruce A., 13/26
Artzt, Russell M., 13/29
ASCC (Automatic Sequence Control Calculator), 1/4, 1/6
ASCII characters, 17/11
ASCII Corporation, 6/12, 12/9
ASCII magazine, 18/3
Ashton, Alan, 13/10-11, 16/15
Ashton-Tate, Inc., 7/9, 13/12, 13/15-16, 13/19, 13/21, 16/14
dBASE II, 7/9, 13/16, 16/14
dBASE III, 13/16
dBASE III PLUS, 13/16
dBASE III PLUS LAN PACK, 13/16
dBASE Mac, 13/16, 13/19
Framework, 13/20-21
Framework II, 13/21
Full Impact spreadsheet, 13/15
ASR-33 terminal. *See* Teletype Corporation
Assembler programming language, 4/19, 7/2, 13/18, 15/8
Also see Microsoft /Prog. Lang.
Assembler/Debugger. *See* Apple Computer/Software
Assistant series of programs.

See IBM and Software Publishing Corporation
Association for Computing Machinery *See* ACM.
Associations, 1/14, 19/17-19
AST Research, Inc., 11/14, 16/15, 17/21, 20/4
Combo Card, 17/19
Premium/286 computer, 11/14
Asymetrix Corporation, 12/4, 19/23
AT Bus. *See* IBM/Miscellaneous
AT&T (American Telephone & Telegraph Corporation),
Apple Computer, 10/5, 10/25
C language, 7/5
Microsoft, 12/1, 12/15
Miscellaneous, 2/12, 12/28, 17/19, 18/2, 19/18, 20/6
Other companies, 8/8, 9/17, 11/26, 16/14
UNIX operating system,
C language, 7/5
IBM, 9/17, 9/21, 12/11
Microsoft, 12/1, 12/15, 12/17, 16/9
Miscellaneous, 2/12, 13/6, 15/10, 19/14, 19/17-18
Other companies, 10/25, 11/12, 11/25-26, 13/4, 13/27, 15/9, 15/11, 16/13
UNIX System V operating system, 9/17
Atanasoff, John V., 1/4-5, 1/8-9
Atari Corporation,
Apple Computer, 5/3, 5/6-9, 5/17, 10/5
Corporate, 11/1, 11/5, 11/14, 16/16
Early developments, 4/17-18
Games, 7/10-11, 19/15
IBM, 9/4
Markets, 4/15, 4/21, 5/15
Software, 13/12, 13/14
Atari 400, 3/14, 4/17
Atari 800, 3/14, 4/17
Atari 1200XL, 11/14
Breakout game, 4/17, 5/7, 7/8
Pong tennis game, 4/17
ATI Technologies Inc., 19/20
Atkinson, Bill, 5/15, 7/5, 10/5, 10/8, 10/15, 10/19-20, 10/22, 19/22
Atlantic Monthly magazine, 1/14
AuctionWeb Internet site, 19/21
AUGMENT. *See* Tymshare
Augmented Research Center. *See* Stanford Research Institute
Autodesk, Inc., 13/23, 16/15
AutoCAD, 13/23
Automatic Sequence Control Calculator. *See* ASCC
A/UX operating system. *See* Apple Computer/Software
Auxiliary storage, 1/7
Avant-Garde Creations, 13/28
Creative Life Dynamic series, 3/28

--B--

Baby Blue card. *See* Xedex
BackOffice software. *See* Microsoft/Applic. Progs.
Backus, John W., 1/13
Baer, Ralph, 2/14
Bailey, David, 17/10
Baker, Al, 13/29
Baker, Bill, 7/7, 7/9, 13/9-10
Balakrisman, Jay, 13/28
Balleisen, Gary, 13/15
Ballmer, Steven (Steve) A., 6/4, 12/1, 12/6, 12/8, 12/11, 12/19, 12/22, 15/8, 16/9-12
Bank Street Writer. *See* Brøderbund Software
Bardeen, John, 1/11
Barksdale, James L., 15/11
Barnaby, Bob, 7/7
Barnes, Susan, 11/11
Barnhart, Dennis, 11/30
Baron, Paul, 2/13
Barrett, Craig R., 16/8
Bartz, Carol, 16/15
Barzilay, Jason, 11/23
BASIC (Beginner's All-purpose Symbolic Instruction Code),
Apple Computer, 5/4, 5/6-8, 5/10, 5/13, 10/10-11, 10/13-14, 10/18-19, 10/21-22
Development of, 2/3, 2/5-6, 7/3-5, 7/12, 13/7
IBM, 4/3, 4/12, 9/4, 9/7
Microsoft, 6/3, 6/5-11, 15/5, 17/17-18
MITS, 4/9-11

Other companies, 4/15, 4/17-20, 7/8, 11/28, 11/3-4, 11/7, 11/10, 11/13, 11/18, 11/20, 11/24, 13/2, 18/3
Also see Microsoft/Progr. Languages
BASIC computer. *See* Sphere Corp
Basic Input/Output System (BIOS), 13/22
Compaq Computer, 11/9
Digital Research, 7/2
IBM, 9/5, 9/7, 9/12, 11/19, 12/14, 13/22-23
Zenith Data Systems, 11/29
BASIC interpreters (Altair and Macintosh). *See* Microsoft /Progr. Lang.
BASIC-80. *See* Microsoft/Progr. Lang.
BASIC-Plus. *See* Digital Equipment/Software
BASICA (Advanced BASIC). *See* Microsoft/Progr. Lang.
Bastian, Bruce, 13/10-11, 16/15
Bates, Roger, 17/23
Bauer company, 12/11, 12/20
Baum, Allen J., 5/1-4, 5/7, 5/10
Bauman, Joe, 9/4, 9/6, 9/23
Bawden, David, 20/2
BBC. *See* British Broadcasting Corporation
BBC BASIC, BBC Model A and B computers. *See* Acorn Computers
BCPL programming language. *See* Xerox/Software
Be Inc., company, 16/14
BeBox computer, 16/14
Beagle Bros. company, 13/20, 18/5
MacroWorks, 13/20
TimeOut modules, 13/20
Bechtolsheim, Andreas, 11/24-25
Beckman Instrument Systems, 5/9
Bedke, Janelle, 13/18
Bell, Jay, 11/17
Bell, Murray, 11/21-22
Bell Telephone Laboratories, 1/3, 1/6, 1/11, 2/4, 2/11, 7/3, 13/7, 20/6
General Purpose Relay Calculator, 1/3
Models III, IV, V and VI computers, 1/3
Bell System Technical Journal, 20/6
Belleville, Robert L., 10/20
Bendix Aviation Corporation, 1/16
G-15 computer, 1/16
Benton Harbor BASIC. *See* Heath Company
Berez, Joel, 13/24-25
Berkeley campus. *See* University of California
Berkeley Softworks, 13/6
GEOS (Graphic Environment Operating System), 13/6
Berkeley UNIX operating system, 11/25
Berners-Lee, Tim, 19/14, 20/3
Bernoulli, Daniel, 17/10
Bernoulli box and disk drive. *See* Iomega
Bernstein, Alex, 1/13
Berry, Clifford E., 1/5
Bezos, Jeffrey, 19/20
Bibliographies, 20/2
Big Blue disk magazine. *See* Softdisk
Bina, Eric, 15/10
BINAC (Binary Automatic Computer), 1/5, 1/8
BIOS. *See* Basic Input/Output System
Bishop, Bob, 7/10
Bit - term origin, 20/6
Bit Pad. *See* Summagraphics
BitBlt procedure, 4/5
BIX on-line service, 19/16
Blank, Marc, 13/24
Blankenbaker, John V., 4/3
Bletchley Park, 1/3-4
Blue Box phone tone generator, 5/3
Blue project. *See* Apple Computer/Software
Blumenthal, Jabe, 12/24
Bob operating system. *See* Microsoft/Oper. Systems
Bobrow, Daniel, 2/12
Boeing Computer Services, 16/8
Boggs, David, 19/13
Boich, Mike, 10/7
Bolt Beranek and Newman Inc., (BB&N), 2/4-5, 2/12-13
Bookshelf. *See* Microsoft/Multimedia
Boone, Gary W., 3/12, 14/8

Boot 13 utility. *See* Apple
Computer/Software
Booth, A. D., 1/9
Borel, Daniel, 19/22
Borland International, Inc.,
Beginning of, 13/8-9
Miscellaneous, 13/19,
15/5, 16/12, 16/14, 16/17
Programs, 12/21, 13/7,
13/15, 13/29
Paradox database, 13/9,
13/19, 16/12
Quattro Pro spreadsheet,
13/9, 13/15 16/12
Sidekick, 13/9, 13/29
Turbo BASIC, 12/21, 13/7
Turbo C, 12/21, 13/7
Turbo Pascal, 12/21, 13/8
Bosack, Leonard, 19/21
Boston Computer Museum. *See*
The Computer Museum
Boston Computer Society, 19/3
Both Barrels game. *See* Sirius
Software
Bowles, Kenneth L., 7/5
Bowman, William, 13/28
BPC microprocessor. *See*
Hewlett-Packard
BPI accounting software, 13/29
Bradley, David J., 9/5, 12/13,
13/22
Bradley, Terry, 13/25
Brainerd, Paul, 13/24
Brattain, Walter, 1/11
Bravo and BravoX word
processors. *See* Xerox/
Software
Breakout game. *See* Atari
Bricklin, Daniel S., 5/14,
7/8, 13/15, 13/29, 19/12
British Broadcasting
Corporation (BBC), 11/13
British Telecom company, 13/27
Brock, Rod, 12/14, 17/18
Brockman, Inc., 19/25
Brockman, John, 19/25
Brøderbund Software, Inc.,
13/12, 13/24-25
Alien Rain, 13/25
Bank Street Writer word
processor, 13/12
Choplifter, 13/25
David's Midnight magic,
13/25
Galactic Empire, 13/25
Galactic Saga series, 13/25
Galactic Trader, 13/25
Lode Runner, 13/25
Print Shop, 13/24
Space Quarks, 13/25
Where in the World is Carmen
Sandiego?, 13/25
Brodie, Richard, 12/25-26
Broedner, Walt, 10/11
Brookhaven National
Laboratory, 1/13
Brother International
Corporation, 17/16, 19/20
Brown, David, 17/9
Brown, Dick, 10/4, 12/6, 19/11
Brown University, 10/4
Browsers, 15/10-11
Bubble jet printing, 17/14
Bubble memory, 11/18
Buchholz, W., 20/6
Budge, Bill, 7/10, 13/26
Budgeco company, 13/26
Raster Blaster game, 13/26
Bug - term origin, 20/6
Bull company, 11/30
Bunnell, David, 6/8, 18/2-3,
18/6-7, 19/4
Burroughs Corporation, 7/6,
17/12
ALGOL, 7/6
Printer, 17/12
Burtis, Don, 6/14
Bus systems, 17/20-21, 20/3
Bush, Vannevar, 1/14-15, 2/9,
20/1
Bushnell, Nolan K., 4/17, 5/9,
7/10
Busicom company, 3/6-8, 4/7
Business Accounting Series
software. *See* Peachtree
Software
Button, Jim, 13/18
ButtonWare company, 13/18
PC File database, 13/18
Bybe, Jim, 4/9
Byte - term origin, 20/6
BYTE magazine,
Beginning of, 18/2-3
Borland, 13/9
Microprocessor, 3/17-18
Miscellaneous, 4/11-12,
17/5, 17/7, 17/23, 19/3,
19/16, 20/6-7
North Star Computers, 17/19
VisiCalc, 7/9
Byte Shop's, 5/6, 5/9, 19/10
Bytec Management Corporation,
11/21, 13/27
Bytec-Comterm, 11/22
Hyperion computer, 11/10,
11/21

Bytesaver board. *See* Cromemco.

--C--

C & E (Coleman & Eubanks) Company, 13/29
C programming language, 4/5, 7/5, 13/7, 13/17, 15/8, 15/13, 19/26
C++ programming language, 13/7, 15/13
C.Itoh Electronics (CIE), Inc., 10/17, 17/14
Cable-TV, 15/15, 16/10
Cache, 8/4, 8/6
CAD (Computer Assisted Drafting), 2/10, 9/17, 11/25, 13/23, 20/4
CAE (Computer Assisted Engineering), 2/10, 11/24
Cahners Publishing, 18/7
Cairo project. *See* Microsoft/ Operating Systems
CalcStar. *See* MicroPro International
Calculators,
  Busicom company, 3/6, 4/6, 4/8
  Early computers, 1/3, 1/7
  Miscellaneous, 3/5, 3/12, 4/2, 7/8, 17/9, 19/1
  Other calculator products, 4/3, 4/8, 4/15, 4/19, 11/23, 11/28
Calculedger spreadsheet, 7/8
Call-A.P.P.L.E. magazine. *See* A.P.P.L.E.
Call Computer company, 5/4
Calma company, 2/10
CAM (Computer Assisted Manufacturing), 11/24
Cambridge University, 1/5, 1/12
Campbell, Gordon, 8/7
Campbell, Rob, 12/10
Campbell, William V., 13/30, 16/2
Canion, Rod, 11/9, 16/3
Cannavino, James A., 9/21-22, 9/24-25, 12/17, 15/8, 16/5-6
Canon Inc., 3/5, 16/15, 17/8, 17/14-15
  LPB-CX laser printer, 17/15
  Pocketronic calculator, 3/5
Cantin, Howard, 5/10
Capellas, Michael D., 16/4
Capps, Steve, 10/20, 14/10
Captain Crunch, 5/3
Carlston, Douglas G., 13/25
Carlston, Gary, 13/25
Carmack, Adrian, 19/22
Carmack, John, 19/21
Carnegie-Mellon University, 13/4
Carr, Robert, 13/21
Cartridge BASIC. *See* Microsoft/Prog. Languages
Cary, Frank T., 9/1
Case, Stephen M., 18/9, 19/15
Cashmere project (Word for Windows). *See* Microsoft/ Application Programs
Cassette BASIC. *See* Microsoft/Progr. Languages
Cassette tape, 5/6, 5/10, 5/12, 11/24, 17/4-5, 17/19
CAT acoustic coupler. *See* Novation
Caulkins, Dave, 19/14
Cayre, Joseph J., 19/22
C-BASIC, 7/4, 11/7, 11/12
CBASIC-86. *See* Digital Research
CBI. *See* Charles Babbage Institute
CBM-8032 computer. *See* Commodore International
CBS company, 19/16
cc:Mail. *See* Lotus Development
CD-ROM (Compact Disk - Read Only Memory), 12/7, 12/28, 15/9, 17/9, 20/2, 20/4-5
  CD-I (Compact Disk - Interactive), 12/7, 20/4
  CD-R (Compact Disk - Recordable), 20/4
  CD-ROM XA standard, 20/4
CD-ROM conference, 12/7, 12/10
CD-ROM division. *See* Microsoft/Multimedia
CD-Write disk. *See* Cytation
CeBIT computer show, 19/5
Celeron microprocessor. *See* Intel/Microprocessors
Central Intelligence Agency. *See* CIA
Central Point software company, 16/14
Centronics Data Computer Corporation, 17/12
  700 series printers, 17/12
  779 printer, 17/12
  Micro-1 printer, 17/12
  Model 101 printer, 17/12
CERES workstation, 13/8
CERN (European Particle Physics Laboratory), 19/14,

20/3
CES. *See* Consumer Electronics Show
CGA (Color Graphics Adapter), 9/5, 11/18, 14/9, 20/5
Challenger series of computers. *See* Ohio Scientific Instruments
Chamberlain, Hal, 4/3, 18/3
Chamberlain, Mark, 4/10-11
Chambers, John T., 19/21
Chang, Dean, 19/23
Chaplin, Charlie characture, 9/6
Charles Babbage Institute (CBI), 19/6-7
Newsletter, 19/6
Charlson, Chester, 17/15
Charney, Howard, 17/18
Chart software. *See* Microsoft/ Application Programs
Cheadle, Edward, 2/11
Cheiky, Mike and Charity, 4/14
Chess project. *See* IBM/ Computers (Personal)
Chicago project (Windows 95). *See* Microsoft/Oper. Systems
Chieftain computer. *See* Smoke Signal Broadcasting
Chips and Technologies, Inc., 8/7, 11/17, 16/7
Choplifter game. *See* Brøderbund Software
CIA (Central Intelligence Agency), 13/17
Cider Press newsletter, 18/5
CIE (Community Information Exchange) Net, 19/13
Circus project (PC AT). *See* IBM/Computers (Personal)
Cirrus project. *See* Microsoft/Applic. Programs
Cisco Systems, Inc., 19/21
CL-9. *See* Cloud-9
Clamshell project (5140 PC Convertible.) *See* IBM/ Computers (Personal)
Claris Corporation, 10/8, 13/20, 13/21
Clark, James H., 11/23, 15/11-12
Clark, Wesley A., 1/15, 2/2, 2/7
Clayton, Dubilier & Rice company, 16/4
Clones, 4/13, 9/20, 11/13, 11/30, 12/3, 13/23
Cloud-9 (CL-9) company, 10/6
Clubs (Computer), 19/2-3
CMC (Corporate Management Committee), *See* IBM/Miscellaneous
CMG@Ventures company, 19/23
CMOS (Complementary Metal Oxide Semiconductor), 3/13
CMP Media Inc., 18/2
COBOL (COmmon Business Oriented Language), 1/12-13, 2/11, 6/2, 6/9
Also see Microsoft/ Programming Languages
Cochran, Michael, 3/12, 14/8
Cocke, John, 8/7, 14/8
CoCo computer. *See* Tandy/Radio Shack
CODASYL (Committee On Data SYstem Languages), 1/13, 2/11
Codd, Edgar F., 13/17
Cole, Ida, 12/6
Coleco Industries, 11/30
Adam computer, 11/30
Coleman, Deborah "Debi" A., 10/7
Coleman, Dennis, 13/29
Coleman, William T., 13/5
Collard, Paul, 17/19
Colligan, John C., 19/23
Color Graphics Adapter. *See* CGA
Colvin, Neil, 13/22
Combo Card. *See* AST Research
Comcast Corporation, 16/11
COMDEX (COMputer Dealers' Exposition),
Beginning of, 19/5
Miscellaneous, 9/21
Product introductions at, 9/11, 11/17, 11/21, 12/17-19, 13/5, 13/7, 13/10-11, 13/14
Softbank purchase of, 19/24
Commodore Business Machines Company, 4/15
Commodore International Inc.,
Apple Computer, 5/8, 5/14
Beginning of, 4/15-16
Later products, 11/4-5
Liquidation of, 16/15
Market, 4/15, 4/21, 5/17, 9/1-2, 11/1
Microsoft, 6/9, 6/11
Miscellaneous, 3/14, 11/14, 11/28, 13/20, 17/20, 19/8, 19/15
3-Plus-1 software, 11/5,

Computerfest, 19/4
ComputerLand, 4/3, 4/13, 4/20, 9/8, 19/11
Computers & Electronics, 18/8
Computervision Corporation, 2/10
ComputerWorld, newspaper, 18/7
Computing-Tabulating-Recording Company, 1/6
Computist magazine, 18/8
Concentric Data Systems, 13/13
Concurrent CP/M-86 and Concurrent DOS. *See* Digital Research
Condor database, 7/9
Connelly, Dan, 20/3
Conner Peripherals, Inc., 17/6, 17/9
Conner, Finis F., 17/7-9
CONSOL operating system. *See* Processor Technology.
Consumer Electronics Show (CES), 5/12, 11/4-5, 11/13, 19/4
Consumer Products Division. *See* Microsoft/Miscellaneous
Context Management Systems, 13/19
  Context MBA, 13/19, 13/21
Continuum company, 12/11
Control Data Corporation (CDC), 2/9, 6/3, 13/5, 17/6, 17/9
  660 computer, 19/9
  Cyber 6400 computer, 6/3
Control Video Corporation, 19/15
Contura computer. *See* Compaq Computer Corporation
Conventions, 19/4
Convergent Technologies, 11/30
  Workslate portable computer, 11/30
Cook, Scott D., 13/30
Cooper, Alan, 7/4
Copland project. *See* Apple Computer/Software
Coprocessors, 8/4-5
Corbató, Fernando J., 2/4
Corbis Corporation, 12/11
Core memory. *See* Magnetic core memory
Corel Corporation, 13/17, 15/9, 15/12, 16/12, 16/15
Corel Systems Corporation, 13/24, 13/27, 16/15
  CorelDRAW, 13/27, 15/9, 16/15
  WordPerfect suite, 15/9
Cornell University Classic Computer Club, 19/10
Cornet, Jean Claude, 3/9
Coronado Corporation, 3/12
Corporate Management Committee (CMC). *See* IBM/Miscellaneous
Corrigan, Robert J., 16/5
Cortland project. *See* Apple Computer/Computers
Corvus Concept computer, 11/15
Couch, John D., 5/15, 10/15
Courtney, Mike, 12/14, 12/21
Cowell, Casey G., 17/19
Cowpland, Michael C. J., 11/21-22, 13/27
CP/M operating systems. *See* Digital Research
CP/NET. *See* Digital Research
CP-DOS. *See* IBM/Software
Crawford, John, 8/3, 14/6
Cray, Seymour, 16/14
Cray Research, Inc., 16/13, 16/16
  Cray-1 computer, 19/8
Cream Soda Computer, 5/2, 5/5
Creative Computing magazine, 18/1, 18/3, 18/9
Creative Labs, Inc., 17/16
Creative Life Dynamic series. *See* Avant-Garde Creations
Creative Technology Ltd., 17/16
  Sound Blaster audio card, 17/16
Cromemco Inc., 4/10, 4/14-15, 5/7, 6/14, 17/16-18
  Bytesaver board, 17/16
  Dazzler machine, 5/7
  Kaleidoscope program, 17/17
  System Zero, One, Two and Three computers, 4/14
  TV Dazzler board, 17/17
  Z-1 computer, 4/14
  Z-2 Computer System, 4/14
  Zilog Z-80 board, 17/17
Crosby, Kip, 19/7
Crow, George, 11/11
Crowther, Will, 7/10
CSMA/CD (Carrier Sense Multiple Access with Collision Detection), 19/13
CSOS (Computer System Operating System). *See* IBM/Software
CT-64 Terminal Kit. *See* Southwest Technical Products
CT-VM video monitor. *See*

Southwest Technical Products
CTC. *See* Computer Terminals Corporation
CTSS (Compatible Time Sharing System). *See* MIT
Culbert, Michael, 14/10
Cutler, David N., 12/17
CW Communications company, 4/3, 18/8
Cyber 6400 computer. *See* Control Data
Cyber Strike game. *See* Sirius Software
Cyrix company, 14/7
Cytation company, 12/7, 20/2
CD-Write disk, 12/7, 20/2

--D--

D.C. Hayes Associates Inc., 17/20
Also see Hayes Corporation
D'Arezzo, James, 9/6, 9/23
Dabney, Ted, 4/17
Daisy Systems company, 11/24
Daniels, Bruce, 13/24
Dartmouth College, 1/3, 2/3-6, 7/3-4, 13/7
ALGOL 30, 2/5
BASIC (Beginner's All-purpose Symbolic Instruction Code), 2/3, 2/6, 7/3-4, 13/5
DARSIMCO (Dartmouth Simplified Code), 2/5
DART (Dartmouth programming language), 2/5
DOPE (Dartmouth Oversimplified Programming Experiment), 2/5
DTSS (Dartmouth Time sharing System), 2/6
SCALP (Self Contained ALGOL Processor), 2/5
Structured BASIC (SBASIC), 13/7
True BASIC, 13/7
Darwin workstation. *See* Sun Microsystems
Data General, 2/9, 4/5, 5/2, 6/3, 13/10, 13/13
Nova 1220, 4/5
Nova computer, 2/9, 5/2, 6/3
Data Machines company, 2/9
Data Processing Management Association. *See* DPMA
Data routers, 19/21
Databases, 7/9, 12/22, 12/26, 13/16-21, 15/5, 16/14
DataMaster. *See* IBM/Computers
Datamation, magazine, 1/14, 18/6
Datanet-30 communications computer. *See* General Electric Company
DataPerfect database. *See* WordPerfect
Datapoint Corporation, 3/7, 3/12, 4/2, 12/1
Datapoint 2200 terminal, 4/2, 19/6
Dataquest information provider, 19/21
DataStar. *See* MicroPro International
Datronic company, 11/28
Davidoff, Monte, 6/6-7
David's Midnight Magic. *See* Brøderbund Software
Davies, Donald, 2/12
Davis, Robert, 19/23
Daytona project (NT). *See* Microsoft/Operating Systems
Dazzler machine. *See* Cromemco
dBASE software. *See* Ashton-Tate
DCS (Digital Computer System), 4/18
Deadline game. *See* Infocom
DEC (Digital Equipment Corporation),
Beginning of, 1/15-16
Compaq purchase of, 16/4
Early systems, 2/8-9, 4/2, 4/4, 4/11, 4/19-20
Later systems (1980's), 11/15-16
Microsoft, 6/2-5, 6/9, 6/13, 12/4, 12/18, 15/2
Miscellaneous, 3/6, 4/1, 4/18, 7/8, 8/8, 10/16, 14/7, 16/14, 18/2, 19/17-18, 20/7
Other companies and organizations, 2/4, 7/1, 16/8
Software, 7/5, 13/8, 13/17
Computers:
2020 minicomputer, 6/13
DECmate II word processing system, 11/15
DECstation (VT78), 4/19-20, 13/8
EduSystem, 4/4
LINC-8 computer, 2/7
LSI-11 microcomputer board, 4/1, 4/11, 4/18,

7/5
LSI-11/23 microcomputer system, 4/11
PDP-1 (Programmed Data Processor - One), 1/16, 2/4, 2/8, 2/14
PDP-4, 2/8
PDP-5, 2/8
PDP-8, 2/8, 2/15, 3/6, 4/1, 4/4, 4/20, 19/6, 19/9
PDP-8/A, 4/4
PDP-8/E, 4/2
PDP-8/I, 2/8
PDP-8/L, 2/8, 6/3
PDP-8/S, 2/8
PDP-10, 6/3-4, 6/6, 7/10, 19/9
PDP-11, 4/2, 4/11, 4/18, 7/5, 11/16
PDP-11/03 microcomputer system, 4/11
Professional 300 Series, 11/16
Models 325 & 350, 11/16
Rainbow 100 series, 11/15-16
Rainbow 100, 11/15
Rainbow 100+, 11/15-16
VAX (Virtual Address Extension) computer, 8/8, 12/17, 13/17
Miscellaneous:
Alpha 21064 microprocessor, 14/7, 15/2
DECUS user group, 6/3
F-11 CPU, 11/16
Omnibus backplane, 4/3
VT52 terminal, 4/20, 10/17
VT78. *See* DECstation
VT100 terminal, 10/17
Software:
DEC BASIC, 6/3, 6/5
DEC BASIC-Plus, 6/6
DECwindows, 13/6
Focal language, 6/9
P/OS operating system, 11/16
RT-11 operating system, 11/16
VMS (Virtual Memory operating System), 12/17
DECUS user group. *See* DEC/Miscellaneous
Delbourg-Delphis, Maryléne, 13/19
Dell Computer Corporation, 9/25, 11/16-18, 16/4-5
Dell, Michael S., 11/16-17, 16/4
Delphi on-line service. *See* General Videotex
Denman, Donn, 10/20, 10/23
Dennis, Jack B., 2/4, 7/2-3
Department of Defense, 2/3, 19/13
Department of Justice (DOJ), 9/1, 9/23, 15/12, 16/9-11, 16/14
Deskpro computers. *See* Compaq Computer
Desktop Management Interface. *See* DMI
Desktop Management Task Force. *See* DMTF
Desktop Manager. *See* Apple Computer/Software
Desktop publishing, 13/23-24, 13/27, 16/15
Desktop Software Division. *See* IBM/Miscellaneous
DESQ and DESQview software. *See* Quarterdeck Office
Develop-65, 68 and 80 software. *See* Microsoft /Miscellaneous
Dhuey, Mike, 10/24
DIF format, 13/14
Digerati: Encounters with the Cyber Elite, 19/25
Digital Computer Newsletter, 1/14
Digital Computer System. *See* Digital Group
Digital Deli book, 17/23, 18/9
Digital Equipment Corporation. *See* DEC
Digital Group Inc., 4/20, 7/4
Digital Computer System, 4/20
Digital Logic Microlab. *See* Southwest Technical Products
Digital Research, Inc. (DRI),
Beginning of, 7/1-2
IBM, 9/6, 9/8, 12/13
Later developments at, 13/2-3, 13/5, 13/7, 13/9, 13/27, 15/9
Microsoft, 6/10, 6/14, 12/12, 12/14, 17/17
Miscellaneous, 7/4, 13/24, 13/29, 20/2
Other companies, 11/6, 13/1

Purchase by Novell, 15/9, 15/12, 16/12
CBASIC-86, 13/7
CD-ROM disk, 20/1
Concurrent CP/M-86, 13/3
Concurrent DOS, 13/3
CP/M (Control Program for Microprocessors),
Development of, 7/1-2
IBM, 9/6, 12,12
Microsoft, 6/10-11, 6/14, 12/1, 12/12, 12/14, 17/17
Miscellaneous, 4/17, 7/3-4, 7/7, 7/9, 7/12, 13/1-2, 13/15-16, 13/27, 19/12
Other companies, 11/6, 11/8, 11/12, 11/15-16, 11/18, 17/19
CP/M-80 operating system, 11/18
CP/M-86 operating system, 9/8, 11/28, 12/13-14, 13/1-2
CP/M Plus, 11/7
CP/NET, 13/27
DR-DOS, 13/3, 15/9, 15/12
GEM (Graphics Environment Manager), 13/5
MP/M (Multi-Programming Monitor), 7/2, 13/27
PL/I compiler, 13/9
PL/M. *See* PL/M
Digital signal processor (DSP). *See* Texas Instruments/Misc.
Digital Video Interactive (DVI). *See* RCA company
Digitizer, 17/22
Dilks, John, 19/4
Direct marketing, 9/25, 11/16, 11/19
Disbrow, Steven, 18/6
DISK BASIC. *See* Microsoft/Progr Lang. and PolyMorphic Systems
Disk drives, 1/9, 17/5-10
Disk II drive. *See* Apple Computer/Accessories
Disks. *See* Floppy disks
Diskworld for the Macintosh. *See* Softdisk
Displays, 17/11-12
Displaywrite word processor and DisplayWriter workstation. *See* IBM
Distributors of software, 19/10-12
DLL (Dynamic Link Libraries), 15/3
DMI (Desktop Management Interface), 19/17
DMTF (Desktop Management Task force), 19/17
Document Type Definition. *See* DTD
Dompier, Steve, 4/13, 19/2
DOPE (Dartmouth Oversimplified Programming Experiment). *See* Dartmouth College
Dorado processor. *See* Xerox/Miscellaneous
DOS (Disk Operating System). *See* Apple Computer/Software and Microsoft/Oper. Systems
DoubleSpace utility. *See* Stac Electronics
Dow Jones and Company, Inc., 19/16
Dow Jones News/Retrieval Service, 9/8, 17/20, 19/16
DPMA (Data Processing Management Association), 1/13
DR-DOS. *See* Digital Research
Dr. Dobb's Journal, 4/11, 5/8, 7/4, 18/2-3, 19/4
DRAM (Dynamic Random Access Memory), 3/6, 4/7, 8/6, 17/3-4
Draper, John, 5/4, 7/7, 13/9
DrawPerfect presentation graphics program. *See* WordPerfect Corporation
DRI. *See* Digital Research
DriveSpace. *See* Microsoft/Application Programs
DTD (Document Type Definition), 20/3
DTSS (Dartmouth Time Sharing System), *See* Dartmouth College
Dubinsky, Donna, 14/12
DuoDisk. *See* Apple Computer/Accessories
DVI (Digital Video Interactive). *See* RCA company
Dynabook concept, 4/5, 7/6, 20/1
Dynalogic Corporation, 11/21
Dynamic Random Access Memory. *See* DRAM
Dynamical Systems Research, Inc., 12/9, 12/16

Mondrian software, 12/16, 13/6
Dysan Corporation, 17/9, 17/22
Dyson, Esther, 19/25

--E--

Eagle Computer company, 11/30
EARS laser printer. *See* Xerox/ Miscellaneous
Eastern Joint Computer Conference, 1/16
Eastman Kodak. *See* Kodak
EasyWriter word processor. *See* Information Unlimited Software
E-BASIC, 7/4
eBay Inc., 19/21
Ebrahimi, Farhad Fred, 13/12
ECHO IV (Electronic Computing Home Operator IV), 2/15
Eckert-Mauchly Computer Corporation, 1/12
Eckert, J. Presper, 1/5, 1/8, 1/11
E-commerce (Electronic-commerce), 15/15
ECS Magazine, 18/2
Edit-80 text editor. *See* Microsoft/Applic. Programs
EDLIN text editor. *See* Microsoft/Applic. programs
EDS (Electronic Data Systems) company, 6/14
EDSAC (Electronic Delay Storage Automatic Calculator), 1/5, 1/8, 1/12
Edu System 20 terminal, 19/1
EduSystem computer. *See* DEC/Computers
EDVAC (Electronic Discrete Variable Automatic Computer), 1/5, 1/8
EdWord word processor, 11/15
EEPROM (Electrically Erasable Programmable Read Only Memory), 17/4
EGA (Enhanced Graphics Adapter), 14/9, 20/5
Eggebrecht, Lewis, 9/4
Egghead Discount Software, 19/22
Egghead, Inc., 19/22
EGO Systems company, 18/6
EIA (Electronic Industries Association), 19/18
EISA (Extended Industry Standard Architecture), 11/11, 17/21
Eklund, Jon, 19/9
Electric Pencil. *See* Michael Shrayer Software
Electrically Erasable Programmable Read Only Memory. *See* EEPROM
Electronic Arts Inc., 13/26
  Music Construction Set, 13/26
  Pinball Construction Set, 13/26
Electronic Data Systems company. *See* EDS
Electronic-mail. *See* E-mail
Electronic Design, 17/11
Electronic Industries Association. *See* EIA
Electronic News, 3/7, 20/9
Electronic Paper spreadsheet. *See* Microsoft/Applic Programs
Electronics magazine, 3/5, 3/12
Electrophotographic printing, 17/15
Electrostatic cathode ray tube memory, 1/5, 1/7-9
Elephant company, 17/22
Ellenby, John, 4/14-15, 11/19
Eller, Marlin, 12/18
Ellison, Lawrence J., 13/17, 16/2
Elly, Carol, 17/19
E-mail (Electronic-mail), 2/11, 12/27, 15/3, 15/11,
eMate (PDA). *See* Apple Computer/Computers
EMS (Expanded Memory Specification), 12/20, 20/4
Enable integrated program, 13/22
Encarta. *See* Microsoft/ Multimedia
Encoder Kit. *See* Southwest Technical Products
Encyclopedia of Computer History, 20/3
Encyclopedia of Microcomputers, 3/18, 20/2-3
Engelbart, Douglas C., 2/9-10, 4/5, 17/23, 20/1
Engineering Research Associates (ERA), 1/9
English, William K., 2/10, 4/6, 17/23
Enhanced BASIC interpreter. *See* Hewlett-Packard /Misc.
Enhanced Graphics Adapter. *See*

EGA
ENIAC (Electronic Numeric Integrator and Calculator), 1/5, 1/8-9, 1/12, 19/9
Enquire program, 19/14
Entry Level Systems (ELS). *See* IBM/Miscellaneous
Entry Systems Division (ESD). *See* IBM/Miscellaneous
EP-101 printing device. *See* Epson America
EPD company, 4/4
  System One computer kit, 4/4
EPIC (Explicitly Parallel Instruction Computing). *See* Intel/Miscellaneous
EPROM (Erasable Programmable Read Only Memory), 17/4
Epson America, Inc.,
    Computers, 11/18-19, 12/21
    Miscellaneous, 17/21
    Printers, 9/8, 11/12, 17/13
  EP-101 printing device, 17/13
  Equity Series, Equity I, II & III, 11/19
  Geneva computer, 11/18
  HX-20 computer, 11/18, 12/21
  HX-40 computer, 11/18
  MX-80 printer, 9/8, 11/12
  MX printers, 17/13
  QX computers, QX-10, QX-11 & QX-16, 11/18
  TX-80 printer, 17/13
Epstein, Robert S., 13/18
Equity Series. *See* Epson America
Erasable Programmable Read Only Memory. *See* EPROM
ESP-1 (Extended Software Package 1), 7/6
Espinosa, Christopher, 5/4, 5/10
Esquire magazine, 5/3
Estridge, Philip D. (Don), 9/3, 9/5, 9/9, 9/13, 9/23-24, 10/3, 12/16
ETH (Eidgenossische Technische Hochschule), 7/5, 13/8
Ethernet network. *See* Xerox/ Miscellaneous
Eubanks, Gordon E., 7/4, 13/29
European Laboratory for Particle Physics. *See* CERN
Evans, Kent, 6/3
Ewing, Marc, 15/10
Excaliber Technologies, 11/30
  Powerstation computer, 11/30
Excel spreadsheet. *See* Microsoft/Applic. Programs
Executer word processor, 7/6
Executive computers. *See* Osborne Computer
Executive Word Processor. *See* MultiMate International
Executive WordPerfect. *See* WordPerfect
Expanded Memory Specification. *See* EMS
Experimenters' Computer System (ECS) magazine, 18/3
Explicitly Parallel Instruction Computing (EPIC). *See* Intel/Misc.
Extended BASIC. *See* Microsoft/ Programming Languages
Extended Benton Harbor BASIC. *See* Heath company
Extended Graphics Array. *See* XGA
Extended Industry Standard Architecture. *See* EISA
Extended Memory Manager. *See* XMM
Extended Memory Specification. *See* XMS
Extended Software Package 1. *See* ESP-1
Exxon corporation, 3/14, 8/8
E-Z Draw utilities. *See* Sirius Software

--F--

Faber, Edward, 19/11
Faggin, Federico, 3/6-8, 3/14, 3/17
Fairchild Camera and Instrument Corporation, 1/11
Fairchild Semiconductor Corporation, 1/11, 3/5-6, 3/15-16, 5/2, 5/9, 8/7, 16/7
Fairs, 19/4
Falcon Technology, 12/3, 12/9
Fano, Robert M., 2/4
FAT (File Allocation Table), 6/11, 12/12, 13/2
Fat Mac computer. *See* Apple Computer/Computers
Federal Trade Commission (FTC), 16/8-9
Feeney, Hal, 3/7
Felsenstein, Lee, 4/13, 11/6, 17/12, 17/18, 19/1-2
Fernandez, William, 5/2, 5/7

Ferranti Ltd., 1/5
  MARK I computer, 1/6
Feurzeig, Wallace, 2/12
File Allocation Table. *See* FAT
File program. *See* Microsoft/Applic. Programs
File server software, 13/28
  Also see Apple Computer/ Software
File System Translators (FST). *See* Apple Computer/Software
Filo, David, 15/14-15
Finder program. *See* Apple Computer/Software
Findley, Robert, 4/7
Fiorina, Carleton S., 16/17
Fire Copter game. *See* Adventure International
First Choice software. *See* Software Publishing Corporation
First West Coast Computer Faire. *See* West Coast Community Faire
Flash memory, 17/4
FLEX computer, 2/11, 4/5
Flight Simulator game. *See* Microsoft/Applic. Programs
Floppy disk drives, 7/1, 17/6-8, 17/19, 17/21
Floppy disks, 17/22, 19/12
Florence, Philip, 12/24
FLOW-MATIC software, 1/12
FMS 80 database, 7/9
Focal. *See* DEC/Software and Microsoft/Progr. Languages
Folsom, Barry James, 11/15
Forbes magazine, 16/8, 16/12, 19/20
Forefront Corporation, 13/21
Forethought, Inc., 12/10, 12/27
Forrestor, Jay W., 1/9
FORTH programming language, 2/12, 13/11
FORTRAN (FORmula TRANslation), 1/13, 2/6, 2/8, 5/1-2, 6/3, 7/5, 10/3, 11/15
  Also see Microsoft/Progr. Languages
FORTRAN-80. *See* Microsoft /Progr. Languages
Fortune magazine, 10/5, 12/6, 19/25
Fox Software, 15/5
  FoxPro database, 15/5
FPB, Model A floating point board. *See* North Star Computers
Framework software. *See* Ashton-Tate
Frank, Richard, 13/15
Frankenberg, Bob, 16/12
Franklin Computer Corporation, 9/13, 10/3, 10/5, 11/13
  Ace 100 computer, 11/13
Frankston, Robert (Bob), 5/14, 7/8, 13/14, 13/29, 19/12
Frassanito, Jack, 4/2
French, Donald H., 4/15
French, Gordon, 19/2
French, Melinda, 16/9
Frenzel, Louis E., 4/3
Friedl, Paul, 4/4
Frohman, Dov, 17/4
FrontPage program. *See* Vermeer Technologies
FST's (File System Translators). *See* Apple Computer/Software
FTC. *See* Federal Trade Commission
Fudge, Don, 13/28
Fujitsu Limited, 17/3
Full BASIC. *See* American National Standards Institute (ANSI)
Full Impact spreadsheet. *See* Ashton-Tate
Fulmer, Howard, 17/19
Funk and Wagnalls Encyclopedia, 12/28
Furukawa, Susumu, 12/9
Fylstra, Daniel, 7/8-9, 7/11, 19/12

--G--

G Subset of PL/I. *See* ANSI
G&G Systems, 4/20
G-15 computer. *See* Bendix Aviation
Galactic games. *See* Brøderbund Software
Galvin Manufacturing Corporation, 3/10
Galvin, Paul V., 3/10
Gamer's Edge disk magazine. *See* Softdisk
Games, 1/13, 2/10, 2/13, 3/9, 4/16, 5/4, 7/10-11, 13/24-26
  Adventure, 6/14, 7/10, 9/7, 13/25
  Adventure Land, 7/10
  Alien Rain. *See* Brøderbund Software

Both Barrels. *See* Sirius Software
Breakout. *See* Atari Corp. and Apple Computer
Choplifter. *See* Brøderbund Software
Computer Space, 7/10
Cyber Strike. *See* Sirius Software
David's Midnight Magic. *See* Brøderbund Software
Deadline. *See* Infocom
Fire Copter. *See* Adventure International
Flight Simulator. *See* Microsoft/Applic. Programs
Galactic Empire, Saga & Trader. *See* Brøderbund Software
King's Quest. *See* Sirius Software
Laser Ball. *See* Adventure International
Lode Runner. *See* Brøderbund Software
Lunar Lander. *See* Apple Computer/Software
Merlin tennis, 1/13
Microchess, 7/11, 19/12
Music Construction Set. *See* Electronic Arts
Mystery House. *See* Sierra On-Line
Odyssey 100. *See* Magnavox
Pac-Man. *See* Namco
Penny Arcade. *See* Apple Computer/Software
Pinball Construction Set. *See* Electronic Arts
Pirate Adventure. *See* Adventure International
Pong tennis. *See* Atari Corporation
Princess. *See* Sierra On-Line
Raster Blaster. *See* Budgeco company
SARGON, 7/11
Space Quarks. *See* Brøderbund Software
Space Wars, 2/14, 7/10
Star Cruiser. *See* Sirius Software
Wizard. *See* Sierra On-Line
Zork. *See* Infocom
Gandalf project (Encarta). *See* Microsoft/Multimedia
Garland, Harry, 4/14, 17/16
Garner, Robert, 8/8
Gartner Group Inc., 19/21
Gassée, Jean-Louis, 10/7, 16/14
Gates, William H.,
1960's and 1970's, 6/1-15;
1980's, 12/1-28;
1990's, 15/15, 16/9-12
Apple Computer, 10/19, 10/23
History of BASIC, 7/3
IBM, 9/21, 9/25, 15/8
Miscellaneous, 4/11, 7/8, 11/2, 18/9, 19/26, 20/1
MITS, 4/10
Gateway 2000, Inc., 9/25, 11/19
Gaudette, Francis (Frank) J., 12/7, 16/9
GBASIC. *See* Microsoft/Progr. Languages
Gebelli, Nasir, 13/25
GEM (Graphics Environment Manager). *See* Digital Research
General Electric (GE) Company, 2/4-6, 6/1, 6/10, 10/5, 10/7, 19/16
Datanet-30 communications computer, 2/5
GE-225 computer, 2/5-6
GE-235 computer, 2/6
GE-645 computer, 2/4
GE-BASIC, 2/6
GEnie on-line service, 19/16
Mark II time sharing system, 6/1
General Ledger program. See Tandy/Radio Shack/Software
General Magic company, 19/22
Magic Cap operating system, 19/22
General Motors Corporation, 2/10, 10/5
General Purpose Relay Calculator. *See* Bell Telephone Laboratories
General Systems Division. *See* IBM/Miscellaneous
General Videotex Corporation, 19/16
Delphi on-line service, 19/16
Geneva computer. *See* Epson America
GEnie on-line service. *See* General Electric Company
Geometry Engine integrated chip. *See* Silicon Graphics
Geophysical Service, Inc.,

(GSI), 3/12
GEOS (Graphic Environment Operating System). *See* Berkeley Softworks
GeoSafari. *See* Microsoft /Applic Programs
German Aeronautical Research Institute, 1/4
Gernelle, Francois, 4/7
Gerstner, Louis V., 16/6
Geschke, Charles M., 13/23
Gibbons, Fred M., 13/18
Gilbert, Paul, 6/4, 20/7
Gill, Timothy E., 13/12
Gimix Ghost computer, 17/21
Glaser, Rob, 12/10
Go Corporation, 14/9
Gold, Tony, 18/6, 19/11
Goldhaber, Nathaniel, 19/19
Golding, Val J., 18/4
Goldman Sachs & Company, 11/17, 12/8
Good Housekeeping magazine, 12/6
Goodhew, Bill, 13/22
Gosling, James, 15/13
Graetz, J. M., 2/14
Grant, Charles, 4/20
Grant, Richard, 2/12
Graphics, 1/12, 2/9-11, 5/15, 10/7, 11/29, 12/17, 12/20, 12/22, 13/21, 13/23, 13/26-27, 20/4
Gray, Stephen B., 2/14, 18/1, 19/1
Grayson, George D., 13/26
Grayson, Paul J., 13/26
Greelish, David A., 19/8
Greelish, Tamara, 19/8
Great Records and Tapes. *See* GRT
Green Book specification, 20/4
Green, Cecil H., 3/12
Green, Wayne, 18/2-5, 18/8
Greenberg, Bob, 6/11, 6/13
Greenberg, Mark, 4/20
Greenia, Mark, 20/3
GRiD Systems Corporation, 11/19, 13/27
  Compass I computer, 11/19, 20/8
  Compass II computer, 11/19
  GRiD Server software, 13/27
  GRiDCase computer, 11/20
  GRiDCase Plus computer, 11/20
  GRiDPad tablet computer, 11/20, 20/8
Grolier company, 20/2
Groupware, 15/6, 15/10
Grove, Andrew S., 3/3, 3/6, 8/6, 16/8, 18/9
GRT (Great Records and Tapes), 6/12, 6/14
GS Works software. *See* Apple Computer/Software
GS+ magazine, 18/6
GS/OS operating system. *See* Apple Computer/Software
GT Interactive Software Corporation, 19/22
GTE Sylvania, 5/1
Guglielmi, Joseph M., 9/22, 15/8, 19/19
GUIDE (Guidance of Users of Integrated Data-processing Equipment), 1/14
Gunji, Akio, 18/3
Gunji, Hiromi, 19/20
Gunter, Tom, 3/11
Gutknecht, Jurg, 13/8
GWBASIC. *See* Microsoft/Progr. Languages
Gypsy text editor. *See* Xerox/Software

--H--

H & R Block Inc., 19/16
H7 to H89 accessories and computers. *See* Heath Company
Haba Systems company, 13/20
HAL-4096 computer, 4/3
Hall, Tom, 19/22
Hallman, Michael R., 16/9
Halpin, James F., 19/10
Haloid Corporation, 17/15
Haltek surplus retailer, 5/2
Hammond, Mike, 11/19
Hanson, Rowland, 12/5
Harbers, Jeff, 12/24
Hard disk drives, 7/1, 17/5, 17/8-9, 20/8
Hardcore Computing magazine, 18/8
Hardcore Computist magazine, 18/8
Hardware, 20/4
Harp, Bob, 17/19
Harp, Lore, 17/19
Harrigan, Sid, 19/11
Harris, James, 11/9
Harvard Graphics program. *See* Software Publishing
Harvard University,
    Bricklin, Daniel, 7/8
    Early computers,1/4, 1/7,

1/12, 19/9, 20/6
Gates, Bill, 6/4-7, 6/9-10
Harvard Business School, 7/7
Mark I computer, 1/4, 1/12, 19/9
Mark II, III & IV computers, 1/4, 20/6
Harvey, Mike, 18/4
Harvey, Will, 13/26
Hassett, Christopher R., 19/24
Haughton, Kenneth E., 17/6
Hawkins, Jeff, 14/12
Hawkins, William "Trip", 10/14, 13/26
Hawley, Jack, 17/23
Hayden Publishing, 18/3
Hayes Corporation and Hayes Microcomputer Products, Inc., 16/15, 16/17, 17/20
80-103A Data Communications Adapter modem, 17/20
Micromodem 100, 17/20
Smartmodem 300, 17/20
Hayes, Dennis C., 17/20
HDOS operating system. *See* Heath Company
Heath Company, 4/18-19, 6/11
Benton Harbor BASIC, 4/18
Extended Benton Harbor BASIC, 4/19
H89 computer, 4/18
H7 floppy disk drive, 4/18
H8 computer, 4/18
H9 video terminal, 4/18
H10 paper tape reader/punch, 4/18
H11 computer, 4/11, 4/18
HDOS operating system, 4/18-19
Heath/Zenith-89, 4/18
Heath-Robinson cryptoanalysis machine, 1/4
Heatherington, Dale, 17/20
Heckel, Paul, 12/22
Hector, Hans-Werner, 19/24
Heiser, Dick, 19/10
Helios disk drive. *See* Processor Technology
Heller, Andy, 14/8
Helmers, Carl T., 18/2-3
Helsinki University, 15/9
Hendrix, Gary, 13/29
Hennessy, John L., 8/8
Henochowicz, Mike, 19/10
Henry, G. Glenn, 9/17
Herbold, Robert J., 16/10
Hercules Computer Technology, 17/18
Hercules Card, 17/18
Hertzfeld, Andy, 10/6, 10/19-20, 19/22
Hewlett, William, 2/8, 5/2
Hewlett-Packard (HP) Company,
Apple Computer, 5/2-6, 5/9-10, 10/8
Beginning of, 2/8-9
Early calculators and computers, 4/3, 4/18
Microprocessors, 3/15, 14/4, 14/6, 16/7
Microsoft, 12/4, 12/10
Miscellaneous, 4/1, 4/19, 11/23, 13/18, 16/12, 16/17, 17/8, 17/21, 19/17-18
Personal computers (1980's), 11/20-21
Printers, 17/14-15
Software, 2/6, 13/6
Calculators and Computers:
Alpha project (HP 3000), 4/3
HP-35 calculator, 5/3
HP 70 series of computers, 11/20
HP-75 Portable computer, 11/20
HP 80 series of computers, 11/20
HP-85, 3/15, 11/20
HP 100 series of computers, 11/20
HP-150 computer, 11/21, 17/8
HP 200 series of computers, 11/20-21
Models 216, 226 and 236, 11/21
HP-2114A minicomputer, 2/9
HP 2116 controller, 2/8
HP 3000 minicomputer, 4/3
HP 9100 series of calculators, 2/9
HP 9100A calculator, 2/9
HP 9800 series of calculators/computers, 4/3, 4/19
HP 9830A calculator, 4/3
HP 9831A computer, 4/19
HP Vectra computer, 11/21
Miscellaneous:
BPC microprocessor, 4/19
Enhanced BASIC Interpreter, 11/20
LaserJet printer, 17/15
NewWave interface program,

10/8, 12/10, 13/6
ThinkJet printer, 17/14
UNIX operating system, 14/6
Higgins, Frank M. (Pete), 12/6, 16/10
High Sierra Proposal, 12/7, 20/4-5
Higinbotham, William, 1/13
Hillman, Dan, 10/3, 10/13
Hinckley, Norton, 4/16
Hinckley-Tandy Leather Company, 4/16
Historical Computer Society, 19/8
Historically Brewed magazine, 19/8
Historical organizations, 19/6-9
Hitachi company, 12/4, 17/8
HITS (Hobbyists' Interchange Tape System), 17/5
Ho, Kwok Yuen., 19/20
Hobbiest computer. *See* Sphere
Hobby computing, 2/14
Hoeffler, Don, 20/9
Hoerni, Jean, 1/11
Hoff, Marcian E. "Ted", 3/3, 3/6-7, 3/17, 20/7
Hoffman, Mark B., 13/18
Holt, Frederick Rodney "Rod", 5/8-10, 10/1-3, 10/19
Home Computer Systems, 12/11
Homebrew Computer Club, 4/11, 5/3-7, 6/7, 17/18, 18/2, 19/2-3
Newsletter, 18/2
Homestead High School, 5/1
Homeword word processor. *See* Sierra On-Line
Honeywell company, 2/9, 6/4, 6/6-7, 8/8
Hoo, Sim Wong, 17/16
Hopp, Dietmar, 19/24
Hopper, Grace M., 1/12, 20/6
Horizon computers. *See* North Star Computers
Horn, Bruce, 10/19-20
Howard, Brian, 5/16, 10/18
Howard, Robert, 17/12
HP company and products. *See* Hewlett- Packard Company
HTML (HyperText Markup Language), 15/12, 19/14, 20/3
HTTP (HyperText Transfer Protocol), 19/14
Huang, Minsiu, 19/23
Huck, Jerry, 14/6
Human Engineered Software (HES), 13/28
Hungarian notation, 12/22
HX-20 and HX-40 computers. *See* Epson America
Hyatt, Gilbert, 3/16-17, 14/8, 20/7
HyperCard. *See* Apple Computer/Software
Hyperion computer. *See* Bytec-Comterm
HyperStudio. *See* Roger Wagner Publishing
HyperTalk programming language. *See* Apple Computer/Software
Hypertext markup language. *See* HTML
Hypertext transfer protocol. *See* HTTP
Hypertext, 1/15, 19/14, 20/1, 20/3, 20/7

--I--

IA-64 microprocessor. *See* Intel/Microprocessors
IAS computer. *See* Institute for Advance Studies
IBM (International Business Machines) Corporation,
1970's, 4/2, 4/4, 4/11, 4/20;
1980's, 9/1-25;
1990's, 14/8-10, 15/8-10, 16/4-7
Apple Computer, 5/1, 5/12, 10/2-3, 10/18, 10/20, 10/24-25, 16/15
Beginning of and early computers, 1/4, 1/6-8, 1/15-16, 2/3-4, 2/8-9, 2/15, 4/2
Digital Research, 13/2-3
Market, 4/1, 4/21, 8/3-7
Microprocessors, 3/10, 14/3
Microsoft, 12/1, 12/3-4, 12/6-7, 12/9-18, 12/21, 12/24, 13/3, 15/1-2, 16/8-9
Miscellaneous, 1/14, 17/5-8, 17/11-14, 17/16, 17/19, 17/21-22, 17/24, 18/8, 19/8, 19/16-18, 20/5-6, 20/9
Other companies, 11/3, 11/7, 11/9-11, 11/16,

13/4, 15/14
PowerPC Alliance, 14/6, 15/7, 16/1, 16/16, 19/19 20/4
Software, 1/12-13, 2/11-12, 7/4, 7/6, 13/5, 13/9, 13/17, 13/22-23, 13/28
Accessories:
23FD floppy disk drive, 17/5
33FD floppy disk drive, 17/6
43FD floppy disk drive, 17/6
53FD floppy disk drive, 17/6
305 RAMAC (Random Access Method of Accounting and Control) system, 1/10
350 Disk Storage System, 1/10
3270 display terminal, 9/15, 17/12
3277 display terminal, 9/15
3340 Disk storage Unit (Winchester), 17/5
Hard disk drives, 17/5
Igar project (33FD), 17/6
Minnow project (23FD), 17/6
Model 2213 printer, 17/12
Model 6640 printer, 17/14
ProPrinter, 17/13
Type 26 keypunch, 17/12
Computers (Mainframe & Mini):
DataMaster. *See* IBM System 23
IBM 405 Accounting Machine, 1/9
IBM 603/4 Electronic Multiplier, 1/7
IBM 608 calculator, 1/7
IBM 610 Auto-Point Computer, 1/15
IBM 700 series, 1/7
IBM 701, 1/7, 1/9
IBM 704, 1/13, 2/3
IBM 705 Model III, 1/7
IBM 709, 2/4
IBM 801 minicomputer, 8/7
IBM 1130, 2/8, 5/1
IBM 1620, 2/8, 2/15, 19/6
IBM 7090 Data Processing System, 1/7, 2/4
IBM Defense Calculator, 1/7
IBM Electronic Data Processing Machine, 1/7
IBM SSEC (Selective Sequence Electronic Calculator), 1/7
IBM Stretch system, 1/7, 20/6
IBM System/3 Model 6, 4/2
IBM System/360, 2/11, 5/1, 7/1, 19/6
IBM System/370, 9/15
Personal Automatic Calculator (PAC), 1/15
Computers (Personal):
3270 PC, 9/15
5100 Portable Computer, 4/4, 4/11, 9/1, 20/8
5110 Portable Computer, 4/12
5120 desktop computer, 9/2, 9/4
5140 PC Convertible, 9/17
5160 Model 588 (PC/XT 370), 9/15
5371 Models 12, 14 & 16, 9/15
9000 Instrument System Computer, 9/11
Acorn prototype (IBM PC), 9/5, 12/12-13
Ambra series, 14/10, 16/5
Chess project (IBM PC), 7/7, 9/5-6, 12/12
Circus project (PC AT), 9/11
Clamshell project (5140 PC Convertible), 9/17
DataMaster. *See* IBM System 23
DisplayWriter workstation, 9/22
IBM System/23 DataMaster, 9/4-5
Olympiad project (PC RT), 9/17
PC (Personal Computer),
Apple Computer, 10/2, 10/9, 10/20
Development of, 9/3-9, 9/23
Intel, 3/10, 8/5
Market, 11/1, 13/1
Microsoft, 10/24, 12/1, 12/5, 12/11-12, 12/15-16, 12/19-21, 12/23-25, 12/27, 13/3, 17/23

Miscellaneous, 9/13-14, 9/18, 17/13, 17/19, 17/21, 17/24, 18/7, 19/9, 19/12, 19/24, 20/4
Other computers compatible with, 11/10-11, 11/16, 11/19, 11/21, 11/30
Software, 13/2, 13/4-6, 13/8-13, 13/16, 13/18, 13/20, 13/22-23, 15/10
PC AT (Advanced Technology),
Development of, 9/10-11, 9/16-17
Microsoft, 12/15, 12/19
Miscellaneous, 8/3, 8/7, 11/19, 17/21, 20/5
PC Convertible. *See* 5140 PC Convertible
PC Junior (PCjr), 9/11, 9/13-14 10/12, 18/7
PC RT workstation, 8/8, 9/17
PC/XT,
Development of, 9/10-12
Microsoft, 12/18, 12/21
Miscellaneous, 9/15, 9/17, 11/19
PC/XT 370, 9/15
PC/XT Model 286, 9/18
Peanut project (PCjr), 9/13
Portable PC, 9/16
POWERstation & POWERserver, 14/8
PS/1 (Personal System/1), 14/8
PS/2 (Personal System/2),
Development of, 9/18-20
Microsoft, 12/17
Miscellaneous, 9/24, 10/24, 17/21, 20/5
PS/2 Model 25, 9/18-19
PS/2 Model 30, 9/18
PS/2 Model 50, 9/18
PS/2 Model 60, 9/18
PS/2 Model 80, 9/18-19
PS/2 Model L40 SX laptop, 14/9
PS/2 Model P70 portable, 9/19
PS/2 Model P75 portable, 14/8
RIOS project (RISC System/6000), 9/17, 14/7
RISC System/6000 workstation, 9/17, 13/8, 14/3, 14/8
SCAMP (Special Computer APL Machine Portable), 4/4, 4/12
ThinkPad notebook, 14/9-10
ValuePoint series, 14/10, 16/5
Miscellaneous:
Almaden Research Laboratory, 14/9
AT Bus, 17/21
Corporate Management Committee (CMC), 9/4-6, 9/11, 12/12, 15/8, 16/5
Data Processing Division, 9/8
Desktop Software division, 9/22
Entry Level Systems (ELS), 9/1
Entry Systems Division (ESD), 9/21, 9/23-24, 12/17
General Systems Division, 4/4
IBM Archives, 19/8
IBM Instruments Inc., 9/11
IBM Personal Computer Company (IBM PC Company), 16/5-6
IBM Product Centers, 9/8
IBM Research Division, 14/9
IBM Scientific Center, 4/3
IBM Technical Newsletter, 1/14
Independent Business Units (IBU's), 9/1, 9/5
Individual Computer Products International (ICPI), 16/5
Information Systems Division, 9/6, 9/15
MCA (Micro Channel Architecture), 9/18-19, 9/24, 12/16, 14/8-9, 17/21
Memory Management Unit (MMU), 9/15
Palm microcontroller, 4/4
PC Bus, 17/21
Personal Computer Group, 9/23-25, 16/6
Personal Systems Group, 16/5
POWER architecture, 14/8, 19/19

PowerPC microprocessor, 14/5, 15/7, 20/4
PowerPC 601 microprocessor, 14/3, 14/6
PowerPC 603 microprocessor, 14/3
Research Division, 14/9
Research Laboratory (San Jose, California), 13/17
RISC central processing unit (CPU), 14/3
ROMP (Research Office products Micro Processor), 8/7-8, 9/17
System R (Relational) group, 13/17
Systems Products Division, 9/9
Tabulating machine, 19/6
Technical Reference manual (IBM PC), 9/9
TrackPoint pointing device, 14/9
Yamato Laboratory (Japan), 14/9

Software:
3270 PC Control Program, 9/15
ABIOS (Advance Basic Input/Output System), 9/20, 12/16
AIX (Advanced Interactive Executive) operating system, 9/17, 14/8, 19/19
Assistant series of programs, 13/18
BIOS (Basic Input/Output System), 9/5, 9/11, 11/10, 11/21, 11/30, 12/13, 13/22-23
Cassette BASIC. *See* Microsoft Corporation/ Programming Languages
Computer System Operating System. *See* CSOS
CP-DOS, 9/19
CSOS (Computer System Operating System), 9/11
Disk BASIC. *See* Microsoft Corporation/Programming Languages
Displaywrite word processor, 9/22, 13/13
Extended BASIC. *See* Microsoft Corporation/ Programming Languages
ICPL (Initial Control Program Load), 17/6
NPL (New Programming Language), 2/11
OfficeVision, 9/20, 9/22, 19/19
OS/2 (Operating System/2), Development of, 9/18-22, 9/24-25, 15/8-9
Microsoft, 12/9, 12/15-17, 12/20, 12/24, 12/27, 13/3, 15/1, 15/4, 16/9
Miscellaneous, 13/12, 13/14, 15/14, 19/19
OS/2 Extended Edition, 9/20-21, 12/16, 15/6
OS/2 Warp, 15/9
PC Network, 9/22, 13/28
PC-DOS, 9/8, 9/12-16, 9/21-22, 12/14-15, 12/21, 13/3
PL/I (Programming Language One), 2/11-12, 7/6, 13/9
Presentation Manager, 9/20-21, 12/10, 12/16-17, 12/20, 12/25, 13/6, 15/9
SAA (Systems Application Architecture), 9/20, 12/16
System R (Relational Database specification), 13/17
TopView user interface, 9/22, 12/9, 12/16, 12/18-19, 13/6
VM/PC (Virtual Machine/Personal Computer), 9/15

IBM Instruments Inc. *See* IBM/Miscellaneous
IBM Personal Computer Company (IBM PC Company). *See* IBM/Miscellaneous
iBook portable computer. *See* Apple Computer/Computers
ICC (International Color Consortium), 19/18
iCOM Microperipherals company, 17/7
Frugal Floppy disk drive, 17/7
Microfloppy disk drive, 17/7
Icons, 2/10-11, 4/5, 10/7, 10/15, 11/29, 12/20, 12/24
ICPL (Initial Control Program Load). *See* IBM/Software

id Software, Inc., 19/22
IEEE (Institute of Electrical & Electronic Engineers), 2/10, 3/7, 8/5, 17/20-21, 17/23
IEEE - Annals of the History of Computing, 18/7
IEEE 696 bus, 17/21
IEEE Computer periodical, 20/7
IEEE Micro periodical, 3/10, 3/18
Igar project. *See* IBM/Accessories
IGES (Initial Graphics Exchange Standard), 20/4
II Computing magazine, 18/6
iMac computer. *See* Apple Computer/Computers
ImageWriter printer. *See* Apple Computer/Accessories
IMP (Interface message processor), 2/13
IMP-8 and IMP-16 microprocessor systems. *See* National Semiconductor
IMS Associates, Inc. (IMSAI), 4/13, 4/19, 19/11
IMSAI Manufacturing Corporation, 4/13, 4/15, 4/20, 7/4, 7/7
IMSAI 8080, 4/13, 7/9, 19/11
VDP-40 computer, 4/20
VDP-80 computer, 4/20
In-A-Vision graphics software. *See* Micrografx
inCider magazine, 18/5
inCider/A+ magazine, 18/5
Independent Business Units (IBU's). *See* IBM/Misc.
Indigo workstation. *See* Silicon Graphics
Individual Computer Products International (ICPI) company. *See* IBM/Misc.
Industry Standard Architecture bus. *See* ISA
Infocom Inc., 13/25
Deadline game, 13/25
Zork I game, 13/24-25
Zork II game, 13/25
Information Appliance Inc., 17/19
SwyftCard, 17/19
Information highway, 15/15
Information Processing Techniques Office (IPTO). *See* ARPA
Information Sciences Inc., (ISI), 6/3-4
Information Systems Division. *See* IBM/Miscellaneous
Information Terminals Corporation (ITC), 17/4-5, 17/22
Information Unlimited Software (IUS), Inc., 7/7, 7/9, 9/8, 13/9-10
EasyWriter word processor, 7/7, 9/8, 13/9, 13/29
WHATSIT? database, 7/9
Informix Software Inc., 13/15, 13/18
WingZ spreadsheet, 13/15
InfoStar word processor. *See* MicroPro International
InfoWorld periodical, 18/8, 20/9
Ingalls, Dan, 4/5
Ingram, Gary, 4/13, 17/17
Initial Graphics Exchange Standard. *See* IGES
Ink jet printers. *See* Printers
Inprise Corporation, 16/17
Input/Output devices, 17/10-11
Institute for Advance Studies (IAS), 1/5, 1/8
IAS computer, 1/5, 1/9
Institute of Electrical & Electronic Engineers. *See* IEEE
Integer BASIC. *See* Apple Computer/Software
Integrated circuit, 1/3, 1/10-11, 2/8, 2/15-16, 3/5-7, 3/11-12, 3/16-18, 4/2-3, 11/23
Integrated programs, 12/26, 13/19-22
Integrated Services Digital Network. *See* ISDN
Intel Corporation,
1970's, 3/3-10, 4/7-8, 4/11, 4/20;
1980's, 8/3-6, 8/9;
1990's, 14/3-7, 16/7-8
IBM, 9/23-24
Memory, 3/5-6, 4/7, 17/3-4, 20/4
Microsoft, 6/11, 12/10, 12/16, 15/2, 17/17
Miscellaneous, 3/11, 3/18, 4/1-2, 5/2, 13/8, 15/9, 16/13, 16/16, 19/8, 19/17, 20/7, 20/9
Other companies, 3/12-15,

4/9-10, 5/4, 5/9, 7/1, 7/6, 11/6
Patent controversy, 3/16-17
Microprocessors:
432. *See* APX 432
1201, 3/7-8
4004, 3/4, 3/6-8, 3/17, 4/7, 7/1
8008,
Development of, 3/8
Miscellaneous, 4/7-8, 6/4, 7/1, 18/3
8080,
Development of, 3/8-9
Microcomputer applic's, 4/9-10, 4/13, 4/17, 4/19, 5/5
Miscellaneous, 3/14, 6/5-6, 6/8, 6/15, 7/5
8085, 3/9, 3/15, 4/11
8086,
Development of, 3/9
Personal computers, 9/6, 9/12, 9/18-19
Miscellaneous, 3/10, 6/12, 6/14-15, 8/3-4, 12/13-14, 17/18
8087 coprocessor, 8/4-5, 9/5
8088, 3/10, 4/20, 8/3-5, 9/5, 9/7-8
80186, 8/3, 9/7
80188, 8/3
80286, 8/3, 8/5, 9/6, 9/8-10
80386DX, 8/3-6, 8/9, 14/7, 16/5
80386SL, 14/3
80386SX, 8/4
80486DX, 8/4, 14/3
80486DX2, 14/4
80486DX4, 14/4
80486SLC, 14/7
80486SX, 14/3
82786 graphics coprocessor, 8/5
Celeron, 14/6
IA-64, 14/5-6
iAPX 432 (Advance Processor Architecture), 3/9, 8/3
Itanium, 14/6
MCS-4 (Micro Computer System 4-bit), 3/4, 3/6
MCS-8 (Micro Computer System 8-Bit), 3/8
Merced project (IA-64), 14/6, 16/13
OverDrive processors, 14/4
P6 (Pentium Pro), 14/5
P54C, 14/4
Pentium, 14/4-5, 14/7, 16/7
Pentium II, 14/5-6
Pentium II Xeon, 14/6
Pentium III, 14/6
Pentium Pro, 14/5
Miscellaneous:
1101 memory chip, 3/6
1103 1K memory chip, 3/6, 17/3
1702 EPROM chip, 17/4
2107 4K memory chip, 17/3
2117 16K memory chip, 17/3
"*intel inside*" logo, 16/7
Above Board specification, 20/4
EPIC (Explicitly Parallel Instruction Computing), 14/6
Intellec 4 & 8 Development Systems, 4/7
MMX technology, 14/5
Museum, 19/8
Operation Crush, 8/5
PCI (Peripheral Component Interface), 16/7
SIM4 simulator board, 4/7
SIM8 simulator board, 4/7
Intelligent agents, 15/15
Intelligent computer. *See* Sphere
Intelligent Systems company, 13/22
Intellimouse. *See* Microsoft/Miscellaneous
Interactive computing, 1/13, 15/15
Interactive Home Systems, 12/11
Interactive Media Division. *See* Microsoft/Miscellaneous
Interface Age magazine, 5/8, 8/2
Interface Group. *See* The Interface Group
Interface Manager (Windows). *See* Microsoft/Oper. Systems
Interface message processor. *See* IMP
Intergalactic Computer Network, 2/13
Intergalactic Digital Research, 7/1
International Business

Machines Corporation. *See* IBM
International Color Consortium. *See* ICC
International Federation for Information Processing (IFIP), 1/14
International Standard Organization. *See* ISO
International Telecommunications Union. *See* ITU
Internet, 2/13, 15/3, 15/8-9, 15/11-13, 16/10-11, 16/13, 19/13-14, 19/20-22, 19/24
Internet Explorer. *See* Microsoft/Applic. Programs
Intersil company, 4/20
  6100 CPU, 4/20
Intuit, Inc., 13/29-30, 16/2, 16/10, 16/15
  Quicken finance program, 13/29, 16/10
  QuickBooks, 16/15
Invention & Technology magazine, 4/2
Inventory Control System. *See* Tandy/Radio Shack/Software
Iomega Corporation, 17/10
  Bernoulli Box, 17/10
  Bernoulli disk drive, 17/10
  Jaz disk drive, 17/10
  Zip disk drive, 17/10
Iowa State College, 1/4
IPTO (Information Processing Techniques Office). *See* ARPA
IRE Transactions on Electronic Computers, 20/6
Iris Associates Inc., 15/10
  Notes communications program, 15/10
IRIS terminal, workstations and IRIS Graphics Library. *See* Silicon Graphics
ISA (Industry Standard Architecture) bus, 17/21
Isaacson, Portia, 19/3
ISDN (Integrated Services Digital Network), 19/14
ISO (International Standard Organization), 20/4
Itanium. *See* Intel/Micro's
ITC. *See* Information Terminals Corporation
ITU (International Telecommunications Union), 20/3
  V.32 standard, 20/3
  V.34 standard, 20/3
Iwatani, Toru, 7/11
IWM (Integrated Woz Machine). *See* Apple Computer/Misc.

--J--

Jacobson, Errol, 19/10
JaM language. *See* Xerox/Software
Janus project (Xerox Star). *See* Xerox/Computers
Java programming language. *See* Sun Microsystems
Javelin Software company, 13/15
  Javelin spreadsheet, 13/15
Jaz disk drive. *See* Iomega
Jazz software. *See* Lotus Development
Jennings, Peter R., 7/11, 19/12
Jewell, Jerry, 13/25
Jini software. *See* Sun Microsystems
Jobs, Steven Paul,
  Apple Computer,
    1970's, 5/9-15.,
    1980's, 10/1-4, 10/6-8, 10/9, 10/15, 10/17-20, 10/24.,
    1990's, 16/3
  Early years, 5/1-9, 19/10
  Microsoft, 12/2-3, 16/11
  Miscellaneous, 18/9, 19/2, 19/23, 20/2
  NeXT, 11/11, 16/2, 16/15, 16/17
  Other companies, 4/17, 7/8, 7/10, 13/24
Johnson, Reynold B., 17/12
Jonsson, J. Erik, 3/12
Joy, William N., 8/8, 11/24-25, 15/13
Jurassic Park movie, 11/23

--K--

K5 and K6 microprocessors. *See* Advanced Micro Devices
K56flex technology, 20/3
Kahn, Philippe, 13/8
Kahn, Robert, 2/13
Kaiman, Art, 12/7
Kaleida company, 15/7-8, 16/5, 19/19
Kaleidoscope program. *See* Cromemco
Kamradt, Alex, 5/4

Kansas City Standard, 17/5
Kaplan, Jerry, 14/10
Kapor, Mitchell D., 7/11, 13/13-14, 13/28-29
Karcher, J. Clarence, 3/12
Kassar, Raymond, 4/17, 11/14
Kawasaki, Guy, 10/7, 13/19
Kay, Alan C., 2/11, 4/5, 7/6, 10/5, 20/1, 20/8
Kay, Andrew, 11/8
Kay, Gary, 4/12
Kaypro Corporation, 11/8
  Kaycomp II portable computer, 11/8
  Kaypro 4, 11/8
  Kaypro 10, 11/8
  Kaypro II portable computer, 11/8
KBD-2 Keyboard. *See* Southwest Technical Products
Kemeny, John G., 2/5-6, 7/3
Kenbak Corporation, 4/3
  Kenbak-1 computer, 4/3
Kentucky Fried Computers, 4/20
Kenyon, Larry, 10/20
Keremedjiev, Barbara and George, 19/6
Keyboards, 17/23
Khalsa, Sat Tara Singh, 13/29
Khosla, Vinod, 11/24-25
Kilburn, T., 1/6
Kilby, Jack St. Claire, 1/11, 3/12
Kildall, Gary A., 6/14, 7/1-2, 7/4, 7/6, 12/12, 16/12, 20/2
Killer application, 7/9
Killian, Joseph, 4/13
kilobaud magazine, 18/3
KIM-1 microcomputer. *See* MOS Technology
Kimsey, James V., 19/15
King, Olin B., 19/24
King's Quest game. *See* Sierra On-Line
Kingston Technology Corporation, 19/22
Klunder, Doug, 12/23-24
Knowledge manipulator, 4/5
Knowledge Navigators, 14/10, 20/1
KnowledgeSet Corporation, 20/2
Knuth, Donald E., 19/26
Kodak company, 17/22, 19/18
Konzen, Neil, 6/14, 12/19, 12/23
Koogle, Timothy, 15/14-15
Krauskopf, Tim, 15/11
Kriya Systems company, 13/29
  Typing Tutor, 13/29
Kuehler, Jack D., 9/25, 16/5-6
Kurtz, Thomas E., 2/5, 7/3
Kyoto Ceramics (Kyocera) company, 11/2, 12/2

--L--

Lakeside Programmers Group, 6/3, 6/8
Lakeside School, 6/1-3, 6/7, 12/9
Lampson, Butler, 4/5, 7/6, 17/15
LAN. *See* Local Area Network
LAN Manager. *See* Microsoft/Applic. Programs
Lancaster, Don, 4/12, 17/11
Lane, Jim, 6/12-13
Languages. *See* Programming Languages
Lanier company, 13/9
Laptop computers,
  IBM, 9/17, 14/9
  Other companies, 11/19-20, 12/2, 12/21
  Term description, 20/8
  Toshiba, 11/27
Large Scale Integration. *See* LSI
Larson, Chris, 6/7, 12/9
Laser 128 computer. *See* Video Technology
Laser Ball game. *See* Adventure International
Laser printers. *See* Printers
LaserJet printer. *See* Hewlett-Packard
LaserWriter printers. *See* Apple Computer/Accessories
Lashlee, Hal, 13/16, 19/12
Lattin, William W., 3/9, 8/3
Lau, Benny, 19/20
Lau, Lee, 19/20
Lawrence Livermore National Laboratory, 19/8
  Computer Museum, 19/9
Lawten, Bob, 9/25, 14/9
LCD (Liquid Crystal Display), 11/16-19, 11/21, 11/26, 14/9
Lear Siegler Inc. (LSI), 17/12
  LSI ADM-1 terminal, 17/12
Learning Research Group. *See* Xerox/ Miscellaneous
Lebling, Dave, 13/24
Leeds, Richard, 12/14
Leff, Robert S., 19/12
Lehtman, Harvey, 10/13
Leininger, Steven, 4/16-17,

7/4
Leng, John, 20/7
Lentz, John L., 1/15
Lerner, Sandra, 19/21
LetterPerfect word processor. *See* WordPerfect
Letwin, Gordon, 4/18, 6/12-14, 12/9, 12/16
Level-I, II and III BASIC. *See* Tandy/Radio Shack/Software
Level II and III BASIC. *See* Microsoft/Prog. Languages
Levin, Michael, 2/12
Levy, Bill, 7/3
Lewin, Dean, 11/11
Lewis, Andrea, 6/13
Lexikon Services company, 20/3
Lexmark International Group, Inc., 16/4
LGP-30 computer. *See* Librascope
Libes, Sol, 19/4
Librascope/General Precision, 1/13
  LGP-30 computer, 1/16, 2/5
Licklider, J.C.R., 2/4, 2/9, 2/13
Liddle, David, 9/22, 11/29, 13/6
Life magazine, 10/2
Lifeboat Associates, 12/15, 18/6, 19/11-12
  Software Bus-86 (SB-86), 12/15, 19/12
Lifetree Software Inc., 13/12
  Volkswriter word processor, 13/12
Light pen, 2/10, 2/14
LIM (Lotus, Intel and Microsoft) specification, 20/4
LINC (Laboratory Instrument Computer). *See* MIT
LINC-8 computer. *See* DEC/Computers
Lincoln Laboratory. *See* MIT
Linnett, Barry, 12/27
Linux operating system, 15/9-10
Liquid crystal display. *See* LCD
Lisa computers and software. *See* Apple Computer
Lissner, Rupert, 13/18, 13/20
Loadstar magazine. *See* Softdisk
Lobo International, 11/22
  Max-80 computer, 11/22
Local-area network (LAN), 9/10, 9/22, 13/27
LocalTalk. *See* Apple Computer/Software
Lockwood, Russ, 7/3
Lode Runner game. *See* Brøderbund Software
Logitech International SA, 19/22
Logo language, 2/12, 13/7 *Also see* Apple Computer /Software
Long Island Computer Association of New York, 19/3
Lopez, Tom, 12/7, 12/10, 20/2
Lord, Geller, Federico and Einstein advertising agency, 9/6
Lorenzen, Lee, 13/5
Lotus Development Corporation,
    Beginning of, 13/13-14
    IBM acquisition of, 16/6
    Microsoft, 12/4, 12/23
    Miscellaneous, 13/15, 13/20, 13/28, 15/10, 19/25, 20/4
  Ami Pro word processor, 13/13
  ccMail, 13/28
  Jazz, 12/24, 13/20-21
  Lotus 1-2-3, 12/23-24, 13/13-15, 13/19, 13/21
  Lotus 1-2-3/3, 13/14
  Lotus Notes, 15/6, 15/10
  Symphony, 13/20
  TR10 project (Lotus 1-2-3), 13/13
  Word Pro wordprocessor, 13/13
Lougheed, Kirk, 19/21
Lowe, William C., 9/1-3, 9/4-6, 9/24, 12/12, 12/16-17
LPB-CX printer. *See* Canon company
LSI (Large Scale Integration), 2/14, 3/5-6, 4/7, 4/20, 17/3, 20/7
LSI-11 microcomputer systems. *See* DEC/Computers
LTE notebook. *See* Compaq Computer
Lucasfilm company, 19/23
Lucent Technologies, 20/3
Luggable portable computer, 9/16, 9/19, 11/11, 20/8
Lunar Lander game. *See* Games
Lutus, Paul, 7/7, 13/12

Lycos Inc., 19/23

--*M*--

M6800 series of microprocessors. *See* Motorola
M-10 computer. *See* Olivetti
MAA. *See* Microcomputer Applications Associates
MACazine magazine, 18/6
MacBusiness Journal, 18/6
MACC. *See* Midwest Affiliation of Computer Clubs
MacGraph. *See* Microsoft /Applic. Programs
MacGregor, Scott, 12/18-19
Mach operating system, 12/17, 13/4, 16/9
Macintosh computers and software. *See* Apple Computer
Macintosh Today magazine, 18/6
Macintosh Word. *See* Microsoft/Applic. Programs
MacMail. *See* Microsoft/Applic. Programs
MaCom company, 4/21
MacPaint and MacProject. *See* Apple Computer/Software
Macro Assembler. *See* Microsoft/Prog. Languages
Macromedia, Inc., 19/23
MacroWorks. *See* Beagle Bros.
MacSketch and MacTerminal. *See* Apple Computer/Software
MacUser magazine, 18/6
MacWeek magazine, 18/6
MacWorld Expo. *See* Apple Computer/Miscellaneous
Macworld magazine, 18/6
MacWrite. *See* Apple Computer/Software
Magazines, 1/14, 18/1-9
Magic Cap operating system. *See* General Magic
Magic Paintbrush. *See* Penguin Software
Magic Wand word processor. *See* Small Business Applications
Magnavox company, 2/14
  Odyssey 100 game, 2/14
Magnetic card reader, 11/20
Magnetic core memory, 1/9, 1/15, 2/7-8, 2/14-15, 3/5, 4/1, 17/3
Magnetic disk storage, 1/10
Magnetic drum storage, 1/7-8, 1/9, 1/13
Magnetic tape, 1/7-8, 1/9, 4/9, 17/4-5, 17/11
Mail program. *See* Microsoft/ Application Programs
MailMerge software. *See* MicroPro International
Main, James, 19/4
Major, Drew, 13/28
Malloy, Tom, 10/16
Management Information Format. *See* MIF
Management Science America Inc., 13/22
Mangham, Jim, 18/5
Mann, Marvin L., 16/4
Manock, Jerrold "Jerry" C., 5/10, 10/19
Manzi, Jim P., 13/14, 16/6
Maples, Michael (Mike) J., 12/10, 16/9-10
Maritz, Paul A., 12/9, 16/10
Mark I, II, III and IV computers. *See* Harvard University
MARK 1 computer. *See* Ferranti
Mark II time sharing system. *See* General Electric
Mark-8 computer, 3/8, 4/8-9, 18/1
Mark-8 Group, 18/1
Markell, Bob, 9/19
Markkula, A. C. "Mike", 5/1, 5/9-11, 5/13, 7/8, 10/1-3, 10/8, 10/15, 16/2
Marquardt, David F., 12/1
Marsh, Robert, 4/13, 17/17, 19/2
Massachusetts Institute of Technology. *See* MIT
Massaro, Donald J., 11/27-28, 13/6, 17/7
MATH-MATIC software, 1/12
Mathews, Bob, 12/23
Mathews, Mark, 12/22
MathPlan spreadsheet. *See* WordPerfect
Matson, Katinka, 19/26
Mattel company, 11/30
  Aquarius computer, 11/30
Mauchly, John W., 1/5, 1/8, 1/11
Mauldin, Michael, 19/23
Max Machine. *See* Commodore International
Max-80 computer. *See* Lobo International
Maxell company, 17/22
Maxtor company, 17/7
Mazner, Marty, 13/21

Mazor, Stan, 3/6-7
MBASIC. *See* Microsoft/Progr. Languages
MC6809 and MC68000 series of microprocessors. *See* Motorola
MCA (Micro Channel Architecture). *See* IBM/Misc.
McCabe, Dan, 12/18
McCarthy, John S., 2/3-4
McCracken, Edward R., 11/23
McCracken, William E., 16/6
McDermott, Eugene, 3/12
McDonald, Marc, 6/9, 6/11, 6/13
McDonnell Douglas Corporation, 2/10
McEwen, Dorothy, 7/1
MCGA (Multi-Color Graphics Array), 20/5
McGraw-Hill, Inc., 11/6, 18/2
McIntosh Laboratories, 5/16, 10/21
McKenna, Regis, 5/9
McNealy, Scott G., 11/24-25, 16/13-14
MCS (Micro Computer System). *See* Intel/ Microprocessors
MDA (Monochrome Display Adapter), 9/7, 20/5
MDL language. *See* MIT
Measday, Tom, 13/28
Media Laboratory. *See* MIT
Medium Scale Integration. *See* MSI
Mega II chip. *See* Apple Computer/Misc.
Melear, Charles, 3/10
Melen, Roger, 4/14, 17/16
Memex device, 1/14-15, 2/9, 20/1
Memorex company, 17/7, 17/22
Memory,
  Early technology, 1/5, 1/8-10, 1/15, 2/14-15
  Intel, 3/5-6, 4/6, 17/3-4
  Miscellaneous, 4/1, 17/3-4, 20/4
Memory Management Unit (MMU). *See* IBM/Miscellaneous
Memory Test Computer (MTC). *See* MIT
Memphis project (Windows 98). *See* Microsoft/Oper. Systems
Mensch, Bill, 8/7
Menu bar and Menus:
  Early development, 2/10
  Apple Computer, 10/15, 10/22
  Microsoft, 12/23
  Xerox, 4/5, 11/29
Merced project. *See* Intel/Microprocessors
Merisel, Inc., 19/11-12
Merlin project (Encarta). *See* Microsoft/Multimedia
Merlin video tennis game. *See* Games
MESA programming language. *See* Xerox/ Software
MessagePad (PDA). *See* Apple Computer/Computers
Metal Oxide Semiconductor. *See* MOS
Metaphor Computer Systems Inc., 13/6
Metaprocessor, 7/11
Metcalfe, Robert, 4/5, 12/15, 17/18, 19/13
Meyer, Dan, 4/12
MGA (Hercules Monochrome Graphics Adapter), 20/5
MIC (Microfloppy Industry Committee), 17/8, 19/18
Michael Shrayer Software company, 7/7
  The Electric Pencil, 7/6-7
  The Electric Pencil II, 7/7
Michels, Douglas L., 13/4
Michels, Larry, 13/3
Micral computer. *See* REE (Recherches et Étude Électroniques)
Micro Channel Architecture (MCA). *See* IBM/Miscellaneous
Micro Computer Inc., 3/16-17, 4/15
Micro magazine, 18/4
Micro-1 printer. *See* Centronics
Micro-8 Computer Users Group, 18/1
Micro-8 Newsletter, 18/1
Micro-Altair computer. *See* PolyMorphic Systems
Micro-Disk system. *See* North Star Computers
Microamerica company, 19/12
Microchess game. *See* Games
Microcomputer - term origin, 3/15, 20/7
Microcomputer Applications Associates (MAA), 7/1-2
Microcomputer Industry Trade Association, 19/18
Microcomputer Periodicals, 18/9

Microcomputing magazine, 18/3
Microfloppy disk drive. *See* iCOM Microperipherals
Microfloppy Industry Committee. *See* MIC
Micrografx Inc., 13/26
  In-A-Vision graphics software, 13/26
Micromainframe computer, 8/3
Micromodem 100. *See* Hayes Microcomputer Products
Micron Electronics, Inc., 16/16
Micron Technology, Inc., 16/16
MicroNET. *See* CompuServe
Micronics Computers, Inc., 19/23
Micropolis Corporation, 17/7
MicroPro International Corporation, 7/7, 11/6, 13/9-10, 16/15
  CalcStar, 13/10
  DataStar, 13/10
  InfoStar, 13/10
  MailMerge, 13/10
  SpellStar, 13/10
  Super-Sort, 7/7
  Word-Master, 7/7, 9/2
  Word-Star, 7/7, 11/6-7, 13/9-10, 13/29
  Word-Star 2000, 13/10
Microprocessor,
  IBM, 14/3
  Intel, 3/6-10, 8/3-6, 14/3-6
  Miscellaneous, 2/14, 2/16, 17/2, 20/2
  Motorola, 3/10-11, 8/6-7, 14/6-7
  Other companies, 3/13-15, 3/18, 8/8-9, 14/7
  Patent controversy, 3/16-17
  RISC, 8/7-8
  Term origin, 20/7
  Texas Instruments, 3/11-13
Microprocessors and Microsystems magazine, 18/3
Micro-Soft partnership, 6/7-8
Microsoft, Inc., 12/1
Microsoft Corporation,
    1960-1970's, 4/11, 6/1-15;
    1980's, 12/1-28, 13/3-5, 13/7-9, 13/11-13, 13/15, 13/17, 13/21, 13/26-29;
    1990's, 15/1-6, 16/8-14
    Apple Computer, 5/11, 10/8, 10/19-20, 10/22-23, 16/1-2
    Commodore, 4/15, 11/4
    Compaq, 11/10-11
    IBM, 9/4-8, 9/10, 9/12, 9/16, 9/18-22, 9/25, 13/3, 15/8
    Miscellaneous, 4/11, 4/17, 7/3, 7/5, 7/8, 7/10, 17/17-18, 17/23-24, 18/8, 19/12, 19/16-18, 19/23, 20/2-4, 20/9
    Netscape, 15/12
    Other companies, 11/7, 11/12, 11/18, 13/2, 13/6, 15/14, 16/14
    Tandy/Radio Shack, 4/16, 7/5, 11/2-3
    User interface, 2/11, 4/6, 13/6
  Application Programs:
    Databases:
      Access, 15/5
      Cirrus project, 12/26
      Omega project, 12/10, 12/26
    Miscellaneous:
      Adventure game, 6/14, 7/8, 9/8, 12/14
      BackOffice, 15/4-5, 15/5
      Chart, 12/6, 12/18, 12/27
      DriveSpace, 16/10
      Exchange, 15/5
      File, 12/6
      Flight Simulator, 13/26
      GeoSafari, 15/3
      Internet Explorer, 15/4, 15/6, 16/2, 16/11
      LAN Manager, 12/27
      MacGraph, 10/20
      MacMail, 12/27
      Mail, 12/27, 15/6
      Microsoft at Work, 15/5
      Money, 15/5
      Mouseworks, 12/26
      MS-NET, 12/27
      Office, 15/4, 15/6, 16/2, 16/11, 16/14
      Office 95, 97 and 2000, 15/4
      Olympic Decathlon game, 12/14
      PCMail, 12/27
      PowerPoint graphics, 12/10, 12/27, 15/1, 15/4-5
      Professional Office, 15/3
      Project, 12/27

Publisher, 12/27
Schedule +, 15/5
Time Manager, 12/14
Typing Tutor, 6/14, 12/14, 13/28
Works for Macintosh, 12/26
Works for PC, 12/27
Spreadsheets:
Electronic Paper, 12/14, 12/22, 13/15
Excel, 10/24, 12/20, 12/23-25, 13/15, 13/21, 15/1, 15/4
Multiplan, 10/20, 10/22, 12/6, 12/18, 12/23, 13/15
Odyssey project (Excel), 12/24
Word Processors:
Cashmere project (Word for Windows), 12/10, 12/26
Edit-80 text editor, 6/15
EDLIN text editor, 12/13
Macintosh Word, 10/23, 12/6, 12/26
Multi-Tool Word, 12/5-6, 12/25, 13/13
Opus project (Word for Windows), 12/26
Word, 12/6-7, 12/18, 12/25-26, 13/9, 13/12-13, 15/5
Word for Windows, 12/11, 12/26, 13/13, 15/5
Miscellaneous:
Applications Division, 12/6, 16/10
ASCII Microsoft, 6/12
Consumer Products Division, 6/14, 12/14, 15/3, 16/11
Develop-65, 68 and 80, 6/9
Headquarters, 12/8
Interactive Media Division, 16/11
Microsoft International, 12/3
Microsoft Network (MSN), 15/3, 16/10-11
Microsoft Press, 12/5
Microsoft Quarterly, 18/8
Microsoft System Journal, 18/8
Mouse, 12/5, 17/24
Intellimouse, 17/24
Natural Keyboard, 17/23
Office of the President, 16/9, 16/11
RamCard, 17/17
Systems Division, 12/6
Tiger system project, 15/6
Z-80 SoftCard, 6/14, 12/1, 17/17
Multimedia:
Bookshelf, 12/7, 12/28
CD-ROM division, 12/7, 12/10, 12/28, 15/6
Encarta, 15/6
Gandalf project (Encarta), 15/6
Merlin project (Encarta), 15/6
MS-CD format, 12/7
Multimedia PC (MPC) standard, 12/28
Multimedia Systems division, 12/10
Operating & Interface Systems:
Advanced DOS, 12/16
Bob, 15/3
Cairo project, 15/1, 15/3, 16/9
Chicago project (Windows 95), 15/3, 16/10
Daytona project (NT), 15/3
Interface Manager, 12/6, 12/18
Memphis project (Windows 98), 15/4
MIDAS project, 6/11
MS-DOS (Disk Operating System),
Computer use of, 11/7, 11/10-11, 11/16, 11/18, 11/20-21, 11/28, 12/4
Development of, 12/13-15, 13/2
Later releases, 12/18-19, 15/1-2, 15/5, 16/10
Miscellaneous, 12/6-7, 12/15, 13/3, 17/24, 19/12, 20/3
Other software, 11/26, 12/19, 12/21, 12/27, 13/28, 15/3-4, 15/9
Seattle Computer Products, 12/10, 12/14, 17/18
MSX system, 12/6
MSX-DOS, 12/6, 12/15

Multi-Tool Interface, 12/22
NT (New Technology), 12/17, 13/4, 15/1-2, 15/4
OS/2. *See* IBM/Software
PC-DOS. *See* IBM/Software
Psycho project (NT), 12/17
SB-86 (Software Bus-86). *See* Lifeboat Associates
Win32, 15/2-3
Windows,
  Apple Computer, 10/8, 10/23, 12/7, 12/10, 16/1, 16/8-9
  Development of, 12/3, 12/6, 12/17-20
  IBM, 9/22, 12/16-17, 15/1, 15/8-9, 16/9
  Later releases, 15/1-4
  Miscellaneous, 11/26, 13/5, 20/3, 20/9
  Other software, 12/25, 12/27, 13/6, 13/11, 13/17, 13/27, 15/5, 15/14
Windows 95, 13/26, 15/3-5, 15/12, 16/10-11
Windows 98, 15/4, 16/12
Windows 286, 12/20
Windows 386, 12/20
Windows CE, 15/4
Windows for Workgroups, 15/2
Windows NT (New Technology), 15/2-4, 16/9
Windows NT Server, 15/4
Windows NT Workstation, 15/4
WINPAD, 15/4, 15/6
XENIX, 11/2, 12/10, 12/12, 12/15, 13/3
XENIX 286, 9/16, 12/15
ZDOS (Zenith), 12/14-15
Programming Languages:
  4K BASIC, 6/6, 7/3
  8K BASIC, 6/6
  6502 BASIC, 6/9-10
  6800 BASIC, 6/8-9
  8080 BASIC, 6/10, 6/12, 17/13
  8086 BASIC, 6/12, 6/14, 13/2, 17/18
  Advanced BASIC (BASICA), 9/8, 12/14
  APL, 6/9, 12/14
  Assembler, 12/13
  BASIC Compiler, 6/13, 6/15
  BASIC interpreter (Altair), 6/5-8
  BASIC interpreter (Macintosh), 10/20, 10/22-23, 12/6, 12/21
  BASIC,
    Apple Computer, 5/12, 12/6, 12/22
    Early developments of, 6/5-13
    IBM, 9/8, 9/12, 12/12-14, 12/21
    Other companies, 4/16, 11/10, 11/18, 12/3-4, 12/21
  BASIC-80, 17/17
  BASICA. *See* Advanced BASIC
  C, 12/21
  Cartridge BASIC, 9/14
  Cassette BASIC, 9/7-8, 9/14
  COBOL, 6/10-11, 6/14, 12/12-13, 12/14, 12/21
  COBOL-80, 6/13-14
  DISK BASIC, 6/9, 9/8, 12/14
  Extended BASIC, 6/8
  Focal, 6/9
  FORTRAN, 6/9-10, 6/12, 6/14, 12/12-13, 12/21
  FORTRAN-80, 6/10, 6/13-14, 7/5
  GBASIC, 17/17
  GWBASIC (Gee Whiz BASIC), 9/13, 12/21
  Level II BASIC (TRS-80), 4/17, 6/12, 7/4
  Level III BASIC (TRS-80), 6/14, 7/5
  Macro Assembler, 6/15
  MBASIC, 11/7-8, 11/12, 17/17
  Pascal, 6/10, 6/13, 9/8, 12/12-13, 12/21, 13/8, 13/12
  PC BASIC, 12/21
  Quick C, 12/21
  Quick Pascal, 12/22, 13/8
  QuickBASIC, 12/21, 13/7
  Stand-alone Disk BASIC, 6/9, 6/11
  Visual BASIC, 15/5
MicroTeck magazine, 18/3
MIDAS project. *See* Microsoft/Oper. Systems
Midwest Affiliation of Computer Clubs (MACC), 19/4

MIF (Management Information Format), 19/17
Mikbug operating system. *See* Motorola
Millard, William H., 4/12, 4/20, 19/11
Mims, Forrest M., 4/8
Miner, Robert N., 13/17
Minicomputer, 2/4, 2/8-9, 4/8, 4/10, 4/18, 11/16
  Term origin, 20/7
Minimal BASIC. *See* ANSI
MiniScribe company, 17/7
MinisPort portable computer. *See* Zenith Data Systems
Minnow project. *See* IBM/Accessories
MIPS Computer Systems, 8/8, 14/7, 15/2, 16/15, 19/17
  R2000 RISC microprocessor, 8/8
  R8000 microprocessor, 14/7
  R10000 microprocessor, 14/7
  Workstation, 15/2
MIT (Massachusetts Institute of Technology),
    Computer graphics, 2/10-11
    Early computers, 1/12, 1/15, 2/2
    Games and other software, 2/13-14, 13/6, 13/24, 15/10
    Magnetic core memory, 1/9
    Miscellaneous, 4/1, 7/8, 7/11, 10/25, 13/13, 17/21, 19/26
    Other companies, 1/7, 1/9, 1/16
    Time sharing, 2/3-5
  ARC (Average Response Computer), 1/15
  CTSS (Compatible Time Sharing System), 2/4
  LINC (Laboratory Instrument Computer), 1/15, 2/2, 2/7
  Lincoln Laboratory, 1/16, 2/7
  MDL language, 13/24
  Media Laboratory, 19/26
  Memory Test Computer (MTC), 1/15
  NuBus, 11/12, 17/21
  Project MAC (Multiple Access Computer), 2/4
  TX-0 computer, 1/15, 2/4, 2/13
  TX-1 and TX-2 computers, 1/15
  Whirlwind computer, 1/5, 1/9, 1/12, 1/15-16
  X Window System, 13/6
Mitel Corporation, 13/27
MITS, Inc.,
    Beginning and end of, 4/8-11
    Microsoft, 6/5-11
    Miscellaneous, 3/8, 3/11, 4/14-15, 18/2, 19/2, 19/4, 19/10, 20/2
    Other companies, 4/13-14, 5/1, 7/7, 17/17-18
  Altair 680, 4/10
  Altair 680b, 3/11, 4/10, 6/8
  Altair 8800,
    Development of, 4/8-11
    Expansion boards, 4/13-14, 17/17-19
    Microsoft, 6/5-7, 7/3
    Miscellaneous, 3/8, 4/13, 5/1, 7/7, 9/1, 17/7, 17/20, 18/2, 18/8, 19/2, 19/9-11, 20/2
  Altair 8800b, 4/10
  Altair BASIC, 4/10, 6/5-7, 7/2
  Altair Bus, 4/10, 17/20
  Computer Notes newsletter, 4/10, 6/7-8, 18/2-3
  Memory boards/cards, 4/10, 6/6
  MITS 816 calculator, 4/8
  PE-8 computer, 4/9
  World Altair Computer Convention (WACC), 6/8, 19/4
Mitsubishi Chemical Company, 17/22
Mitsubishi Kasei, 17/22
MMX technology. *See* Intel/Miscellaneous
Model 100 computer. *See* Tandy/Radio Shack/Computers
Model 101 printer. *See* Centronics
Models 33 and 35 teletypes. *See* Teletype Corporation
Modem, 9/14, 17/19-20, 19/17, 20/4
Modula programming language, 13/7
Modula-2 programming language, 13/8
Molnar, Charles E., 1/15, 2/7
Mondrian software, *See* Dynamical Systems Research
Money magazine, 12/6

Monochrome Display Adapter. *See* MDA
Monochrome Graphics Adapter. *See* MGA
Moore School of Electrical Engineering, 1/5
Moore, Charles H., 2/12
Moore, Dave, 12/23
Moore, Fred, 18/2, 19/2
Moore, Gordon E., 3/3, 3/5-6, 16/8
Moore's Law, 3/5
Moore, Rob, 10/13
Morgan, Charles R., 2/12
Morgan, James J., 11/14
Morgridge, John P., 19/21
Morill, Lyall, 7/9
Morrow's Micro Decisions, 11/30
Morse, Stephen P., 3/9
MOS (Metal Oxide Semiconductor), 17/3
MOS Technology, Inc.,
  Apple Computer, 5/4, 5/14
  Beginning of, 3/13-14
  Commodore purchase of, 4/15, 5/8
  Miscellaneous, 3/18, 4/12, 6/9, 7/11, 11/28
  KIM-1 (Keyboard Input Monitor-1), 4/12, 7/11
  MOS 6501, 3/13, 4/13
  MOS 6502, 3/13-14, 3/18, 4/12, 4/14-16, 5/4, 10/12-13, 11/25
  MOS 65C02, 10/11
Mosaic browser, 15/10-12
Mosaic Communications Corporation, 15/11
  Mosaic Navigator, 15/11
Mostek company, 3/15, 4/18, 8/8
Motherboards, 16/7, 19/17
Motif software. See OSF
Motorola, Inc.,
  1970's, 3/9-11, 4/8, 4/10, 4/20, 5/4;
  1980's, 8/6-7;
  1990's, 14/6-7
  Microprocessor application, 4/8-9, 4/13, 5/15-16, 9/11, 11/2, 11/12, 11/14, 16/13, 17/21
  Miscellaneous, 3/13, 3/15-16, 6/12, 7/1, 10/8, 11/28, 19/9
  PowerPC Alliance, 14/3, 14/10, 15/7, 16/1, 16/5, 19/19, 20/4
  68008 microprocessor, 3/11
  68882 math coprocessor, 10/8
  88000 RISC microprocessor, 8/6
  M6809E microprocessor, 5/15
  MC6800 MPU (Micro Processor Unit), 3/9, 3/10, 3/13, 4/7, 4/10, 4/12, 4/14, 5/4, 11/25, 17/21
  MC6801 microprocessor, 3/11
  MC6809 microprocessor, 3/11, 10/18, 11/22
  MC68000 microprocessor, 3/11, 4/20, 5/15, 6/12, 8/6, 9/6, 10/15, 17/11
  MC68010 microprocessor, 8/6
  MC68020 microprocessor, 8/6, 17/16
  MC68030 microprocessor, 8/6, 10/8
  MC68040 microprocessor, 8/7
  MC68060 microprocessor, 14/6
  Mikbug operating system, 4/12, 7/1
  MPC601 microprocessor, 14/6-7
  MPU (Micro Processor Unit), 3/9
  Museum of Electronics, 19/9
  PowerPC 601 microprocessor, 14/6
  PowerPC 603 microprocessor, 14/7
  PowerPC 604 microprocessor, 14/7
  PowerPC 620 microprocessor, 14/7
  PowerPC 750 G3 microprocessor, 14/9, 14/11
  PowerPC microprocessor, 14/5, 15/8, 16/1, 20/5
  Versabus bus standard, 9/11
Mott, Tim, 7/6
Mouse,
  Also see. Trackball
  Apple Computer, 10/12, 10/14-15, 10/21-22
  Development of, 2/10, 17/23-24
  Microsoft, 12/5, 12/17, 12/19, 12/22, 12/25-26
  Miscellaneous, 2/3, 2/13, 11/19, 13/5, 19/22
  Xerox, 4/5, 5/14, 11/29
Mouse Systems company, 17/24

Mouseworks. *See* Microsoft/ Application Programs
Moussouris, John, 8/8
MP/M (Multi-Programming Monitor). *See* Digital Research
MPC601 microprocessor. *See* Motorola
MPU (Micro Processor Unit). *See* Motorola
MS-CD format. *See* Microsoft/Multimedia
MS-DOS. *See* Microsoft/ Operating Systems
MS-NET. *See* Microsoft/ Application Programs
MSI (Medium Scale Integration), 3/5, 4/3, 11/20, 11/29
MSX hardware, 12/16
MSX-DOS. *See* Microsoft/ Operating Systems
MTC (Memory Test Computer). *See* MIT
Muka, Steve, 17/19
Multi-Color Graphics Array. *See* MCGA
Multi-Tool Interface. *See* Microsoft/Operating Systems
Multi-Tool Word. *See* Microsoft/Applic. Programs
Multics (Multiplexed Information and Computing Service), 2/4
MultiFinder operating system. *See* Apple Computer/Software
MultiMate International Corporation, 13/12
  Advantage Professional Word Processor, 13/12
  Executive Word Processor, 13/12
  Professional Word Processor, 13/12
Multimedia,
  Microsoft, 12/7, 12/10, 12/28, 15/6
  Miscellaneous, 15/9, 15/15, 19/23, 20/2
  PowerPC Alliance, 15/7-8, 16/5, 19/19
Multimedia PC (MPC). *See* Microsoft/Multimedia
Multimedia Systems division. *See* Microsoft/Multimedia
Multiplan spreadsheet. *See* Microsoft/Applic. Programs
MultiSpeed portable computer. *See* NEC
Multitasking, 8/3, 9/16, 9/19, 9/22, 10/25, 12/19
Multitech International Corporation, 11/13
Mundie, Craig, 16/9
Murto, William, 11/9
Museums *See*:
  Intel Corporation/Misc.
  Motorola Museum of Electronics
  National Museum of American History (Smithsonian)
  The American Computer Museum
  The Computer Museum (Boston)
Music Construction Set game. *See* Electronic Arts
MX printers. *See* Epson America
Myhrvold, Nathan P., 12/9, 12/17, 16/8, 16/10
Mystery House game. *See* Sierra On-Line

--N--

N M Electronics, 3/5
N.V. Philips. *See* Philips
Namco Limited, 7/11
  Pac-Man game, 7/11
National AppleWorks Users Group (NAUG), 13/20, 18/5
  AppleWorks Forum newsletter, 18/5
National Bureau of Standards (NBS), 1/10
National Cash Register Company, 1/6, 1/13, 6/9, 6/11, 17/4
  8200 Terminal, 6/11
National Center for Supercomputing Applications (NCSA), 15/10-11
National Computer Conference (NCC), 4/16, 6/14, 7/9, 10/10, 11/3, 11/29, 12/5, 13/14, 17/18, 19/5
National Museum of American History. *See* Smithsonian Institution
National Physical Laboratory (NPL), 2/13
National Radio Institute (NRI), 4/3
  NRI 832 computer kit, 4/3
National Science Foundation (NSF), 19/13
  NSFnet, 19/13
National Semiconductor Corporation, 3/9, 3/15,

5/9, 6/11, 8/7, 10/14, 16/1
16032 microprocessor, 8/7
32032 microprocessor, 8/7
IMP-8 microprocessor systems, 3/15
IMP-16 microprocessor systems, 3/15
Pace microprocessor, 3/9, 3/15
SC/MP microprocessor, 3/15
Super-Pace microprocessor, 3/15
Natural Keyboard. *See* Microsoft/Miscellaneous
Naval Postgraduate School, 7/1, 7/4
Navigator browser. *See* Netscape Communications
NBC television network, 16/11
NCC. *See* National Computer Conference
N-channel Metal Oxide Semiconductor. *See* NMOS
NCSA. *See* National Center for Supercomputing Applications
Near Letter Quality (NLQ), 17/13
NEC (Nippon Electric Company), 6/13, 11/19, 11/22, 12/2, 12/28, 16/16, 17/14, 17/21
PC-100 computer, 11/22
PC-8200 Portable Computer, 12/2
MutiSpeed portable computer, 11/22
UltraLite portable computer, 11/22, 20/8
V-30 microprocessor, 11/19, 11/22
NED (New Editor), 7/7
Negroponte, Nicholas, 18/9, 19/26
Nelson, Theodor H., 18/2, 20/1, 20/7
Netscape Communications Corporation, 15/11-13, 16/17, 19/15
Communicator Professional, 15/12
Netscape Navigator, 15/11-12
NetWare. See Novell
Network computers, 11/25
Networks, 11/25, 12/27, 13/27, 15/15, 16/10, 16/13, 19/13-14, 19/21
Newell, Martin, 13/23
Newman, M. H. A., 1/4, 2/4, 2/9-10
Newton MessagePad (PDA). *See* Apple Computer/Computers
NewWave interface program. *See* Hewlett-Packard Company
NexGen company, 14/7, 16/16
NeXT Computer, Inc., 9/21, 10/7, 11/11-12, 15/7, 16/15-16
NeXT computer, 11/12, 13/4, 17/10, 19/14
NeXT laser printer, 11/12
NeXTSTEP operating system, 9/21, 11/12, 12/11, 13/4, 15/7
NeXT Software, Inc., 15/7, 16/2, 16/17
Nibble - term origin, 20/8
Nibble Mac magazine, 18/4
Nibble magazine, 18/4, 20/6
Nippon Electric Company. *See* NEC
Nishi, Kazuhiko (Kay), 6/12-13, 11/2, 12/2, 12/6, 12/9, 18/3
NLQ. *See* Near Letter Quality
NLS (On-line system). *See* Stanford Research Institute
NMOS (N-channel Metal Oxide Semiconductor), 3/7
Noble, David L., 17/6
Non-Linear Systems company, 11/8
Noorda, Raymond, 13/27, 16/12
North Star Computers, Inc., 4/20, 5/12, 11/21, 17/7, 17/19, 19/12
FPB, Model A floating point board, 17/19
Horizon-I, 4/20
Horizon-II, 4/20
Micro-Disk system, 17/7
Peripheral boards, 17/19
Northrop Aircraft, Inc., 1/8
Norton, David, 17/10
Notebook computer,
Apple Computer, 14/11
Compaq, 11/11
Epson, 11/18
IBM, 14/9
Miscellaneous, 4/5, 14/7, 16/17
NEC, 11/22
Tandy/Radio Shack, 11/2
Term description, 20/8
Notes. *See* Iris Associates and Lotus Development Corporation

Nova computers. *See* Data General
Noval Inc., 4/19
  Noval 760 computer, 4/19
Novation, Inc., 17/20
  CAT acoustic coupler, 17/20
Novell Data Systems, 13/27
Novell, Inc., 13/27, 15/9, 15/12, 16/8, 16/12-13, 19/17
  NetWare, 13/27
  Novell DOS, 15/9, 15/12
  PerfectOffice, 16/12
Noyce, Robert N., 1/11, 3/3, 3/5, 8/6, 16/7, 19/8, 20/7
NPL (New Programming Language). *See* IBM/Software
NRI 832 computer kit. *See* National Radio Institute
NSFnet. *See* National Science Foundation
NT (New Technology). *See* Microsoft/Operating Systems
NuBus architecture. *See* Apple Computer/Misc. and MIT
N.V. Philips company. *See* Philips

--O--

O2 workstation. *See* Silicon Graphics
Oak project (Java). *See* Sun Microsystems
Oates, Edward, 13/17
Oberon programming language, 13/8
Object Linking and Embedding. *See* OLE
O'Connor, Dave, 9/13
Odyssey 100 game. *See* Magnavox
Odyssey book, 14/10, 20/1
Odyssey project (Excel). *See* Microsoft/Applic. Programs
Office Products Division. *See* Xerox/Miscellaneous
Office program. *See* Microsoft/Applic. Programs
OfficeVision program. *See* IBM/Software
Ogdin, Jerry, 17/5
Ohio Scientific Instruments (OSI), 4/14-15, 4/21
  Challenger series of computers, 4/21
  OS 65 operating system, 4/21
  OSI 300 computer training board, 4/14
  OSI 400 Superboard computer, 4/14
Oki, Scott D., 12/3
Okidata company, 17/14
OLE (Object Linking and Embedding), 10/25, 13/8, 15/1-2, 15/4
Olivetti company, 12/2, 12/28, 17/21
  M-10 computer, 12/2
Olsen, Kenneth H., 1/15-16, 11/15, 16/14
Olympiad project (PC RT). *See* IBM/Computers (Personal)
Olympic Decathlon. *See* Microsoft/Applic. Programs
Omega database project. *See* Microsoft/Applic. Programs
Omidyar, Pierre, 19/21
Omnibus backplane. *See* DEC/Miscellaneous
On Disk Monthly magazine. *See* Softdisk
On-line services, 19/15-17
On-Line Systems company, 13/25
Onyx Computer company, 11/25
Opel, John R., 9/4, 9/23, 10/2
Open bus architecture, 4/8, 9/3, 10/20
Open Look software. *See* UNIX International
Open Software Foundation. *See* OSF
Open-Apple newsletter, 18/5
Open-source software, 15/9
Operating systems, 7/1-3, 9/19-21, 12/15-20, 13/1-6, 13/18, 15/7, 15/11-13, 16/9
Operation Crush. *See* Intel Corporation/Miscellaneous
Optical disk drives, 11/12, 17/10
Opus project (Word for Windows). *See* Microsoft/Application Programs
Oracle Corporation, 13/17, 15/13, 16/2
  Oracle relational database, 13/17
Orange Book specification, 20/4
Orange Computer, 11/14
O'Rear, Bob, 6/11-14, 12/13
OS 8 and 9 operating systems. *See* Apple Computer/Software
OS-65 operating system. *See* Ohio Scientific Instruments
OS/2 (Operating System/2). *See* IBM/Software

Osborne Computer Corporation, 11/1, 11/6-7
Executive portable computer, 11/7
Executive II portable computer, 11/7
Osborne 1 portable computer, 11/6-7, 12/23, 19/2, 19/9, 20/8
Osborne, Adam, 11/6, 12/18, 19/2, 20/1
OSF (Open Software Foundation), 9/21, 11/26, 12/11, 13/6, 19/18
Motif software, 13/6, 14/8
OSI computers. *See* Ohio Scientific Instruments
Ovation integrated program, 13/22
OverDrive processors. *See* Intel Corporation /Microprocessors
Owens, Don, 13/10-11
Oyama, Terry, 10/19
Ozzie, Raymond, 13/20, 15/10

--P--

P6 and P54 microprocessors. *See* Intel/Microprocessors
PAC (Personal Automatic Calculator), 1/13
Pac-Man game. *See* Namco
Pace microprocessor. *See* National Semiconductor
Packard Bell Electronics, Inc., 11/22-23, 16/16
Packard Bell NEC, Inc., 16/16-17
Packard, David, 2/8
Packet communications, 2/13
Page Description Language. *See* PDL
Page, John D., 13/18
Page, Rich, 10/5, 10/14, 11/11
PageMaker software. *See* Aldus
Palladin Software company, 13/15
Palm microcontroller. *See* IBM/Miscellaneous
Palmer, Robert, B., 16/14
PalmPilot PDA. *See* U.S. Robotics
Palo Alto Research Center (PARC). *See* Xerox/Misc.
Palo Alto Scientific Center. *See* IBM/Miscellaneous
Paper tape, 1/15, 4/18, 17/4, 17/11
Paperback Software International, 11/7, 13/15
VP-Planner, 13/15
Papert, Seymour, 2/12
Paradox database. *See* Borland International
Parasitic Engineering, 17/19
Peripheral boards, 17/19
PARC (Palo Alto Research Center). *See* Xerox/Misc.
Parkinson, Joseph, 16/16
Parkinson, Ward, 16/16
Parsons, Keith, 7/4
Pascal programming language,
Also see Microsoft/Prog. Languages and Apple Computer/Software
Apple Computer, 5/12, 7/3, 10/10, 10/21
Development of, 7/5
Miscellaneous, 4/5, 11/15, 13/8, 13/18
Patch, Glenn E., 18/7
Paterson, Tim, 6/11, 6/14, 12/3, 12/6, 12/9, 12/12-13, 12/15, 13/1-2, 17/17-18
Patterson, David, 8/8
Patterson, James, 17/9
Paul Allen Group, 19/23
PC BASIC. See Microsoft /Programming Languages
PC Bus. *See* IBM/Miscellaneous
PC computers. *See* IBM/Computers (Personal)
PC Computing magazine, 18/7
PC File database. *See* ButtonWare
PC Forum conference, 19/26
PC Magazine, 18/6
PC Network. *See* IBM/Software
PC Transporter card. *See* Applied Engineering
PC Week magazine, 18/7
PC World Communications Inc., 18/7
PC World magazine, 12/25, 18/6-7
PC's Limited company, 11/16, 11/18
Also see Dell Computer
PC-8200 computer. *See* NEC
PC-DOS. *See* IBM/Software
PCC Newsletter. *See* People's Computer Company
PCI (Peripheral Component Interconnect). *See* Intel/ Miscellaneous
PCjr magazine, 18/7

PCMail. *See* Microsoft/Applic. Programs
PCNET (Personal Computer NETwork), 19/14
PDA (Personal Digital Assistant), 14/10, 14/12, 19/22
PDL (Page Description Language), 10/6, 13/17
PDP computers. *See* DEC/Computers
PE-8 computer. *See* MITS
Peachtree Software, Inc., 9/8, 13/22
  Business Accounting Series, 13/22
  Peachtree Complete, 13/22
Peacock, H. B., 3/12
Peanut project (PCjr). *See* IBM/Computers (Personal)
Peddle, Charles H. "Chuck", 3/10, 3/13, 4/12, 4/15, 5/14, 11/28
P-Edit editor. *See* WordPerfect
Peelings II magazine, 18/5
Pelczarski, Mark, 13/26, 18/3
Pen computing, 10/8, 14/9-10
Penguin Software company, 13/26
  Magic Paintbrush graphics utility, 13/26
Penny Arcade game. *See* Apple Computer/Software
Pennywhistle modem, 17/19
Pentium microprocessors. *See* Intel/Microprocessors
People's Computer Company, 18/1, 19/2, 20/1
  PCC Newsletter, 7/4, 18/1, 19/1
  People's Computers magazine, 18/1
People magazine, 12/6
PepsiCola company, 10/3
PerfectCalc, 11/8, 11/13
PerfectFiler, 11/8, 11/13
PerfectOffice. *See* Novell
PerfectSpeller, 11/8, 11/13
PerfectWriter, 11/8, 11/12
Peripheral cards, 11/14, 17/16-19
Perot, Ross H., 6/15, 11/12
Personal Automatic Calculator (PAC). *See* IBM/Computers
Personal computer,
  - term origin, 20/7
  - the "first," 1/16, 2/2, 2/7, 4/2, 4/7
Personal Computer Fair and Exposition, 19/5
Personal Computer Group. *See* IBM/Miscellaneous
Personal computing,
  Hobby and amateur computing, 2/14-16
  Miscellaneous, 1/3, 1/16, 2/3, 4/1, 4/6, 4/10, 4/21, 17/4, 18/1
  MITS Altair, 4/8
  Time sharing, 2/5
  Xerox, 4/20
Personal Computing magazine, 18/3
Personal Computing 76 show, 4/13, 5/7, 19/4
Personal Computing Show!, 19/4
Personal Digital Assistant. *See* PDA
Personal Filing System. *See* Software Publishing Corp.
Personal Pearl database, 11/7
Personal Software, Inc.,
  Also see VisiCorp
  Apple Computer, 5/14
  Founding of, 19/12
  IBM, 9/6, 9/8
  Other software, 7/11, 13/25, 13/28-29
  VisiCalc development, 7/8-9
  VisiCorp, 13/5, 13/14
Personal Systems Group. *See* IBM/Miscellaneous
Personality Module. *See* Processor Technology
PERT (Program Evaluation and Review Technique), 10/17
Pertec Computer Corporation, 4/11, 6/10-11, 17/7
PET computers. *See* Commodore
Peter Norton Computing, 16/14
Peters, Chris, 12/2
Peterson, Rich, 4/14
Peterson, W.E. Pete, 13/11, 16/15
Pfeiffer, Eckhard, 16/3-4
pfs software. *See* Software Publishing
PGA (Professional Graphics Array), 20/5
Philips N.V. company, 12/7, 12/28, 17/9, 20/4
Phoenix project (Apple IIGS). *See* Apple Computer
Phoenix Technologies Ltd., 13/22
Pico magazine, 18/8

Pinball Construction Set game. *See* Electronic Arts
Pineapple computer, 11/14
Pink project. *See* Apple Computer/Software
Pinpoint Publishing, 13/20
  Pinpoint Desk Accessories, 13/20
PIP (Programmable Integrated Processor). *See* Signetics
Pirate Adventure game. *See* Adventure International
Pitman, Doug, 16/16
Pittman, Tom, 7/4
Pixar Animation Studios, 19/23
  Toy Story film, 19/23
PL/I (Programming Language/One). *See* IBM/Software
  PL/I "G" subset. *See* ANSI
  PL/I compiler. *See* Digital Research
PL/M (Programming Language for Microcomputers), 7/1-2, 7/6, 13/9
Planar process, 1/11
PlanPerfect spreadsheet. *See* WordPerfect Corporation
Platt, Lew, 16/17
Plattner, Hasso, 19/24
PLUS/4 computer. *See* Commodore
Pocketronic calculator. *See* Canon
Pohlman, William B., 3/9
PointCast Inc., 19/24
  PointCast Network, 19/24
PolyMorphic Systems, 4/14-15, 7/4
  Disk BASIC, 4/14
  Micro-Altair computer, 4/14
  POLY 88 computer, 4/14
  System 8813, 4/14
Pong tennis game. *See* Atari Corporation
Poor, Vic, 4/2
Popular Computing magazine, 18/3
Popular Electronics magazine, 4/8, 4/13, 6/5, 17/5, 17/11, 17/19, 18/1, 18/8
Popular Mechanics magazine, 2/15
Popular Science magazine, 3/18
Porat, Marc, 19/22
Portal, 15/13, 19/22
Portable computers,
  Apple Computer, 10/12, 10/26, 14/11
  Compaq, 11/9-11, 12/3, 12/21, 13/22
  IBM, 4/4, 4/10, 9/1, 9/25, 14/8
  Other companies, 11/6, 11/8, 11/12, 11/19-22, 11/30, 13/11, 17/8,
  Tandy/Radio Shack, 11/2, 12/2
  Term definition, 20/8
  Texas Instruments, 11/26
P/OS operating system. *See* DEC/Software
POSIX operating system, 15/2
PostScript. *See* Adobe Systems
Potter, Dave, 1/13
Power architecture. *See* IBM/Miscellaneous
PowerOpen environment, 19/19
PowerPC Alliance, 14/3, 14/6, 14/9, 15/7-8, 16/1, 16/5, 19/19
PowerPC microprocessor. *See* IBM/Misc. and Motorola
PowerPoint graphics. *See* Microsoft/Applic. Programs
Powerstation computer. *See* Excaliber Technologies
POWERstation and POWERserver. *See* IBM/Computers (Personal)
PPS microprocessors. *See* Rockwell
Premium/286 computer. *See* AST Research
Prentice-Hall Inc., 19/5
Presentation Manager. *See* IBM/Software
Princess game. *See* On-Line Systems
Princeton University, 1/8
Print Shop software. *See* Brøderbund Software
Printers, 17/12-16
  Ink jet, 17/14
  Laser, 11/11, 12/25, 13/27, 17/15-16
  Thermal, 11/20, 17/16
  Wire matrix, 17/12-14
Processor Technology Corporation, 4/10, 4/13-15, 5/7, 6/12, 7/3-4, 17/17-18, 19/2, 19/10
  3P+S (Parallel + Serial) board, 17/18
  CONSOL operating system, 4/13
  Helios disk drive, 4/14
  Memory expansion boards,

17/17-18
Personality Module, 4/13
PT-DOS operating system, 4/14, 7/3
Sol-10 Terminal Computer, 4/13, 5/8, 6/10, 19/2, 19/4, 19/10
Sol-20, 4/13
SOLOS operating system, 4/14
VDM-1 (Video Display Module) board, 17/7, 17/18
PRODAC IV computer. *See* Westinghouse
Prodigy Services Company, 19/16
ProDOS (Professional Disk Operating System). *See* Apple Computer/Software
Professional Graphics Array. *See* PGA
Professional Office. *See* Microsoft/Applic. Programs
Professional Word Processor. *See* MultiMate International
ProFile hard disk. *See* Apple Computer/Accessories
Profit Plan, 11/8
Program Evaluation and Review Technique. *See* PERT
Programming languages, 1/12-13, 2/11, 7/3-6, 12/21, 13/7-9, 15/12
Project MAC (Multiple Access Computer). *See* MIT
Project program. *See* Microsoft/Applic. Programs
ProLinea and ProSignia computers. *See* Compaq Computer
ProPrinter. *See* IBM /Accessories
PS/1 & 2 computers. *See* IBM/Computers (Personal)
Psycho project (NT). *See* Microsoft/Operating Systems
p-System. *See* SofTech Microsystems
PT-DOS operating system. *See* Processor Technology
Publish It! *See* Timeworks
Publisher program. *See* Microsoft/Applic. Programs
PUP (PARC Universal Packet). *See* Xerox/Miscellaneous

--Q--

Q & A integrated program. *See* Symantec
QDOS (Quick and Dirty Operating System). *See* Seattle Computer Products
QL (Quantum Leap) computer. *See* Sinclair Research
Q-Link on-line service. *See* Quantum Computer Services
QST magazine, 4/7, 18/1
Quadram company, 17/19
Quadboard, 17/19
Quantum Corporation, 17/7, 17/9
Quantum Computer Services Inc., 19/15
Q-Link on-line service, 19/15
Quark, Inc., 13/12, 13/24
QuarkXPress desktop publishing program, 13/24
Word Juggler, 13/12
Quarterdeck Office Systems, 12/18, 13/5
DESQ windowing system, 12/18, 13/5
DESQview, 13/5
Quasar project (Visi On). *See* VisiCorp
Quattro Pro spreadsheet. *See* Borland
Quick Pascal. *See* Microsoft/ Programming Languages
QuickBASIC and Quick C. *See* Microsoft/Programming Languages
QuickBooks. *See* Intuit
QuickDraw. *See* Apple Computer/Software
Quicken. *See* Intuit
QuickFile database, *See* Apple Computer/Software
Quinn, Peter, 10/11-12
Qume company, 9/14
Qume disk drive, 9/13
Qume printer, 10/17
Qureshey, Safi, 11/14
QX computers. *See* Epson America

--R--

R/1, R/2 and R/3 software. *See* SAP
R2000, R8000 and R10000 RISC microprocessors. *See* MIPS
R2E. *See* REE (Realisations Études Électroniques)
Rabinow, Jacob, 1/10
Raburn, Vern, 6/12, 6/14, 12/1, 12/4, 19/23

Radio Shack. *See* Tandy/Radio Shack
Radio-Electronics magazine, 4/8-9, 17/11, 18/1
Raikes, Jeffrey (Jeff) S., 10/11, 12/2, 16/11
Rainbow 100 computer. *See* DEC/Computers
RAM (Random Access Memory), 17/3
Rambo project (Apple IIGS). *See* Apple Computer/Computers
RamCard. *See* Microsoft/Misc.
RAND Corporation, 2/13
Random Access Memory. *See* RAM
Rashid, Richard (Rick), 13/4, 16/9
Raskin, Jef, 5/15-16, 10/18-19, 10/21, 10/24, 17/19
Raster Blaster game. *See* Budgeco
Ratliff, Wayne C., 7/9, 13/16
Rattner, Justin, 8/3
Ray, Phil, 4/2, 4/13, 13/12
RCA company, 3/15, 12/7
  1802 microprocessor, 3/15
  1804 microprocessor, 3/15
  DVI (Digital Video Interactive), 12/7, 12/10
Read Only Memory. *See* ROM
Reader's Digest organization, 19/16
Real Estate program. *See* Tandy/Radio Shack/Software
Recherches et Études Électroniques. *See* REE
Red Book specification, 20/4
Red Hat Software Inc., 15/10
Reduced Instruction Set Computing. *See* RISC
REE (Realisations Études Électroniques), 4/7
  Micral computer, 3/8, 4/7
  R2E, 4/6
Reed College, 5/3
Reference sources, 20/3
Regis McKenna agency, 5/9-10
Relational database, 13/17-18
Relational Software Inc. (RSI), 13/17
Relay technology, 1/3-4
Release 1.0 newsletter, 19/25
Remala, Rao, 12/18, 12/20
Research and Engineering (The Magazine of Datamation), 1/14
Resource One, 19/1
Retailers, 19/10-12
Reuters news service, 15/15
Rhapsody project. *See* Apple Computer/Software
Ribardiére, Laurent, 13/19
Rickard, Jay, 10/13
Ricoh Company, Ltd., 6/14
Riddle, Michael, 13/23
Rider, Ronald, 17/15, 17/23
Ringewald, Erich, 10/25, 16/14
RIOS project (RISC System/6000). *See* IBM/Computers (Personal)
RISC (Reduced Instruction Set Computing),
  Apple Computer, 16/1, 16/5
  Development of by IBM, 8/7-8
  IBM, 9/17, 13/8, 14/3, 14/8
  Microsoft, 12/17
  Miscellaneous, 8/8, 14/6, 19/17
  Sun, 8/8, 11/26
RISC-I microprocessor, 8/8
Ritchie, Dennis M., 2/12, 7/5
RJR Nabisco Inc., 16/6
Roark, Raleigh, 12/5, 12/7
Roberts, H. Edward, 4/8-11, 4/14, 6/5-6, 6/8, 20/2
Roberts, Lawrence G., 2/13
Robitaille, Roger, 18/3
Robwin company, 19/12
Roche, Gus, 4/2
Rock, Arthur, 3/6, 5/14, 7/8, 9/13
Rockwell International, 3/16, 3/18, 20/3
  PPS-4 microprocessor, 3/16
  PPS-8 microprocessor, 3/16
Roger Wagner Publishing company, 13/30
  HyperStudio, 13/30
Roizen, Heidi and Peter, 13/15
Rollins, Kevin B., 16/4
Rolm Systems, 16/5
ROM (Read Only Memory), 17/3
ROM magazine, 18/3
Romeo, John, 19/22
ROMP (Research Office products MicroProcessor). *See* IBM/Miscellaneous
Rosen Electronics Letter, 19/25
Rosen, Benjamin M., 11/9, 13/13, 16/4, 19/25
Rosenfeld, Eric, 7/11
Rosing, Wayne, 10/15
Rotenberg, Jonathan, 19/3
Rothmueller, Ken, 5/15, 10/15

Routers. *See* Data routers
Rowley, John, 13/2
RS-232C standard, 9/7, 19/18
RSBASIC. *See* Tandy/Radio Shack/Software
RT-11 operating system. *See* DEC/Software
Rubinstein, Seymour, 7/7
Ruby project. *See* Compaq Computer
Russell, Stephen, 2/14

--*S*--

S-100 Bus, 4/10, 4/20, 6/14, 13/1, 17/18, 17/20-21, 19/9
SA-400 and SA-900 diskette drives. *See* Shugart & Associates
SAA (Systems Application Architecture). *See* IBM/Software
Sachs, Jonathan M., 13/13
Sackman, Robert, 11/25
Safeguard Scientifics company, 13/27
SAGE (Semi Automatic Ground Environment), 1/7, 1/9, 1/12, 1/16
Sakoman, Steve, 10/8, 14/10
Salsberg, Arthur P., 18/8
Samna company, 13/13
Sams, Jack, 9/4, 12/11-12
Samson, Peter, 2/14
Samuel, Arthur L., 1/13
Sandberg-Diment, Erik, 18/3
Sandel, Alex, 11/23
Sander-Cederlof, Bob, 18/5
Sander, Wendell, 10/9
Sanders Associates, 2/14
Sanders, Walter Jeremiah (Jerry), 3/15
Santa Cruz Operation (SCO), Inc., 12/10, 12/15, 13/4, 13/15, 19/17
  SCO Professional spreadsheet, 13/15
SAP AG (Systems Applications Products), 19/24
  R/1 and R/2 accounting systems software, 19/24
  R/3 system software, 19/24
Sara project (Apple III). *See* Apple Computer/Computers
Sargent, Murray, 12/20
SARGON chess program. *See* Games
Sarubbi, Joseph, 9/6, 9/12, 9/23-24
Satellite Software International (SSI), 13/10-11
  Also see WordPerfect Corporation
Satz, Greg, 19/21
SB-86 (Software Bus-86). *See* Lifeboat Associates
SBASIC (Dartmouth Structured BASIC). *See* Dartmouth College
SBASIC (Kaypro), 11/8
SC/MP microprocessor. *See* National Semiconductor
Scalable Processor Architecture (SPARC). *See* Sun Microsystems
SCALP (Self Contained ALGOL Processor). *See* Dartmouth College
SCAMP (Special Computer APL Machine Portable). *See* IBM/Computers (Personal)
SCCS Interface newsletter. *See* Southern California Computer Society
SCDP. *See* Software Consultation Design and Production
Scelbi Computer Consulting, Inc., 4/7-8
  Scelbi-8B, 4/8
  Scelbi-8H, 3/8, 4/7
Schedule + program. *See* Microsoft/Applic. Programs
Schlumberger Ltd., 4/18
Schmidt, Eric, 16/12
Schriber, Gene, 3/11
SCI Systems, Inc., 19/24
Scientific American magazine, 4/3, 18/8
Scientific Control Systems (SCS), 2/9
Scientific Data Systems (SDS), 2/9, 4/4
SCO Professional spreadsheet. *See* Santa Cruz Operation
Scott, Bruce, 13/17
Scott, Michael M., 5/9, 5/11, 7/7, 10/2, 10/6, 10/15, 10/19, 11/16, 12/3, 12/14-15
Scribe printer. *See* Apple Computer/Accessories
SCSI (Small Computer System Interface), 17/21
Sculley, John C., 10/3-8, 10/9, 10/23, 12/7, 14/10, 16/1, 16/5, 20/1

Seagate Technology, Inc., 9/12, 17/7-9
ST-506 hard drive, 17/9
Sears Roebuck and Company, 9/8, 19/16
Seattle Computer Products, Inc., 6/11, 6/14, 12/3, 12/9, 12/12-14, 13/1-2, 17/17-18
8086 microprocessor card, 6/12, 12/13, 13/1
86-DOS operating system, 12/13-14, 13/2
Memory cards, 17/18
QDOS (Quick and Dirty Operating System), 6/11, 12/12, 13/1-2, 17/18
Seiko company, 17/13
Selective Sequence Electronic Calculator (SSEC). *See* IBM/Computers
Selector database, 7/9
Selker, Ted, 14/9
SEMATECH, Inc., 8/6
Sequoia Capital, 5/9, 5/14, 15/14, 19/21
Series 70, 80, 100 and 200 computers. *See* Hewlett-Packard
Servers, 9/25, 11/11, 13/18, 13/27, 14/6, 15/5, 15/12, 16/13-14, 19/21
Set-top terminals, 15/6, 16/11
Seuss, C. David, 13/28
Sevin-Rosen Partners, 11/9, 13/13
Sevin, L.J., 3/15, 13/13
SGML (Standard Generalized Markup Language), 20/3
Shannon, Claude E., 1/13, 20/6
Shapiro, Fred R., 20/6
SHARE user group, 1/14, 2/11
Shareware, 13/16, 19/22
Shaw, Greg, 17/18
Shih, Stan, 11/13
Shima, Masatoshi, 3/8, 3/14
Shirley, Jon A., 12/5-6, 12/9, 12/11, 16/8
Shockley, William B., 1/11
Shows, 19/4-5
Shrayer, Michael, 7/6-7
Shugart Associates, 5/12, 7/1-2, 11/28, 17/7-8, 17/22
5.25-inch floppy disk drive, 17/6
SA-400 diskette drive, 17/7
SA-900 diskette drive, 17/7
Shugart Technology, 17/8
Shugart, Alan F., 17/7-8
Sidekick program, *See* Borland
Sideways program, 13/16
Siemans AG company, 3/15
Sierra On-Line Inc., 13/13, 13/25
Homeword word processor, 13/13
King's Quest game, 13/25
Mystery House game, 13/25
Princess and Wizard games, 13/25
Sigma V time computer. *See* Xerox/Computers
Signetics Corporation, 3/16, 5/2
2650 processor, 3/16
Programmable Integrated Processor (PIP), 3/16
SIGPC. *See* Association for Computing Machinery (ACM)
Silentype printer. *See* Apple Computer/Accessories
Silicon Graphics, Inc. (SGI), 11/23, 14/7, 14/11, 15/11, 16/13, 16/15-16, 19/18
Geometry Engine integrated chip, 11/23
Indigo workstation, 14/11
IRIS 1000 3-D terminal, 11/23
IRIS 1400 3-D workstation, 11/23
IRIS Graphics Library, 11/23
O2 workstation, 14/11
Silicon Valley -term origin, 20/9
Silver, David, 13/28
Silverberg, Brad A., 16/8
SIM4 and 8 simulator boards. *See* Intel/ Miscellaneous
Simonyi, Charles, 4/5, 7/6, 12/9, 12/18, 12/22-23, 12/25
Sinclair Radionics company, 11/23
Sinclair Research Ltd., 11/23-24, 11/27
QL (Quantum Leap) computer, 11/24
ZX80 computer, 11/24
ZX81 computer, 11/24, 11/27
Sinclair, Clive M., 11/23
Singer, Hal, 18/1
Singleton, Henry E., 5/14, 9/13
Sippl, Roger, 13/18
Sirius Software, 13/25
Both Barrels game, 13/25-26

Cyber Strike game, 13/26
E-Z Draw utilities, 13/25
Star Cruiser game, 13/26
Sirius Systems Technology, 11/28
Sirius 1 computer, 11/28
Sketchpad, 2/10
Skiwriter word processor, 11/18
Skoll, Jeff, 19/21
SLOT (Scanned Laser Output Terminal) printer. *See* Xerox/Miscellaneous
Small Business Applications, Inc., 7/7
Magic Wand word processor, 7/7
Small Computer System Interface. *See* SCSI
Small Scale Integration. *See* SSI
Smalltalk language. *See* Xerox/Software
Smart integrated program, 13/22
Smartmodem 300. *See* Hayes Microcomputer Products
Smith, Burrell C., 5/16, 10/18-19, 17/16
Smith, Marshall, 11/5
Smith, Mary Carol, 13/28
Smith, Steve, 6/13
Smithsonian Institution, 19/6, 19/9
National Museum of American History, 19/9
Smoke Signal Broadcasting company, 17/21
Chieftain computer, 17/21
Soft Warehouse Inc., 19/10
Softalk Publishing, Inc., 18/4
Softalk, magazine, 18/4-5
Softbank COMDEX Inc., 19/5
Softbank Corporation, 15/15, 19/5, 19/24
SoftCard. *See* Microsoft/Misc.
Softdisk Inc., 18/5, 18/7-8
Big Blue Disk magazine, 18/7
Diskworld for the Macintosh magazine, 18/5
Gamer's Edge magazine, 18/7
Loadstar magazine, 18/5, 18/8
On Disk Monthly magazine, 18/5
Softdisk GS magazine, 18/5
Softdisk magazine, 18/5
SofTech Microsystems, Inc., 9/8, 13/4 13/8
UCSD p-System, 9/8, 11/7, 13/4, 13/19
SoftImage, Inc., 16/10
Softkey International, Inc., 16/15
Softkey Software Products, 16/15
Softsel Computer Products, 19/12
SoftSide, magazine, 18/3
Software, 1/12-14, 2/11-13, 7/1-11, 13/1-29
Software & Information Industry Association, 19/19
Software Arts, Inc., 7/8-9, 13/14-15, 13/29, 19/12
TK!Solver program, 13/29
Software Bus-86. *See* Lifeboat Associates
Software Consultation Design and Production (SCDP) company, 7/9
Vulcan data base, 7/9, 9/2, 13/16
Software Development Laboratories, 13/17
Software Garden, Inc., 13/15
Software Package One (SP-1), 7/6
Software Plus company, 13/16, 19/12
Software Publishers Association (SPA), 19/19
Software Publishing and Research Co. (S.P.A.R.C.), 18/4
Software Publishing Corporation (SPC), 13/13, 13/15, 13/18, 13/21
Assistant series of programs (IBM), 13/17
First Choice, 13/21
Personal Filing System, 13/18
pfs:File, 13/18
pfs:Graph, 13/15, 13/18
pfs:Plan, 13/18
pfs:Report, 13/18
pfs:Word, 13/18
pfs:Write, 13/13
Software Publishing, Inc., 13/27
Harvard Graphics, 13/27
Sol computers. *See* Processor Technology
Solid State Software command module. *See* Texas

Instruments
Solid State Speech synthesizer. *See* Texas Instruments
Solomon, Leslie (Les), 4/9, 4/13, 18/8
SOLOS operating system. *See* Processor Technology
Son, Masayoshi, 15/15, 19/24
Sony Corporation, 10/17, 10/21, 17/8, 17/9, 17/22, 20/4
Soon, Chay Kwong, 17/16
Sophisticated Operating System (SOS). *See* Apple Computer/ Software
Sorcim, 11/6, 13/15
  SuperCalc spreadsheet, 11/6-7, 13/15, 13/29
  SuperCalc3 spreadsheet, 13/15
  SuperCalc5 spreadsheet, 13/15
Sound Blaster audio card. *See* Creative Technology
Source. *See* Telecomputing Corporation of America
Southern California Computer Society (SCCS), 7/6, 18/2, 19/3
  SCCS Interface newsletter, 18/2
Southwest Technical Products Corporation (SwTPC), 4/12, 4/15, 6/12, 7/4, 11/24, 17/21, 19/10
  Digital Logic Microlab, 4/12
  KBD-2 Keyboard and Encoder Kit, 4/12
  SS-50 bus, 4/12, 17/21
  SwTPC 6800 Computer System, 3/11, 4/12, 7/4, 17/21
  SwTPC 6809, 11/24
  SwTPC BASIC, 7/4
SPA. *See* Software Publishers Association
Space Craft, Inc., 19/24
Space Quarks game. *See* Brøderbund Software
Space Wars game. *See* Games
SPARC (Scalable Processor Architecture) microprocessor and workstation. *See* Sun Microsystems
Sparks, H. L., 9/6, 9/23, 11/9
Spatial navigation, 20/1
Specifications, 20/4
Spectrum Information Technologies, 16/1
SpellStar. *See* MicroPro International
Sphere Corporation, 3/11, 4/12
  BASIC computer, 4/12
  Hobbiest computer, 4/12
  Intelligent computer, 4/12
  System 340 computer, 4/12
Spindler, Michael "Mike" H., 10/7, 16/1
Spinnaker Software Corporation, 13/28, 16/15
Sporck, Charles E., 3/15
Spracklen, Dan and Kathe, 7/11
Spreadsheets, 7/8-9, 11/30, 12/22-24, 12/26, 13/13-15, 13/20-21
Springer-Verlag company, 18/8
Sprite graphics, 11/5
Spyglass, Inc., 15/11, 15/12
SQL (Structured Query Language), 13/17
SR-60A personal computer/calculator. *See* Texas Instruments
SRAM (Static Random Access Memory), 17/3
SRI. *See* Stanford Research Institute
SS-50 bus. *See* Southwest Technical Products Corporation
SSEC (Selective Sequence Electronic Calculator). *See* IBM/Computers
SSI (Small Scale Integration), 3/5
SSI Software company, 13/11 *Also see* WordPerfect Corporation
SSI* software. *See* WordPerfect Corporation
ST-506 hard drive. *See* Seagate Technology
Stac Electronics, 16/10
  DoubleSpace utility, 16/10
Stand-alone Disk BASIC. *See* Microsoft/Programming Languages
Standard Generalized Markup Language. *See* SGML
Standards, 20/4
Stanford Linear Accelerator Center (SLAC), 5/1, 19/2
Stanford Research Institute (SRI), 2/3, 2/9, 2/11, 17/23
  ARC (Augmented Research

Center), 2/10-11
NLS (On-line System), 2/10-11
Stanford University,
Apple Computer historical collection, 19/9
Association with, 4/14, 11/23, 11/25, 13/21, 15/14, 17/16, 19/21-22
Miscellaneous, 20/9
RISC microprocessor, 8/8
Software, 7/4, 7/6
Stanley, Robert C., 20/2
Star computer. *See* Xerox/Computers
Star Craft company, 13/25
Star Cruiser game. *See* Sirius Software
Star Division, 16/13
StarOffice, 16/13
StarPortal, 16/14
Starkweather, Gary, 17/15
Static Random Access Memory. *See* SRAM
Statistical Analysis program. *See* Tandy/Radio Shack/ Software
Stephensen, John, 4/14
Stephenson, Robert, M., 16/6
Stibitz, George, 1/3
Stiskin, Nahum, 12/5
Storage devices, 1/9-10, 17/4-10
Strachey, Christopher, 2/3
Stritter, Skip, 8/8
Strom, Terence, M., 19/22
Strong, David, 12/5
Stroustrup, Bjarne, 13/7
Structured BASIC. *See* Dartmouth College
Structured programming, 13/6
Structured Query Language. *See* SQL
Styleware company, 13/21
subLogic company, 13/26
Subnotebook computer, 20/8
Suding, Robert, 4/20
Sugiura, Go, 17/12
Summagraphics Corporation, 17/22
Bit pad, 17/22
Sun, David, 19/22
Sun Microsystems, Inc.,
Beginning of, 11/24-26
Java, 15/13
Miscellaneous, 9/17, 13/8, 16/13-15, 16/17, 19/15, 19/18
RISC microprocessor, 8/8, 14/7
386i personal computer, 11/26, 16/13
Darwin workstation, 16/13
Java programming language, 15/12-13, 16/17
Jini software, 15/13
Oak project (Java), 15/13
Solaris operating system, 11/26, 16/13
SPARC (Scalable Processor Architecture) microprocessor, 8/8, 11/26, 16/13
SPARCstation 1, 8/8, 11/26, 13/8
Sun-1 workstation, 11/25
Sun-2 workstation, 11/25, 19/9
Sun-3 workstation, 11/25-26
SunOS operating system, 11/25
Sun UNIX, 11/26
SuperSPARC microprocessor, 14/7, 16/13
The Network is the Computer slogan, 11/25
UltraSPARC microprocessor, 14/7, 16/13
SunSoft company, 19/17
Super VGA, 20/5
Super-Pace. *See* National Semiconductor
Super-Sort. *See* MicroPro International
SuperCalc spreadsheets. *See* Sorcim
SuperPET 9000 series. *See* Commodore
Superscalar processor, 14/3-4, 4/7
SuperSPARC microprocessor. *See* Sun Microsystems
Sutherland, Ivan E., 2/10, 4/5
Sutherland, James F., 2/15
SWEET16, software, 7/11
Swingle, Richard, 14/11
Switcher program. *See* Apple Computer/Software
SwTPC. *See* Southwest Technical Products Corporation
SwyftCard. *See* Information Appliance
SX-64 computer. *See* Commodore
Sybase, Inc., 13/18
Sydnes, Bill, 9/4, 9/13
Symantec Corporation, 13/29,

Tele-Communications, Inc., 16/10
Telecomputing Corporation of America, 19/16
  The Source, 9/8, 17/20, 19/16-17
Teledyne Inc., 5/14, 9/13, 11/22-23
Teleprinter, 2/9, 17/10
Teletype Corporation, 10/17, 17/11
  ASR-33 terminal, 6/1, 10/17, 17/11
  Models 33 and 35, 17/11
  Teletype terminal, 2/7, 2/10, 2/15, 5/4, 5/10, 6/1, 10/14, 17/4, 17/10-11, 19/1
Teletypewriter, 2/4
Terminology origins, 20/6-9
Terrell, Paul, 5/6, 5/9, 19/10
Tesler, Lawrence G. "Larry", 4/6, 5/15, 7/6, 10/14-15, 14/10
Tevanian, Avidis, 13/4, 15/7
Texas Instruments, Inc.,
    Integrated circuit, 1/11
    Microprocessor, 3/7, 3/11-13, 3/16-17, 8/8, 14/8
    Miscellaneous, 3/5, 4/2, 5/17, 6/10, 11/1, 11/9
    Personal computers, 4/18-19, 11/27, 16/17
  Computers:
    SR-60A personal computer /calculator, 4/19
    TI CC40 computer, 11/27
    TI-99/4 computer, 3/13, 4/19, 11/27
    TI-99/4A computer, 11/27, 13/7
    TIPC (Texas Instruments Professional Computer), 11/27
  Miscellaneous:
    Solid State Software command module, 4/19
    Solid State Speech synthesizer, 4/19
    TI 9980 microprocessor, 3/13
    TI BASIC, 11/27
    TMS320 Digital signal processor (DSP), 8/8
    TMS1000 microprocessor, 3/12
    TMS9000 series of microprocessors, 3/13
    TMS9900 microprocessor, 3/13, 4/19, 6/11
Text editor, 4/19
Thacker, Charles P. "Chuck", 4/5
The American Computer Museum Ltd., 19/6
The Apple II Review magazine, 18/6
The Computer Hobbyist newsletter, 18/1
The Computer Museum, 2/15, 4/3, 19/7
The Computer Museum History Center, 19/7-8
The Electric Pencil. *See* Michael Shrayer Software company
The Interface Group Inc., 19/5
The Network is the Computer slogan. *See* Sun Microsystems
The Reactive Engine thesis, 4/5
The Road Ahead book, 15/15, 20/1
The Source. *See* Telecomputing Corporation of America
Thermal printers. *See* Printers
Thi, Truong Trong, 4/7
ThinkJet printer. See Hewlett-Packard
ThinkPad computer. *See* IBM/Computers (Personal)
Thoman, G. Richard, 16/6
Thompson, Kenneth L., 2/12
Thomson-CSF company, 8/8
TI-99/4 and TI-99/4A computers. *See* Texas Instruments
Tiger system project. *See* Microsoft/Miscellaneous
Time magazine, 10/2, 12/6, 18/9, 19/20
Time Manager. *See* Microsoft/ Application Programs
Time sharing,
  1960's, 2/3-6, 2/9, 2/12, 2/16
  Miscellaneous, 4/1, 6/1-2, 19/1, 19/15
TimeOut modules. *See* Beagle Bros.
Timeworks company, 13/24
  Publish It!, 13/24
Timex Corporation, 11/24, 11/27
  Timex/Sinclair 1000

computer, 11/24, 11/27
Tiny BASIC, 4/17, 7/4, 18/2
Tiny Troll program, 7/11, 13/13, 13/28
TIPC computer. *See* Texas Instruments
Titus, Jonathan A., 4/8
TK!Solver program. *See* Software Arts
T/Maker spreadsheet, 13/15
TMS processors. *See* Texas Instruments/Miscellaneous
Tokyo Shibaura Electric Company. *See* Toshiba
Tom Swift Terminal, 19/2
Tomash, Erwin, 19/6
Tommervik, Al, 18/4-5
Tommervik, Margot, 18/5
Toolbox. *See* Apple Computer/ Software
Topfer, Morton L., 16/4
TopView user interface. *See* IBM/Software
Torode, John, 7/1
Torvalds, Linus, 15/9
Toshiba Corporation, 11/27, 17/3-4, 17/16
  T3100 laptop computer, 11/27
Towne, James C., 12/2-3, 12/5
Toy Story film. *See* Pixar
TR10 project (Lotus 1-2-3). *See* Lotus Development Corporation
Trackball pointing device, 10/26
TRACKPOINT pointing device. *See* IBM/Miscellaneous
TRADIC (TRAnsistor Digital Computer), 1/6
Traf-O-Data company, 6/4-5
Tramel Technology Limited (TTL), 11/5, 11/14
Tramiel, Jack, 4/15, 11/5, 11/14
Transistor, 1/5, 1/10-11, 2/14-15, 3/5, 3/9-10, 3/12
Transport Control Protocol/ Interface Program. *See* TCP/IP
Transportable computer, 20/8
Transwarp card. *See* Applied Engineering
Trenton Computer Festival, 19/4
Tribble, Guy "Bud" L., 10/18-20, 11/12
Tripp, Dr. Robert, 18/4
TRS computers. *See* Tandy/Radio Shack/Computers
TRSDOS operating system. *See* Tandy/Radio Shack/Software
True BASIC. *See* Dartmouth College
TrueImage fonts, 12/11, 12/20
TrueType fonts, 12/20, 15/2
TRW company, 6/4
Tschira, Klaus, 19/24
TSR (Terminate and Stay Resident), 13/29
Tu, John, 19/22
Tuckey, John W., 20/6
Turbo BASIC, Turbo C and Turbo Pascal. *See* Borland International
TurboMac computer. *See* Apple Computer/Computers
TV Dazzler board. *See* Cromemco
TV Typewriter, 17/11, 19/2
  TVT-1 prototype, 17/11
Twiggy floppy disk drive. *See* Apple Computer/ Accessories
TX-0, TX-1 and TX-2 computers. *See* MIT
TX-80 printer. *See* Epson America
Tymshare, Inc., 2/11
  AUGMENT system, 2/11
Typing Tutor. *See* Kriya Systems and Microsoft /Application Programs

--U--

U.S. Army Ordnance Department, 1/5
U.S. Department of Commerce, 19/13
U.S. Government, 15/2, 17/4, 19/13
U.S. Patent and Trademark Office, 3/16-17, 14/8, 15/3
U.S. Robotics Corporation, 14/12, 16/17, 17/18-20, 20/3
  PalmPilot PDA, 14/12
  X2 technology, 20/3
U.S. Venture Partners, 11/25
UCSD p-System. *See* SofTech Microsystems
UCSD Pascal, 7/5, 9/8, 13/4, 13/8
Uiterwyk, Robert, 7/4
Ultimax computer. *See* Commodore
UltraLite portable computer. *See* NEC

Ultraportable computer, 20/8
UltraSPARC microprocessor. *See* Sun Microsystems
Ungermann, Ralph, 3/14
UniDisk. *See* Apple Computer/ Accessories
United Technologies, 3/15, 8/8
UNIVAC (UNIVersal Automatic Computer), 1/5, 1/8, 1/12, 19/9
Universal Resource Locator. *See* URL
University of California, 5/3, 7/5, 8/8, 10/3, 10/7, 11/25, 12/21, 12/23, 13/8
University of Arizona, 12/20
University of Colorado, 5/3, 12/3
University of Delaware, 11/6
University of Illinois, 15/10-11
University of Manchester, 1/6, 1/9
University of Minnesota, 19/6
University of Pennsylvania, 1/5, 10/4
University of Texas, 11/16
University of Utah, 2/11, 7/10, 13/10, 13/23
University of Washington, 6/3-5, 7/1
UNIX International, 13/6
  Open Look software, 13/6
UNIX operating system. *See* AT&T
Ural II computer, 12/22
URL (Universal Resource Locator), 19/14
UseNet (Users NETwork), 19/14
User interface:
  Early developments, 2/3, 2/9-11
  Apple Computer, 5/14, 10/15, 10/18, 10/22, 12/18, 13/5, 13/21
  IBM, 9/17, 9/20-22, 12/10, 12/16, 15/4, 15/8
  Microsoft, 12/3, 12/5, 12/7, 12/17-18, 12/20-22, 15/1, 15/3
  Other developments, 2/10-11, 13/5-6, 13/11, 13/13, 15/10
  Xerox, 4/4-5, 10/2, 11/29
USI International company, 13/28

--V--

V.32 and V.34 standards. *See* ITU
V-30 microprocessor. *See* NEC
Valentine, Donald T., 5/9, 5/14, 15/14, 19/21
ValuePoint computers. *See* IBM/Computers (Personal)
Van Natta, Bruce, 4/13
Vaporware - term origin, 12/19, 20/9
Varian Associates, 5/2
VAX computer. *See* DEC/Computers
VDM-1 (Video Display Module) board. *See* Processor Technology
VDM-1 Video Display Terminal, 17/12
VDP computers. *See* IMS Associates
Vector Graphic, 17/19
  Peripheral boards, 17/19
Veit, Stanley, 5/7-8 18/7, 18/9, 19/10
Venrock Associates, 5/14, 7/8
Ventura Software Inc., 13/24
  Ventura Publisher program, 13/24
Verbatim Corporation, 17/5, 17/22
Vermeer Technologies Inc., 16/11
  FrontPage program, 16/11
Versabus bus standard. *See* Motorola
VESA (Video Electronics Standards Association), 17/21, 19/20
  VL bus, 17/21
Vezza, Albert, 13/24
VGA (Video Graphics Array), 9/18, 14/9, 20/5
VIC BASIC, VIC-20, VICMODEM and VICTERM. *See* Commodore
Victor Business Products, 11/28
Victor Technologies Inc., 11/27, 12/4
  Victor 9000 computer, 11/28
Video standards, 20/5
Video Electronics Standards Association. *See* VESA
Video games, 1/11, 2/14, 4/16, 7/8, 11/14
Video Graphics Array. *See* VGA
Video Interface Chip. *See* Commodore
Video Technology, 11/14

128EX/2 computer, 11/14
Laser 128 computer, 11/13
Vieth, Kathy, 14/9
ViewPoint system. *See* Xerox/Computers
Viruses, 20/2
VisiCorp, 12/3, 12/17-18, 13/5, 13/14-15, 13/28, 19/12
Quasar project (VisiOn), 13/5
VisiCalc (Visible Calculator),
Apple Computer, 5/14, 5/17, 10/1
Development of, 7/8-9
IBM Personal Computer, 9/2, 9/6, 9/8
Microsoft, 12/22-23
Miscellaneous, 7/12, 13/5, 13/28, 19/12
VisiCorp, 13/14-15
VisiCalc Advanced Version, 13/14
VisiCalc III, 10/11
VisiGraph, 13/5
VisiOn graphic interface, 12/3, 12/17-18, 13/5, 13/15
VisiPlot program, 7/11, 13/13, 13/28-29
VisiTrend program, 7/11, 13/13, 13/28
VisiWord word processor, 13/5
Visio Corporation, 16/12
Visual BASIC. *See* Microsoft/ Programming Languages
VL bus. *See* VESA
VLSI (Very Large Scale Integration), 11/12
VM/PC (Virtual Machine/Personal Computer). *See* IBM/Software
VMS operating systems. *See* DEC/Software
Volition Systems company, 13/8
Volkswriter word processor. *See* Lifetree Software
von Meister, William, 19/16
von Neumann, John, 1/12
VP-Planner spreadsheet. *See* Paperback Software International
VT52, VT78 and VT100 computers and terminals. *See* DEC /Miscellaneous
Vulcan data base. *See* Software Consultation Design and Production

--W--

W65C816. *See* Western Design Center
Wa, Ng Kai, 17/16
Wadsworth, Nat, 4/7
Wagman, David, 19/12
Waitt, Norm, 11/19
Waitt, Theodore W., 11/19
Walker, John, 13/23, 16/15
Wallace, Bob, 6/13
Wang Laboratories, Inc., 4/13, 12/24, 13/9, 19/25
Wang, An, 19/25
Wang, Charles B., 13/29
Wang, Li-Chen, 7/4
Warner Communications Inc., 4/17, 11/5, 11/14
Warnock, John E., 13/23
Warren, Jim, 5/10, 18/2, 19/4
Watson Jr., Thomas J., 9/1
Watson Sr., Thomas J., 1/6
Wayne, Ron, 5/6-7, 7/6, 10/13, 13/10, 18/2-4
Web. *See* WWW
WebTV Networks, Inc., 16/11
Weiland, Richard, 6/3, 6/8-9, 6/11, 12/27
Weise, Dave, 12/10, 12/20
Weishaar, Tom, 18/5
Weiss, Eric A., 18/8
Weiss, Larry, 13/9
Wellenreuther, Claus, 19/24
WESCON show, 3/14, 5/5
West Coast Computer Faire,
Apple Computer, 5/10-11, 20/2
Founding of, 19/4-5
Miscellaneous, 7/7, 19/13-14
Other product introductions, 7/9, 11/6, 12/1, 13/12, 13/25
Western Design Center, 8/7, 10/3, 10/14
65802 microprocessor, 8/7
65816 microprocessor, 8/7, 10/3
W65C816 microprocessor, 8/7, 10/14
Western Digital Corporation, 17/7, 17/9
Western Joint Computer Conference, 1/10
Westfield, Bill, 19/21
Westinghouse company, 2/15, 17/4

PRODAC IV computer, 2/15
What-You-See-Is-What-You-Get. *See* WYSIWYG
WHATSIT? database. *See* Information Unlimited Software
Where in the World is Carmen Sandiego? game. *See* Brøderbund
Whirlwind computer. *See* MIT
White, Phillip E., 13/18
Whitman, Meg, 19/21
Wigginton, Randy, 5/4, 5/10-12, 6/10, 10/19-20, 10/22, 13/15
Wilber, Mike, 19/13
Wilcox, Brian, 4/14
Wilkes, Maurice V., 1/5
Wilkie, Dan, 9/6, 9/23-24
Wilkins, Jeffrey, 19/15
Williams, Don, 12/26
Williams, F. C., 1/5, 1/9
Williams, Ken, 13/25
Williams, Roberta, 13/25
Williams, S. B., 1/3
Wilson, Camila, 13/12
Win32. *See* Microsoft/Operating Systems
Winblad, Ann, 19/26
Winchester hard disk drive, 9/12-13, 10/14, 17/5, 17/8-9, 20/9
  term origin, 20/9
Window Manager. *See* Apple Computer/Software
Windows:
  Early developments, 2/10-11
  Apple Computer, 10/7, 10/15, 10/22, 15/8
  IBM, 9/15, 9/22
  Microsoft, 9/21, 10/8, 10/21-22, 12/17-20, 12/23-27, 15/1-6, 15/8, 15/12, 20/3, 20/9
  Xerox, 4/5, 11/29
  Also see Microsoft/Operating Systems
Windows World Exposition Conference, 15/1
WingZ spreadsheet. *See* Informix
Winkless, Nelson, 18/3
WINPAD. *See* Microsoft /Operating Systems
Wintel term, 20/9
Wire matrix printers. *See* Printers
Wired magazine, 18/9, 19/26
Wirth, Niklaus, 7/5, 13/8
Wise, Mike, 4/12
Wizard game. *See* Sierra On-Line
Wong, Albert, 11/14
Wong, Harvey, 19/23
Wood, Marla, 6/13
Wood, Steve (Microsoft manager), 6/9, 6/11, 6/13, 12/1
Wood, Steve (Microsoft programmer), 12/18
Woods, Don, 7/10
Word Juggler word processor. *See* Quark
Word Plus word processor 11/8
Word Pro. *See* Lotus Development
Word processors, 7/6-7, 11/5, 11/15, 11/18, 12/7, 12/22, 12/25-27, 13/9-13, 13/20-21, 13/29, 19/24
Word, word processing program. *See* Microsoft/Application Programs
Word-Master and Word-Star. *See* MicroPro International
WordPerfect Corporation, 13/9-12, 15/14, 16/12, 16/15
  DataPerfect database, 13/11
  DrawPerfect presentation graphics program, 15/14
  Executive WordPerfect, 13/11
  LetterPerfect word processor, 15/14
  MathPlan spreadsheet, 13/11
  P-Edit editor, 13/10-11
  PlanPerfect spreadsheet, 13/11
  SSI*Data database, 13/11
  SSI*Forth programming language, 13/11
  SSI*Legal, 13/11
  SSI*WP word processor, 13/10-11
  WordPerfect for Windows, 15/14
  WordPerfect Office, 13/11, 16/12-13
  WordPerfect word processor, 12/26, 13/9-12, 15/12, 15/14
WordStar International, 13/10, 16/15
  Also see MicroPro International
Works programs. *See*

Microsoft/Applic. Programs
Workslate portable computer. *See* Convergent Technologies
Workstation, 8/6, 9/11, 11/21, 11/24-26, 11/29, 14/6-8, 14/10, 16/13-14
World Altair Computer Convention. *See* MITS
World Wide Web. *See* WWW
WorldCom company, 19/16
Wozniak, Stephen G. "Woz",
  Apple Computer 1970's, 5/1-13., 1980's, 10/1-4, 10/8, 10/13, 10/21
  Homebrew Computer Club, 19/2
  Miscellaneous, 7/11, 20/2
  Other companies, 4/1, 4/17, 6/10
WWW (World Wide Web), 15/11, 15/14, 16/14, 19/14, 19/23, 20/3
Wynn-Williams, C. E., 1/4
Wyse company, 17/21
WYSIWYG (What-You-See-Is-What-You-Get), 4/5, 7/6, 10/15, 10/21, 11/29, 12/25, 13/23

--X--

X Window System. *See* MIT
X2 technology. *See* U.S. Robotics
Xanadu data management system, 20/1, 20/7
Xedex Corporation, 17/19
  Baby Blue card, 17/19
XENIX. *See* Microsoft/Operating Systems
Xerox Corporation,
  1970's, 4/4-6, 4/14-15, 4/20;
  1980's, 11/28-30
  Apple Computer, 5/15-16, 10/1, 10/5, 10/8, 10/14-16, 16/1
  IBM, 9/22, 9/24
  Laser printer, 17/15
  Microsoft, 12/18, 12/22, 12/25
  Miscellaneous, 2/11, 4/1, 6/3, 11/19, 17/7, 17/18, 19/13
  Software, 7/6, 13/21, 13/23-24
  Star computer, 4/20, 17/23
 Computers:
  820 personal computer, 11/29
  6085 "ViewPoint" workstation, 11/29
  8010 "Star" Information System, 4/20, 11/29, 17/22
  Alto computer,
    Development of, 4/4-6, 4/14-15, 4/20-21
    Miscellaneous, 4/1, 10/14, 11/29-30, 17/15, 17/23, 19/9, 19/13
    Software, 7/6, 12/22
  Alto II computer, 4/13, 11/19
  Alto III computer, 4/15
  Janus project (Xerox Star), 4/20
  Sigma V, 6/3
  Star system. *See* 8010 "Star" Information System
  ViewPoint system. *See* 6085 workstation
 Miscellaneous:
  7000 copier, 17/15
  9700 laser printing system, 17/15
  Advanced Systems Division (ASD), 4/20
  Dorado processor, 11/30
  EARS laser printer, 17/15
  Ethernet network, 4/5, 11/29, 17/18, 19/13
  Learning Research Group (LRG), 7/6
  MESA processor, 11/29
  Office Products Division, 11/27
  PARC (Palo Alto Research Center),
    Apple Computer, 5/15-16, 10/1, 10/5-6, 10/9, 10/15-16, 16/1
    Beginning of, 4/4-6, 4/14
    Microsoft, 12/18, 12/22, 12/25
    Miscellaneous, 2/11, 4/20, 7/6, 11/30, 17/15, 17/23, 19/13
    Other companies, 9/22, 11/19, 13/21, 13/23, 17/18
    Star Computer, 4/20, 11/29
  SLOT (Scanned Laser Output Terminal) printer, 17/15
  Systems Development

Division (SDD), 4/20, 11/28
Xerox Data Systems, 6/3

Software:
BCPL programming language, 4/5, 7/6
Bravo word processor, 4/5, 7/6, 12/22, 12/25
BravoX word processor, 7/6
Gypsy text editor, 7/6
JaM language, 13/23
MESA programming language, 4/5, 7/6
PUP (PARC Universal Packet), 4/6
Smalltalk language, 4/5, 7/6

XGA (Extended Graphics Array), 14/9, 20/5
Xidex company, 17/22
XMM (eXtended Memory Manager), 20/5
XMS (eXtended Memory Specification), 20/4
XPL language, 7/6
XYZ Corporation, 19/11

--Y--

Yahoo! Inc., 15/14-15
Yahoo! search engine, 15/14
Yale University, 13/13
Yang, Jerry, 15/14-15
Yates, John E., 2/4
Yates, William, 4/9, 6/5
Yee, Min S., 12/5
Yellow Book specification, 20/4
Yocam, Delbert "Del" W., 10/7
Young, Bob, 15/10
Yuen, Thomas C., 11/14

--Z--

Z-1, Z-2 and Z-3 computers, 1/4
Also see Cromemco
Z-80 SoftCard. *See* Microsoft/Miscellaneous
Zaltair microcomputer, 20/2
Zander, Ed, 11/26, 16/13
Zappacosta, Pierluigi, 19/22
ZDOS (Zenith). *See* Microsoft/ Operating Systems
Zenith Data Systems (ZDS), 11/30, 16/16
MinisPort portable computer, 11/30
Zenith Radio Corporation, 4/18, 11/30, 12/16, 12/28, 17/21
Z-89 computer, 4/18
ZDOS. *See* Microsoft /Operating Systems
Zeos company, 16/16
Ziff-Davis Publishing Company, 18/2, 18/5-7, 19/21, 19/24
Zilog Inc.,
Beginning of, 3/14
Intel, 3/15, 8/5
Miscellaneous, 3/8-9, 6/12, 7/2, 8/7
Products using microprocessor, 4/14, 4/16, 4/18-19, 11/8, 11/23, 11/28, 17/17
Z-80, 3/8, 3/14-15, 4/14, 4/15, 4/17, 6/13, 7/2, 8/7
Z-80A, 3/14, 4/16, 11/7, 11/14, 11/23, 11/28
Z-280, 8/7
Z-8000, 3/14-15, 6/12, 8/6
Z-80000, 8/7
Zip disk drive. *See* Iomega
Zork games. *See* Infocom
Zuse, Konrad, 1/3-4
ZX80 and ZX81 computers. *See* Sinclair Research

Blank page.